Dynamics and Relativity

Dynamics and Relativity

W.D. McComb

Department of Physics and Astronomy
University of Edinburgh

OXFORD
UNIVERSITY PRESS

UNIVERSITY PRESS

Great Clarendon Street, Oxford ox2 6DP

OxfordUniversity Press is a department of the University of Oxford.
It furthers the University's objective of excellence in research, scholarship,
and education by publishing worldwide in

Oxford New York

Athens Auckland Bangkok Bogotá Buenos Aires Calcutta
Cape Town Chennai Dar es Salaam Delhi Florence Hong Kong Istanbul
Karachi Kuala Lumpur Madrid Melbourne Mexico City Mumbai
Nairobi Paris São Paulo Singapore Taipei Tokyo Toronto Warsaw

with associated companies in Berlin Ibadan

Oxford is a registered trade mark of Oxford University Press

Published in the United States
by Oxford University Press Inc., New York

A catalogue record for this book is available from the British Library

Library of Congress Cataloging in Publication Data
(Data available)

ISBN 0 19 850112 9 (Pbk)

Typeset by Newgen Imaging Systems (P) Ltd., Chennai, India

Printed in Great Britain
on acid-free paper by Bath Press, Bath, Avon

For Patricia, Emma and Rachel

Preface

Relativity, as a subject, has a special aura which it shares only with quantum mechanics. Together, these two branches of physics have made possible a revolution in scientific thought and they are rightly regarded as being among mankind's greatest intellectual achievements. Yet they are plagued by two persistent myths. The first is that they are difficult subjects; and given the strange or even bizarre nature of many of their concepts (when judged against the test of everyday experience) this is not altogether surprising. The second is that they have somehow made all of classical physics obsolete.

Taking the question of difficulty first, we should note that the sheer speed with which these ideas were adopted suggests that they are not really very difficult to understand; a suggestion supported by the fact that both subjects have been taught successfully to many generations of students all over the world. Indeed, on a personal note, the present writer, when a student, found both these subjects much easier to understand (and much more interesting) than certain classical subjects, such as thermodynamics.

The second myth, that relativity and quantum mechanics have replaced classical physics, arises from what is really no more than a superficial judgement. But it is worth tackling, nevertheless, as an understanding of the precise relationship of modern to classical physics is surely crucial to an understanding of physics itself.

As we shall not be concerned with quantum mechanics in this book, we will not pursue this aspect. Here we shall be concentrating on relativity, and for this subject we believe that a good way of tackling both myths is to emphasise the inevitability of the theory of special relativity and (without selling Einstein short!) the fact that it was not particularly new. We may do this by taking advantage of hindsight and treating classical mechanics as if it were a form of relativity, which we shall call Galilean relativity.[1] This gives us the freedom to formulate a particular problem in one frame of reference and solve it in another, where it takes a simpler form. Then, we may transform our solution back to the first frame. This method of treating problems—which is absolutely central to special relativity—can therefore become familiar in the context of classical dynamics, and this should facilitate the transition to special relativity.

[1] Of course the subject of classical mechanics contains a great deal of material which requires exposition without much if any emphasis on the choice of reference frame. Thus we cannot expect to give a fully relativistic treatment where this would be at odds with the pedagogic requirements of the subject. Nevertheless, where it seems natural to do so, we shall bring in the concept of transformations between frames of reference; even if occasionally, there is some danger of this seeming a little contrived.

In order to emphasize this approach, we have divided the book into two parts. The first part consists of Chapters 1–10, and deals with Newton's laws, which are presented here in the context of Galilean relativity. The second part deals with the theory of special relativity, as enunciated by Albert Einstein in the year 1905. We may sum up our reasons for presenting our material in this particular way, as follows:

1. We want to make it clear that special relativity did not come about in some miraculous way, but was part of the normal process of development in physics.
2. We also want to make it clear that classical mechanics did not become obsolete with the coming of special relativity.
3. We can overcome much of the supposed difficulty of special relativity if we establish the right habits of thought before we meet the strange new concepts, such as time-dilation or length-contraction, which one encounters in Einstein's theory.

In the first part of the book, we shall make Galilean transformations between inertial frames of reference (these terms are defined in Chapter 1) and employ Newton's laws to describe the real world. We shall refer to Newton's laws as being 'Galilean invariant' and emphasize that they give us a good theoretical description of a range of phenomena ranging from the motion of planets to the statistical mechanics of molecules. At the same time, we must bear in mind that this classical picture breaks down when we attempt to describe the motion of moving bodies at speeds which are an appreciable fraction of the speed of light.

Of course, this is something which we now know because of special relativity. But, historically the difficulty with the classical picture arose because it could not accommodate the propagation of electromagnetic waves. The equations describing the propagation of electromagnetic waves (known as Maxwell's equations) are not Galilean invariant.

In the second part of the book, we explicitly recognize this problem, so now we make Lorentz transformations between inertial reference frames. We do this because the equations of electromagnetic wave propagation *are* Lorentz invariant, hence we are forced to modify Newton's laws such that the equations of dynamics also become Lorentz invariant. We note, and shall enlarge on this later, that for speeds much less than the speed of light *in vacuo*, the Lorentz transformation reduces to the Galilean transformation. As we shall see, this implies that for speeds $v \ll c$, Newton's laws are restored to us.

In both parts of the book, we shall lay emphasis wherever possible on the use of conservation laws. It is generally accepted (in both modern and classical physics) that these are truly fundamental, as they are a direct consequence of the symmetries of the physical problem under consideration. In contrast, laws of motion—be they Newton's or Einstein's—are inherently phenomenological and can be overturned at any time by a contrary observation.

Within the limits of our present treatment, special relativity is covered in Chapters 11–14, and an introduction to general relativity is given in

Chapter 15. The last chapter may seem over-ambitious to some— particularly as it makes an attempt to give an account of the Einstein field equations, the ground for this having been prepared by a treatment of the Newtonian field equations in Chapter 1. However, it should be borne in mind that many physics students are intensely curious about this topic, yet do not get as far as taking a formal course on general relativity or relativistic gravitation. In my view, any self-respecting physicist has the social duty of being able to talk knowledgeably at parties on the subject of black holes. If nothing else, Chapter 15 may help!

It may be of interest to know that this book arose out of a lecture course at Edinburgh University, which in a mere eighteen lectures tried to teach Galilean and special relativity in an integrated way. Just by way of adding to the pedagogical challenge, the course was also given to a joint audience of second-year mathematical physicists and mathematicians, on the one hand, along with the third-year mainstream physicists on the other. Of course the two different 'years' had to be tutored and examined separately. Nevertheless, the joint lectures seemed to go down quite well and, if judged by the overall performance in examinations, quite successfully. However, one thing that I learned from the experience was that my colleagues have strong feelings on the subject of relativistic mass! Ultimately I was convinced by their arguments that one should do without it, only to find that virtually all students came to the course well-versed in its use. Evidently the world is a less perfect place than we would like it to be. So, as a compromise I have discussed the subject in an appendix. That way, the reader can make up his or her own mind on the utility or otherwise of the concept.

Acknowledgements

Lastly, it is a pleasure to turn to the subject of acknowledgements. The inspiration for this book (and some of the exemplary material) came from my teaching experience on the Mathematical Physics 2 course at Edinburgh. Over the years, I have learned much from both students and colleagues in MP2. In particular I am happy to acknowledge helpful and stimulating discussions with Peter Higgs, Peter Osborne, Brian Pendleton and Lance Vick. It is also a pleasure to thank my research students Gary Fullerton, Craig Johnston, Adrian Hunter, Anthony Quinn and Alistair Young, each of whom read numerous chapters of the book and pointed out errors and possible improvements.

W.D. McC.
Edinburgh
October 1998

Contents

Galilean relativity

1

In this book we assume that the reader has attended an elementary course in mechanics, although the treatment presented here is intended to be reasonably complete. With this in mind, the main purpose of this chapter is first to revise some concepts from mechanics and, second, to indicate how we need to be more precise about certain of these concepts (for instance, frames of reference) in order to pursue a relativistic treatment of the subject. As the chapter is somewhat miscellaneous in character, it is also a convenient place to put topics which will be useful later. In this category comes the Newtonian field equation for gravity, which will turn out to be useful when we discuss general relativity in the final chapter of the book.

1.1 Brief reminder of Newton's laws

Let us begin with a reminder of Newton's laws of motion. These may be stated as follows:

N1: A body continues in a state of rest or uniform motion unless acted upon by a force.

N2: The rate at which the momentum of a body changes with time is proportional to the force acting on that body, and is in the direction of the force.

N3: To every action (on a body) there is an equal and opposite reaction.

Here, by a body, we mean a point mass or particle. Later, we shall consider extended bodies of finite size. We should note that these so-called laws are just a set of hypotheses which appear to be in accordance with everyday experience. In the second part of this book, we shall find that Newton's first and third laws go over without modification into the theory of special relativity. However, despite the fact that N1 is just in some ways a special case of N2, and does not need to be changed, it will turn out that N2 needs substantial modification. We shall just note, in passing, the important distinction that N1 is *qualitative*, whereas N2 is *quantitative*.

As the second law plays such a central role in classical mechanics, we shall look at it now from two points of view. First, as defining the concepts of force and mass; and second, as governing the motion of a particle which is subject to a given force.

1.1.1 *Newton's second law as a definition of the concepts of force and mass*

Many of our physical concepts arise as intuitive ideas which are then refined in the process of studying physics. For instance, we may think of temperature as being the 'degree of hotness of a body'; or of mass as the 'amount of matter in a body'. In the case of force, our intuitive ideas have something to do with the effort required of our muscles when we move a body (including, of course, our own bodies!). In order to make this intuitive idea more quantitative and systematic, we invoke N2, as this gives us a definition of force in terms of its effects.

Consider a particle of mass m and velocity $\mathbf{v}$. It has, by definition, linear momentum, $\mathbf{p} = m\mathbf{v}$. From N2, any force $\mathbf{F}$ acting on the particle will produce a change in its momentum; thus:

$$\mathbf{F} \propto \frac{d\mathbf{p}}{dt} = k\frac{d\mathbf{p}}{dt}, \tag{1.1}$$

where k is a constant of proportionality. If we choose a system of units such that $k = 1$, then we have :

$$\mathbf{F} = \frac{d\mathbf{p}}{dt}. \tag{1.2}$$

This step in fact defines the unit of force for a given set of units. For instance, in SI units, equation (1.2) gives the force in **newtons**.

If the mass, m, is constant, we may write Newton's second law in its most familiar form, by substituting $\mathbf{p} = m\mathbf{v}$, and writing (1.2) as

$$\mathbf{F} = m\frac{d\mathbf{v}}{dt}. \tag{1.3}$$

Or, in words, *force = mass × acceleration*. Later, in Chapter 8, we shall consider situations where m is not constant but depends on the time t.

Thus for a given body, we can measure its acceleration (assuming only that we can readily measure length and time), and hence the force acting on it, in terms of its mass. As the basis of all measurement is comparison with a standard, we can set up a scale of forces in this way for a given mass. Then equation (1.3) may be used to compare various masses with the standard mass and this is, in principle, our way of defining and measuring the mass of a body.

Alternatively, some books on mechanics suggest that N3 is the basis of the measurement of mass. Certainly this can be the case in high-energy particle physics, where collisions are routinely used to establish properties of particles, including their mass. However, the idea that we assess the mass of a can of baked beans by colliding it with a brass weight[1] and measuring their subsequent trajectories is clearly not in the realm of practicality, although, to be strictly correct, we should note that

[1] Once upon a time, shopkeepers used brass weights for the purpose of weighing out commodities such as sugar or flour.

N3 plays its part in nearly every aspect of mechanics (including the everyday operation of weighing out a quantity of flour or sugar!). We shall defer discussion of the use of weight as a measure of mass until Section 1.4.2.

1.1.2 *Newton's second law as an equation of motion*

If we turn equation (1.2) round, then Newton's second law may be interpreted as an equation of motion for a particle of mass m, thus

$$\frac{\mathrm{d}}{\mathrm{d}t}(m\mathbf{v}) = \mathbf{F}, \tag{1.4}$$

where the force $\mathbf{F}$ is given. Then, if we are also given the initial conditions of the particle, we may integrate both sides of the above equation with respect to time in order to obtain the velocity of the particle at any time. A further integration with respect to time then gives the position of the particle at any time, and the problem is solved, at least in principle. This is known as the *initial value problem*; or sometimes as the *Cauchy problem*.

As an example, let us consider a particle of mass m, which is constrained to move along the x-axis, subject to a force $F(t)$, in the x-direction. We shall take the initial conditions to be given by $x = x_0$ and $v = v_0$ at the initial time which in turn we take to be $t = t_0$.

From Newton's second law, we have

$$m\frac{\mathrm{d}v}{\mathrm{d}t} = F(t). \tag{1.5}$$

(Note that we have assumed here that the mass of the particle is constant, so that we may take m to the left of the differential coefficient.) Then, integrating with respect to time, and using a dummy variable of integration t', we may write the equation for the particle velocity at any time as

$$v(t') \mid_{t_0}^{t} = \frac{1}{m} \int_{t_0}^{t} F(t') \, \mathrm{d}t', \tag{1.6}$$

where we have also divided both sides by m. (For those unfamiliar with this way of using dummy variables in order to express indefinite integrals in the form of definite integrals, a short account of the method is given in Appendix A.) Using the initial conditions, we evaluate the integral on the left-hand side as

$$v(t) - v_0 = \frac{1}{m} \int_{t_0}^{t} F(t') \, \mathrm{d}t', \tag{1.7}$$

or, rearranging,

$$v(t) = v_0 + \frac{1}{m} \int_{t_0}^{t} F(t') \, \mathrm{d}t'. \tag{1.8}$$

Integrating again (and remembering that $v = \mathrm{d}x/\mathrm{d}t$) we obtain

$$x(t')\,|_{t_0}^{t} = \int_{t_0}^{t} v_0 \mathrm{d}t' + \frac{1}{m} \int_{t_0}^{t} \mathrm{d}t' \left\{ \int_{t_0}^{t'} F(t'')\,\mathrm{d}t'' \right\}, \qquad (1.9)$$

where we have introduced a second dummy variable of integration, $\mathrm{d}t''$, and we have put curly brackets around the innermost integral in the last term, in order to emphasize that this is just a function of the variable t'.

Again, using the initial conditions, we can evaluate the integral on the left-hand side, along with the first integral on the right-hand side. Then, with a little rearrangement, the expression for the distance travelled by the particle from the initial time t_0 to the current time t is given by

$$x(t) = x_0 + v_0(t - t_0) + \frac{1}{m} \int_{t_0}^{t} \mathrm{d}t' \int_{t_0}^{t'} F(t'')\,\mathrm{d}t'', \qquad (1.10)$$

where we have dropped the curly brackets, as the meaning of this term should now be clear. Providing we are given the force in an explicit fashion and that it is integrable in this way, the above procedure may be regarded as the general solution to the problem.

1.2 The nature of force

As we have seen, force as a rigorous concept in physics is defined using N2. Now we consider the various kinds of forces which we shall encounter throughout this book. We begin with a very special class of forces which are of prime importance in mechanics: the conservative forces, which are associated with the conservation of energy.

1.2.1 Conservative forces

If a force $\mathbf{F}$ is applied to a particle, and as a result the particle moves from $\mathbf{x}$ to $\mathbf{x} + \mathrm{d}\mathbf{x}$, the amount of work $\mathrm{d}W$ done by the force in this displacement $\mathrm{d}\mathbf{x}$ is defined to be

$$\mathrm{d}W = \mathbf{F} \cdot \mathrm{d}\mathbf{x}.$$

The total work done in moving the particle over a finite path length between points 1 and 2 is given by the path integral

$$W = \int_{1}^{2} \mathbf{F} \cdot \mathrm{d}\mathbf{x}, \qquad (1.11)$$

and for certain forces this amount of work depends only on the positions of the two points and not on the path taken from one to another. Such forces are known as **conservative forces**. This is because they conserve energy.

We should note an important corollary of this property. If we take the path integral to begin and end at the same point then the total work done must be zero. In other words, the path integral of the force round any closed circuit must be zero for a conservative force, or:

$$\oint \mathbf{F} \cdot d\mathbf{x} = 0, \tag{1.12}$$

where the circle on the integral sign indicates that the integration is round a closed path. It follows from Stokes' theorem in vector calculus that the above equation may be written as

$$\nabla \times \mathbf{F} = 0. \tag{1.13}$$

Then, it is usual to argue that, as the curl of a gradient always vanishes, $\mathbf{F}$ may be written as the gradient of some scalar U (say), so that we may write

$$\mathbf{F} = -\nabla U, \tag{1.14}$$

where the minus sign is conventional. The quantity denoted by U is generally referred to in vector calculus as the **potential** and in the particular case of mechanics as the **potential energy**.

We shall return to this topic in more detail in Chapter 2, but at this point it may be helpful to envisage a specific example. Suppose one throws a ball vertically into the air and then catches it again. If we do this experiment in a vacuum (on the Moon, say), then the ball will return with exactly the same speed with which it left. However, if we do the experiment on Earth, then air resistance will slow the ball, with the amount of the reduction in speed depending on the path taken. In this case, some of the energy of the ball is dissipated in heating up the air around it.

It is probably quite obvious that dissipative forces such as friction between solids or air resistance cannot be conservative forces. We shall discuss dissipative forces in Section 1.2.3.

1.2.2 The fundamental forces

The fundamental forces of physics may be listed as follows:

- electromagnetic forces;
- gravity;
- strong nuclear force;
- weak nuclear force.

We are not concerned with the nuclear forces in this book and indeed will really only be concerned with gravity for the most part. However, it is of interest at this stage to discuss the electromagnetic forces to some extent.

The electrostatic, magnetostatic and gravitational forces have in common the property that they are all conservative. Indeed, they all have the form of an inverse-square law, when one considers a point source which is located in an environment possessing spherical symmetry. Taking the case of the electrostatic force as an example, we have Coulomb's law for the force between point charges q and q' in the form

$$\mathbf{F} = -\frac{qq'}{4\pi\epsilon_0 x^2}\hat{\mathbf{x}}, \qquad (1.15)$$

where $\mathbf{x}$ is the displacement of one charge from the other, ϵ_0 is the permittivity of a vacuum, and the 'hat' symbol denotes a unit vector. If we introduce the electric field $\mathbf{E}$ due to the charge q', then the expression for the force on charge q may be rewritten as

$$\mathbf{F} = q\mathbf{E}, \qquad (1.16)$$

where the form of the electric field due to the charge q' may be inferred from a comparison of the two equations. Evidently the field is the force per unit charge.

Exactly analogous laws hold for magnetostatics and gravity, although in the latter case the force can only be attractive. However, when we consider the case of a moving charge, then the magnetic force takes an interesting form. Consider a charge q moving with velocity $\mathbf{v}$ in a region where there is an electric field $\mathbf{E}$ and a magnetic induction $\mathbf{B}$. The total force acting on the moving charge is given by

$$\mathbf{F} = q\mathbf{E} + q(\mathbf{v} \times \mathbf{B}). \qquad (1.17)$$

This is known as the Lorentz force. The fact that a vector product of the velocity and the magnetic induction is involved has an important consequence. From the properties of vector products, we can deduce that the magnetic component of $\mathbf{F}$ is perpendicular to the velocity $\mathbf{v}$ and hence the magnetic field cannot do work on the moving charge.

Evidently, unlike in the electrostatic or gravitational cases, the magnetic field cannot be obtained as the gradient of a scalar field, as in equation (1.14). Instead, it has to be obtained as the curl of a vector potential. That is, instead of a form like (1.14) we must write:

$$\mathbf{B} = \nabla \times \mathbf{A}, \qquad (1.18)$$

where $\mathbf{A}$ is known as the **vector potential**. We put this brief mention of the vector potential in partly for completeness and partly because it will be helpful when we wish to refer to it again at the end of Chapter 14.

It may also be noted that, unlike the case of the electric field, there is no component of acceleration in the direction of the magnetic induction $\mathbf{B}$. Accordingly, the role of the magnetic field is to provide a constraint to the motion of a charged particle. A corresponding case in mechanics would be to have a particle moving in a circle when tied by a string to a fixed point. That is, the effect of the magnetic field may be seen as being analogous to the tension in the string which constrains the particle to move in a circle.

1.2.3 *Forces encountered in applications*

In this section we consider how forces that arise in practice—particularly in macroscopic systems—can seem quite different from the fundamental forces which we have just been discussing. As we shall see later, problems posed in Chapter 3 on central forces may involve forces which are quite different from the inverse-square form. Also, there is the matter of resistive forces such as friction. In fact, all such forces—however varied and apparently dissimilar—can be interpreted in terms of the basic fundamental forces.

Let us continue to take our examples from electrostatics. If we equate the expressions for the force given by equations (1.15) and (1.16), we may write[2] an expression for the field due to electric charge q' as

$$\mathbf{E} = -\frac{q'}{x^2}\hat{\mathbf{x}}.$$

Then, if we integrate this expression from some arbitrarily chosen position (remember this is a field associated with a conservative force and $\mathbf{E} = \nabla\phi$) we get an expression for the electrostatic potential ϕ. In this case, we choose to integrate in from infinity, where the potential is zero, to get the result

$$\phi = -\frac{q'}{x}. \tag{1.19}$$

Thus, the classic result for one of the fundamental inverse-square law forces is a potential which depends inversely on the distance. This, of course, applies to gravity and magnetostatics as well.

However, we must remind ourselves that this result applies only to a point source where there is spherical symmetry. In practice this is often not the case. For instance, the field due to a long cylindrical distribution of charge falls off with the inverse power of the radial distance, while the corresponding potential falls off logarithmically.

We may illustrate this point by examining the simplest deviation from the single-particle case: a dipole formed from charges $\pm q$ separated by $\mathbf{dl}$, as shown in Figure 1.1.

In this case, each separate charge will have a potential varying like (1.19). From the principle of superposition, we may therefore write the potential at the point P due to the pair of charges as

$$\phi = q\left(\frac{1}{x_+} - \frac{1}{x_-}\right). \tag{1.20}$$

If we assume the fiction that the displacement $\mathbf{dl}$ is generated by a motion of the test point P, then from vector calculus, we may relate the first inverse to the second by taking the gradient, thus:

$$\frac{1}{x_+} = \frac{1}{x_-} - \mathbf{dl}\cdot\left(\nabla\frac{1}{x}\right) = \frac{1}{x_-} + \frac{\mathbf{dl}\cdot\hat{\mathbf{x}}}{x^2}.$$

[2]For the sake of simplicity, we put the prefactor on the right-hand side of equation (1.15) equal to unity.

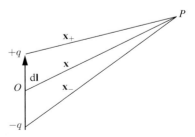

Fig. 1.1 Sketch for working out the potential of a dipole at the point P.

Then, substituting this result into equation (1.20), we immediately obtain

$$\phi = -q\frac{d\mathbf{l} \cdot \hat{\mathbf{x}}}{x^2}. \qquad (1.21)$$

Thus the potential of a dipole decreases as the inverse square of the distance.

One can get virtually any power law for the potential (and hence for the force) by choosing an appropriate distribution of charges. Similar conclusions apply to gravitational potentials due to distributions of mass.

Let us turn now to the question of resistive forces. Take the example of a particle moving through air. The resistive force will depend on the speed of the particle and the amount of work done against it will depend on the total distance travelled. However, such a force only has meaning at a macroscopic level, where we interpret the loss of the particle's kinetic energy in terms of the heating up of the gas through which it is passing. In contrast, at the microscopic level, the interactions of the molecules of the gas are mediated by the fundamental forces. At this level, there has merely been a transfer of kinetic energy from a directed motion to a random motion. To sum up, the idea of dissipation of energy is really only a value judgement imposed at the macroscopic level!

1.3 Force of gravity

From a study of Kepler's laws of planetary motion (see Chapter 3), Newton came to the conclusion that there was a gravitational force between any two bodies of mass m and m' which was given by

$$\mathbf{F} = -G\frac{mm'}{x^2}\hat{\mathbf{x}}, \qquad (1.22)$$

where $\mathbf{x}$ is the displacement of one mass from the other and G is a constant. In arriving at this result, he had to rely on his own invention of calculus. In order to illustrate this process, we shall now consider how to work out the gravitational potential due to a massive spherical shell. However, first we repeat the steps taken in going from equation (1.18) to (1.19) for the electrostatic case. That is, we integrate both sides of equation (1.22) with respect to x, integrating in from infinity where the potential is zero, to give for the gravitational potential U due to mass m'

$$U = -\frac{Gm'}{x}. \qquad (1.23)$$

Referring to Figure 1.2, we consider a spherical shell of radius r and with a surface mass density of σ per unit area. We want to know the potential at a point P which is a distance x from the centre of the shell. If we take as our elementary area dA the ring defined by the line segment $rd\theta$, which subtends the angle $d\theta$ at the origin, and by a revolution of this

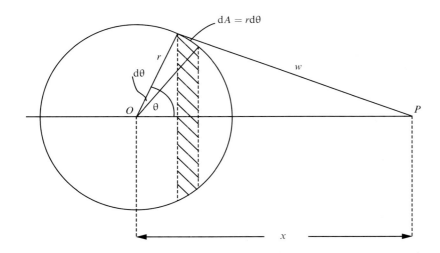

Fig. 1.2 Gravitational potential due to a spherical shell of uniform density.

line element through an angle of 2π round the x-axis, we may write

$$dA = 2\pi r \sin\theta \times r\,d\theta.$$

Then if we let the distance from the elementary area to the test point P be w, and substitute $\sigma dA = m'$ in equation (1.23), then Newton's law of gravitation gives the potential at P as

$$dU = -\frac{G\sigma\,dA}{w} = -\frac{2\pi G\sigma r^2 \sin\theta\,d\theta}{w}, \tag{1.24}$$

where the last step follows from substitution for dA.

In order to generate the complete spherical shell, we have to integrate the variable θ over the range from 0 to π. At the same time, we have to remember that w is a function of θ and in fact it is better to change this into an integration with respect to w. We do this as follows: from the figure, and using the cosine law, we may write w as

$$w^2 = x^2 + r^2 - 2xr\cos\theta. \tag{1.25}$$

Differentiating both sides, cancelling the factors of 2 and the minus signs, we obtain

$$w\,dw = xr\sin\theta\,d\theta, \tag{1.26}$$

and hence from (1.24), the elementary potential is

$$dU(x) = -\frac{2\pi G\sigma r\,dw}{x}. \tag{1.27}$$

For the case where the point P is outside the spherical shell, as θ varies from 0 to π, w varies from $x - r$ to $x + r$. So the total potential at the test point due to the mass shell is

$$U(x) = -\frac{2\pi G\sigma r}{x}\int_{x-r}^{x+r} dw = -\frac{4\pi G\sigma r^2}{x}. \tag{1.28}$$

If we introduce the total mass of the shell $m = 4\pi r^2 \sigma$, then this result takes the form

$$U(x) = -\frac{Gm}{x}, \tag{1.29}$$

indicating that the shell behaves as if it were a point mass, with all its mass concentrated at its centre. It is left as an exercise for the reader to show that, for the case where the test point is inside the shell, the potential is constant.

Lastly, we should comment on the axiomatic nature of the 'law of gravitation'. Although the law was arrived at by Newton in a purely phenomenological way—that is, by trying to explain various astronomical observations of the solar system—its real significance is that it was postulated as a **universal law of gravitation**, applying just as much to the aprocryphal apple falling to the ground as to the relationships between the Sun and the planets. This suggests that the law should be given the status of an axiom. That is, we should proceed on the basis that we assume it to be true and then try to explain all relevant physical phenomena from it. Then, there is no reason to enquire further into its status until we find that its use leads to an incorrect prediction about the physical universe.

1.3.1 Newtonian field equations for gravity

The classical field equations are probably best known in the context of electromagnetism, but the static forms—Laplace's and Poisson's equations—can also be applied to the case of gravity, in order to deal with problems where there is a continuous but non-uniform distribution of matter. We shall derive Poisson's equation here by adapting the usual derivation from electrostatics. We do this by introducing the gravitational field $\mathbf{g}(\mathbf{x})$, defined such that Newton's law in the form of equation (1.22) may be rewritten as

$$\mathbf{F} = m\mathbf{g}(\mathbf{x}). \tag{1.30}$$

That is, $\mathbf{g}(\mathbf{x})$ is gravitational force per unit mass, and by comparison of this defining relation with (1.22), we see that it takes the form

$$\mathbf{g} = -G\frac{m'}{x^2}\hat{\mathbf{x}}, \tag{1.31}$$

as the gravitational field due to a point mass m'.

We begin by considering a surface S which encloses a volume V within which there are point masses $m_1, m_2, \ldots, m_n, \ldots$; the gravitational analogue of Gauss's theorem in electrostatics takes the form

$$\int_S \mathbf{g} \cdot \mathbf{n}\, dS = -4\pi G \sum_i m_i. \tag{1.32}$$

This theorem is easily verified for the case of a single point mass m_1, as follows. Referring to Figure 1.3, we may put

$$\mathbf{g} \cdot d\mathbf{S} = g \cos \theta \, dS = \frac{m_1 G \cos \theta \, dS}{x^2}, \tag{1.33}$$

where the last step follows when we substitute from Newton's law for g. However, the solid angle[3] subtended by the element of surface dS at the point O is

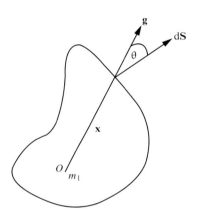

[3]Solid angles are discussed in Appendix B.

$$d\Omega = \frac{\cos \theta \, dS}{x^2},$$

and hence

$$\int_S \mathbf{g} \cdot \mathbf{n} \, dS = -G m_1 \int d\Omega = -4\pi G m_1. \tag{1.34}$$

From the principle of superposition, this result may be extended to any number of point charges and hence Gauss's theorem is verified.

Now consider the field due to a continuous distribution of mass within V with density $\rho(\mathbf{x})$. In these circumstances, the generalization of Gauss's theorem becomes

$$\int_S \mathbf{g} \cdot \mathbf{n} \, dS = -4\pi G \int_V \rho(\mathbf{x}) \, dV. \tag{1.35}$$

Fig. 1.3 Volume V enclosed by surface S for the proof of Gauss's theorem.

We may use the divergence theorem in order to turn the left-hand side into a volume integral

$$\int_V \nabla \cdot \mathbf{g} \, dV = -4\pi G \int_V \rho(\mathbf{x}) \, dV. \tag{1.36}$$

This equality must hold for arbitrary volumes and surfaces and hence the integrands must themselves be equal:

$$\nabla \cdot \mathbf{g} = -4\pi G \rho(\mathbf{x}). \tag{1.37}$$

As gravity is a conservative force, the associated field is derivable from a potential, thus

$$\mathbf{g} = -\nabla U, \tag{1.38}$$

where U is known as the gravitational potential. With this form substituted for $\mathbf{g}$, equation (1.37) may be written as

$$\nabla^2 U = 4\pi G \rho(\mathbf{x}), \tag{1.39}$$

which is the Newtonian field equation for the gravitational potential.

Having derived this general equation, we shall show how it can be put to use in order to calculate the properties of stars, as these may be regarded as self-gravitating fluids.

1.3.2 Self-gravitating fluids

We shall consider stellar atmospheres as an important example of self-gravitating fluids. As a first step, we obtain the condition of **hydrostatic equilibrium** for a fluid at rest.

Consider a fluid *at rest* in a volume V, enclosed by a surface of area S. Introduce the *force per unit volume* $\mathbf{f}(\mathbf{x})$ and examine the effect of the pressure $p(\mathbf{x})$ on an element dV, with surface area dS.

The force on the elementary volume is $\mathbf{f}(\mathbf{x})\,dV$ and this is due to pressure acting inwards at the surface which results in a force $p(\mathbf{x})\,d\mathbf{S}$ (recall that pressure has dimensions of force per unit area). Equating these two expressions for the force on the elementary area yields

$$\mathbf{f}\,dV = p\,d\mathbf{s},$$

as the condition for equilibrium; and for the whole system:

$$\int_V \mathbf{f}\,dV = \int_S p\,d\mathbf{S} = \int_V \nabla p\,dV,$$

where we have again used a version of the divergence theorem of vector calculus for the last step.

Hence the equilibrium equation becomes

$$\int_V \mathbf{f}\,dV = \int_V \nabla p\,dV,$$

and as this holds for arbitrary volumes V, equating the two integrands gives us the condition

$$\mathbf{f} = \nabla p, \tag{1.40}$$

which is the equation of hydrostatic equilibrium. In words, this may be written as:

$$\text{force/unit volume} = \text{pressure gradient.}$$

This equation can be further written in terms of the force per unit mass $\mathcal{F} = \mathbf{f}/\rho$, with the result

$$\rho\mathcal{F} = \nabla p. \tag{1.41}$$

Now let us turn to a description of **self-gravitating fluids**. We may draw an analogy with a cloud of electrostatic charge of given density. As we have seen, the gravitational field is derivable from a scalar potential, as in equation (1.38), and this potential is a solution of Poisson's equation, which is given by (1.39). In equilibrium, the fluid must also satisfy (1.41) in the form

$$\nabla p = \rho\mathbf{g}(\mathbf{x}). \tag{1.42}$$

Substituting from (1.38) into (1.42) yields

$$\nabla p = \rho(-\nabla U),$$

and, with some rearrangement,

$$\nabla U = -\frac{1}{\rho} \nabla p. \tag{1.43}$$

Now write (1.39)—our general form of Newtonian field equation—as

$$\nabla^2 U = \nabla \cdot (\nabla U) = 4\pi G\rho.$$

Then, further substituting for ∇U from (1.43), we obtain the Newtonian field equation for the specific case of a self-gravitating fluid as

$$\nabla \cdot \left(\frac{1}{\rho} \nabla p\right) = -4\pi G\rho. \tag{1.44}$$

We shall restrict our attention to situations where the mass of fluid is spherically symmetric. In this case p and ρ depend only on $|\mathbf{x}| = r$ (say), and so equation (1.44) becomes

$$\frac{1}{r^2} \frac{\partial}{\partial r} \left(\frac{r^2}{\rho} \frac{\partial p}{\partial r}\right) = -4\pi G\rho. \tag{1.45}$$

This equation can be solved for the pressure in terms of the density and we shall consider some examples in the next section.

1.3.3 Example: *Radial variation of pressure and density in model stars*

A usual model for a star which is made up of hot gases is to assume a **polytropic** equation of state in the form $p = k\rho^\Gamma$, where Γ is called the polytropic exponent and is written as $\Gamma = (n+1)/n$ for integer n.

Problem A: Obtain an expression for the radial variation of the pressure in an incompressible star.

Problem B: Solve for the radial variation of the density in a star when $n = 1$.

Problem A: An incompressible star   In this case $\Gamma = 0$ and $\rho = \rho_0$ which is constant. We tackle this problem as follows. Substitute for ρ in the Poisson equation, as given by (1.45):

$$\frac{\mathrm{d}}{\mathrm{d}r} \left(\frac{r^2}{\rho_0} \frac{\mathrm{d}p}{\mathrm{d}r}\right) = -4\pi Gr^2 \rho_0.$$

Integrate twice:

$$\Rightarrow r^2 \frac{\mathrm{d}p}{\mathrm{d}r} = -\frac{4\pi}{3} \rho_0^2 r^3 G + C_1;$$

$$\Rightarrow p(r) = -\frac{2\pi}{3} G\rho_0^2 r^2 + \frac{C_1}{r} + C_2.$$

We use the boundary conditions to fix values for the constants of integration C_1 and C_2.

- At $r = 0$ we have the condition that $p(0)$ should be finite, therefore $C_1 = 0$.
- At $r = R$ (the surface) we must have $p(R) = 0$, hence the second constant of integration is given by:

$$C_2 = \frac{2\pi}{3} G\rho_0^2 R^2.$$

Substituting these results back into the expression for the pressure yields:

$$p(r) = \frac{2\pi}{3} G\rho_0^2 (R^2 - r^2),$$

as required.

Problem B: The case where $n = 1$ We have $n = 1$, and hence $\Gamma = 2$. For this case the equation of state is $p = k\rho^2$ and differentiating both sides of this relationship gives us

$$\frac{dp}{dr} = 2k\rho\rho',$$

where a dash indicates differentiation with respect to r.

Again, our starting point is equation (1.45) and this time we substitute for dp/dr to obtain

$$\frac{d}{dr}\left(\frac{r^2}{\rho} \cdot 2k\rho\rho'\right) = -4\pi Gr^2\rho.$$

Trick: treat $r\rho$ as the dependent variable, so that the equation may be written as

$$2kr(r\rho)'' = -4\pi Gr(r\rho),$$

or, with some rearrangement,

$$(r\rho)'' + (2\pi G/k)(r\rho) = 0.$$

Now put

$$\omega^2 = \frac{2\pi G}{k},$$

to obtain

$$(r\rho) = A\sin(\omega r + \varepsilon),$$

as the standard solution to the equation of simple harmonic motion.

Again we use the boundary conditions to fit the constants of integration. At the centre, we take $\rho(0)$ to be finite, therefore we must have $\varepsilon = 0$. This leaves us with

$$(r\rho) = A\sin(\omega r).$$

Then, for small r we can replace the sine function by its argument, so that $(r\rho) \to A\omega r$ and hence $A = \rho_0/\omega$. Substituting back for A then gives:

$$r\rho = \frac{\rho_0}{\omega} \sin \omega r,$$

and with some rearrangement

$$\rho(r) = \rho_0 \left(\frac{\sin \omega r}{\omega r} \right),$$

as required.

1.4 The nature of mass

We have previously mentioned our intuitive definition of the mass of a body as being the amount of matter in that body. In order to take this helpful, but imprecise, concept into the realm of physics, we have to proceed along two quite different paths. First we have to reconcile this intuitive idea of the meaning of mass with Newton's laws of motion—this leads us to the concept of **inertial mass**. Second, we have to interpret our intuitive idea in terms of Newton's law of universal gravitation—this leads us to the concept of **gravitational mass**.

1.4.1 Inertial mass

For this section, let us restrict our attention to situations where the amount of matter in a body is constant. Then we can write N2 in the form:

$$\mathbf{F} = m\ddot{\mathbf{x}}. \tag{1.46}$$

Although it is quite usual to refer to the scalar coefficient m in this relation as the mass, strictly speaking, what we detect when we apply a force to a body is its **inertia**.

It is helpful to be clear about this, as later on we shall consider the rotational motion of a body fixed at a point. Then we shall find that the mass of the body is, in itself, not an immediately relevant quantity. Instead, we shall find that the relevant quantity is the **moment of inertia**. For the present, therefore, we shall adopt the usual compromise and refer to m in equation (1.46) as the **inertial mass**.

In order to determine the inertial mass of a body, we compare it to the inertial mass of some standard or reference body, which we take to be unity. Then we can use N2 to establish the inertial mass of some specimen particle m_{spec} in terms of its acceleration $\ddot{\mathbf{x}}_{\text{spec}}$. Then it is just a matter of measuring the accelerations produced by the same force when applied to each of the two particles.

However, we shall defer until a later stage the question of how one ensures that the force is of the same magnitude in both instances. We can evade this question for the moment by (despite our previous strictures on the subject) simply invoking N3. If there is some interaction between the two particles (e.g. they collide), then by N3 the forces acting on the two bodies are equal and opposite. Or, in the form of an equation:

$$m_{\text{spec}} \ddot{\mathbf{x}}_{\text{spec}} = -\ddot{\mathbf{x}}_{\text{ref}}, \tag{1.47}$$

where we have taken the reference mass to be unity. Thus the inertial mass of the specimen particle is given by the ratio

$$m_{\text{spec}} = -\frac{\ddot{\mathbf{x}}_{\text{ref}}}{\ddot{\mathbf{x}}_{\text{spec}}}, \tag{1.48}$$

in units of the reference mass.

1.4.2 Gravitational mass

Our second approach to a more quantitative definition of the mass of a body is through Newton's law of universal gravitation. From (1.22) we have this as

$$\mathbf{F} = -G\frac{mm'}{x^2}\hat{\mathbf{x}},$$

giving the force between two bodies separated by a distance x. In this equation, the masses of the two bodies are by definition their **gravitational masses**. It is not trivially obvious that the gravitational mass and the inertial mass of a body are the same thing. We shall return to this point in the next section. Here we deal with two topics which are relevant to the concept of gravitational mass.

First, there is the concept of the **weight** of a particle. This is the force exerted on a particle by the Earth. Let us take the Earth to be a sphere of mass M. Then the gravitational force on a particle of mass m, a distance r from the centre of the Earth, is

$$F = \frac{GMm}{r^2}, \tag{1.49}$$

where the force is directed along a radius towards the centre of the Earth.

We define the weight by:

$$\text{weight} = mg = \frac{GMm}{r^2}, \tag{1.50}$$

where

$$g = \frac{GM}{r^2}. \tag{1.51}$$

We may summarize this result in words by noting that, if the weight of a particle of mass m is mg, it follows that its acceleration would be g, if the particle were allowed to fall freely to Earth.

Evidently from (1.51) the gravitational acceleration depends on the value of r; but in most experiments the variation in height above the ground is likely to be very small compared to the radius of the Earth. So for most purposes we are justified in treating r and hence g as constant. However, we should bear in mind that the Earth is not a perfect sphere, so that the gravitational acceleration measured at sea level varies with latitude, with a variation of about one half of one percent between the equator and the North Pole.

We conclude this discussion of gravitational mass by pointing out that a rigorous distinction can be drawn in principle between two possible forms, viz.,

- **passive gravitational mass** of a particle, which is a measure of its reaction to a gravitational field;
- **active gravitational mass** of a particle, which is a measure of its strength as the source of a gravitational field.

If we postulate that these two kinds of mass are not equal, then we break the symmetry between m and m' in equation (1.22). However, if we consider the interaction between two bodies, it follows from N3 that the first body cannot have a greater effect on the second than the second has on the first. Hence it is really quite clear that active and passive gravitational mass must be the same.

1.4.3 *The principle of equivalence*

The equality of inertial and gravitational mass has long been accepted in physics. Its justification has been the pragmatic one that it has been experimentally verified, beginning with Galileo's famous experiments (although, sadly, it seems that the Leaning Tower of Pisa may not have been involved!). We now examine the implications of this equality.

Just for the moment, let us assume that the equality does *not* hold. We denote a particle's inertial mass by m_I and its gravitational mass by m_G and consider the particle to be falling freely towards Earth. As usual, we ignore air resistance. Then, equating the expressions for the forces acting on the particle, as given by N2 and Newton's law of gravitation, we obtain an expression for the gravitational acceleration of the particle, thus:

$$\ddot{\mathbf{x}} = -\frac{m_G}{m_I}\frac{GM\hat{\mathbf{x}}}{x^2}, \tag{1.52}$$

where M is the mass of the Earth. It follows at once from this expression that if the inertial mass is equal to the gravitational mass then we have $(m_G/m_I) = 1$ and the gravitational acceleration of the particle does not depend on the particle's mass. Of course, this is the underlying justification for the introduction of the concept of weight in the previous section. Accepting the experimental verification that the inertial and gravitational masses are equal is equivalent to choosing the gravitational constant, thus:

$$G = 6.67 \times 10^{-11}\,\mathrm{N\,m^2\,kg^{-2}}. \tag{1.53}$$

If we go further, and assert that the two kinds of mass are actually the same, then we have the formal statement:

The principle of equivalence states that inertial mass equals gravitational mass.

This was put forward by Einstein as a fundamental postulate leading to the general theory of relativity. We shall discuss this further in Chapter 15.

1.5 The space-time continuum

When formulating and solving problems in classical mechanics, we tend to take the concepts of space and time somewhat for granted. We assume that both space and time are continuous and that we can measure intervals of either as precisely as we like. In practice of course we realize that our measurement accuracy is limited by the instruments available to us; but, as these grow ever more refined, we retain the idea that space and time are infinitely divisible (or, what is the same thing, continuous) and that our measurements tend towards realizing—if never quite reaching—the mathematical idealization just mentioned.

However, in the context of classical mechanics, the two concepts of space and time are seen as quite distinct. On the one hand, space is regarded as essentially Euclidean in character and indeed as possessing such geometrical properties for every part of the universe. We talk about the positions of points as being represented by triads of numbers, which are their coordinates when referred to orthogonal coordinate axes in three dimensions. Although Newton stated that the movement of bodies was always specified in relation to the positions and motions of other bodies, he nevertheless believed that this was all within an absolute frame of reference which spanned the entire universe.

On the other hand, time was treated on a different footing in the Galilean–Newtonian system. Unlike space, which is static, time was seen as dynamic and something, which as the poet[4] said, was 'like an ever-rolling stream'. Yet, time was also seen as being, like space, universal in character. Of course if an event happens in the vicinity of Proxima Centauri—the nearest star to the solar system—then it will be some time before the light signal telling us of that event reaches us on Earth. In the context of classical mechanics, we simply interpret this as the news of the event taking time to reach us. However, if two events (separated by one hour, say) happened at Proxima Centauri, then we should expect the associated light signals to arrive on Earth separated by one hour.

All of these very natural and hence intuitively obvious ideas are subject to some revision in the context of both special and general relativity. In order to avoid unnecessary confusion, we shall have to be very precise, even to the point of seeming pedantic, on the question of how these quantities are measured and recorded. We deal with these aspects in the next two sections.

[4]Isaac Watts 1674–1748.

1.5.1 *Measurement of length and time*

We have pointed out that all measurements involve a process of comparison. We compare lengths to some standard length and intervals of time to some standard interval of time. Note, however, the form of words used. We take it for granted that a length is not an absolute length, but rather a distance between two points. In other words, it is natural to think of a length rather than an interval of length, and to speak of an interval of length would seem unnecessarily pedantic.

In the case of time, this is not quite so obvious. In everyday, colloquial speech the question 'What is the time?', with its implications of absoluteness and uniqueness, is heard and answered without thought. Thus when we come to physics, we tend to recognize the need for greater precision in the way we express ourselves. Thus the expression 'time interval' is quite usual in physics, but the expression 'length interval' is not. Admittedly one tends to talk about a 'displacement' but this is not quite the same thing, as it implies that something has actually moved from one place to another.

In fact, just as one does when considering how to calculate experimental error, one should recognize that each and every measurement involves two comparisons. If we wish to measure the length of the distance between two points then we must align both of the points with the scale marks on a measuring rod. There is no escaping this! Even when a measuring instrument has a 'zero', that zero has to be set by a process of comparison.

Let us now extend these ideas to the problem of measuring the length of a moving rod. The question that we have in mind is this: does it matter at what time we make the measurement of the position of each end of the rod on our measuring scale? Obviously, if the rod is at rest with respect to the measuring scale, then one may make the two comparisons at different times. However, if the rod is moving with speed v parallel to the measuring scale, then the time at which each comparison is made does matter. If the time difference between our two comparisons is Δt, then (in classical mechanics) our measurement of the length of the moving rod will be subject to a systematic error of an amount $\pm v\Delta t$, the sign depending on whether we do the measurement of the position of the front of the moving rod first or second.

This might seem like rather a trivial and obvious point; but, later on when we find that in special relativity a moving rod has its length contracted in the direction of motion, it is important to appreciate the need to make simultaneous measurements in any frame other than the reference frame in which the rod is stationary.

Lastly, we should consider who actually makes the measurements. We shall follow standard practice and refer to the person who makes the measurements as 'the observer'. In the interests of precision, we should normally avoid colloquial speech, and always draw a clear distinction between 'seeing' and 'observing'. So far as we are concerned in relativity, the latter term means 'measuring'. Indeed, so crucial is this point that we shall give a formal definition, as follows:[5]

Definition: an **observer** is an experimenter, who is equipped to make measurements of length and time on bodies which are moving, or at rest, relative to his own frame of reference.

It is important to realize that an observer normally has no more information than can be obtained by his or her measurements. The only exceptions to this rule are when information is given as part of a set exercise and we are told that the observer knows some particular fact or relationship between quantities.

We close this section with two further definitions which will be of great importance.

Definition: the **rest frame** of a body or an observer is the frame of reference in which the body or observer is at rest. It is also referred to as the **comoving frame**.

Definition: proper measurements are measurements made on a body which are in the rest frame of that body. Thus, for example, we speak of **proper length** and **proper time**.

1.5.2 *Ideal measuring rods and ideal clocks*

In principle, our standards for the measurement of length and time are based on the wavelength of light of a given colour and the vibrations of atoms—the so-called atomic clock. We therefore can think of our observer as being at rest in his own reference frame, and provided with modern optical and radar methods of measuring velocities, lengths and times. We shall only be more specific about these when it helps a particular discussion to be so.

However, once we encounter special relativity, we shall learn that moving clocks run slow and that moving rods contract in the direction of their motion. We may take account of this effect mathematically and the way in which we do so will be one of the important topics of this book. But, the question then arises: are such measuring instruments affected by acceleration?

In fact the experimental evidence seems to suggest that the fundamental measurements of length and time are unaffected by acceleration, at least to quite a high degree of accuracy. Moreover, in the context of both Galilean and special relativity, acceleration is an absolute and not a relative quantity. This implies that if there were an effect, one could compensate accordingly with some kind of feedback loop, and in effect devise a measuring instrument which would be unaffected by acceleration.

Considerations of this kind have been formalized in the concept of **ideal clocks** and **ideal measuring rods**, which are postulated not to depend on acceleration. A corollary of this idea is the principle that an accelerated observer makes the same local time and distance measurements as an inertial observer momentarily comoving with him. This principle underlies the introduction of the instantaneous comoving frame in Section 1.9.

1.6 Relativity of motion in Newtonian mechanics

1.6.1 Inertial frames of reference

In order to make the above statements, we need to have a way of specifying the position of the particle. That is, we need a frame of reference. In effect, this just means a coordinate system. We shall often refer to such a system as a 'frame'. Of course a coordinate system is really a mathematical abstraction; but in physics, as in everyday life, we will normally choose our frame to be appropriate to the situation under consideration, so that it forms a basis for making and analysing measurements. For example, a surveyor would choose his frame of reference to be fixed to the surface of the Earth, with the origin of coordinates chosen to be at sea level; while a physicist who is interested in (say) how the properties of a fixed volume of gas depend on the behaviour of the constituent molecules would naturally choose a coordinate system fixed to the container. In practice, one would expect that a frame could be chosen without much difficulty; but once we start to consider relativity, we have to be more fastidious. In particular, we must use a frame of reference which does not of itself add anything to the dynamics of the system which we are studying. In order to ensure this, we restrict our choice to that class of coordinate systems in which a body obeys Newton's first law. As the property of a particle which satisfies N1 is known as *inertia*, so a frame of reference in which such a state can be observed is known as an *inertial frame*. This is such an important idea that we now highlight it as an important definition.

Definition: A frame of reference in which a body satisfies Newton's first law is known as an **inertial frame**.

In order to understand the significance of this definition, and how it restricts our choice of frame, it may be helpful to first consider the question of what is *not* an inertial frame. For instance, it is obvious that a rotating coordinate system would not be an inertial frame, as a freely moving particle would appear to travel in a curved path in such a system, and the existence of a curved path implies the existence of a force. So this brings us to a fundamental difficulty. We have to consider the question of how such inertial frames may be realized in practice. After all, when we talk of a coordinate system fixed to the Earth, it is necessary to remember that the Earth rotates and inevitably the result is that we have a rotating system. However, if we want to measure the period of (say) a simple pendulum,[6] then a frame fixed to the surface of the Earth is a perfectly adequate inertial frame, with the effects of the Earth's rotation being quite undetectable, as indeed they are for most of us for most of the time! Let us sum up our position on this important issue. For the purposes of this book, when we use the term inertial frame, we have in mind a mathematical abstraction, which meets the above definition exactly. In general, we shall not worry unduly about how such frames may be realized in practice.

[6]Of course what we have in mind here is a small pendulum with length of order of one metre. In Section 7.2.6, we shall discuss Foucault's pendulum, which is used to demonstrate the rotation of the Earth.

1.6.2 Inertial frames in standard configuration

Let us consider two frames, S and S', as illustrated in Figure 1.4. The frames have the following properties:

(a) They are identical.

(b) S' moves relative to S with speed V.

(c) The motion of the origin of S' is along the x-axis of S.

(d) x and x' are co-axial.

(e) At $t = t' = 0$, the origins O and O' coincide.

This set-up is called the **standard configuration**. It can be extended to any number of frames, $S', S'', S''', \ldots$, with origins moving along the x-axis with speeds $V, V', V'', \ldots$, in S. That is, we may speak of two frames as being in standard configuration or of any number of frames as being in standard configuration. Lastly, from now on we shall restrict our attention to problems which are essentially two dimensional, in that there is no significant variation in the z-direction and accordingly, for simplicity, we shall often omit the z-axis from our coordinate systems.

1.6.3 Galilean transformation

In order to specify an event, we have to say *where* it happened and *when* it happened. Alternatively, we can say that in order to specify an *event*, we specify the position of a point P (say) in the space-time continuum. Then we may consider how the event appears from two different inertial frames of reference, S and S'. As always, from now on, we take the two frames to be in standard configuration. Obviously, there is only the one event, but its coordinates will be different in two different frames of reference, thus:

- in frame S, $P \equiv$ the point (x, y, z, t);
- in frame S', $P \equiv$ the point (x', y', z', t).

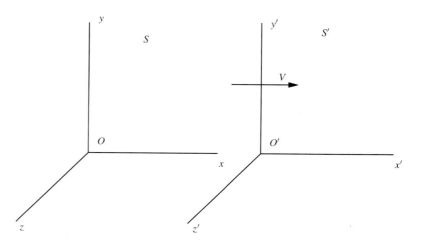

Fig. 1.4 Two inertial frames in standard configuration.

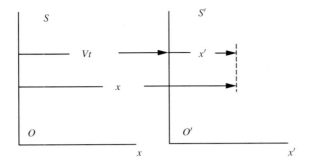

Fig. 1.5 Definition sketch for the Galilean transformation between two frames in standard configuration.

Note that we have implicitly assumed the invariability of time; or that $t = t'$. It seems intuitively obvious that time should be the same for observers in different frames and hardly worth mentioning. Yet, when we encounter special relativity in the second part of this book, we shall find that this is not actually true. This will be only one of the strange things that we shall discover that special relativity tells us about the universe in which we live.

Evidently, in standard configuration, the y and z coordinates of an event are unchanged by the motion of S' relative to S, as this involves motion purely in the x-direction. It follows that these coordinates are related by: $y = y'$ and $z = z'$. But for x and x', the relationship between the two coordinates has to take into account that the two origins coincided at $t = 0$, and thereafter separated at a speed V. Referring to Figure 1.5, we see that the origin of S' is located at $x = Vt$ at time t, and hence the x-direction coordinates are related by

$$Vt + x' = x,$$

or rearranging to make an expression for x'

$$x' = x - Vt.$$

Thus, collecting all our results for coordinate transformations between the two frames, we may summarize the relationship between the two observations of the same event as measured in S and S' as follows:

$$x' = x - Vt, \tag{1.54}$$
$$y' = y, z' = z, t' = t.$$

These are known as the standard Galilean transformations. Note that we give only the first one an equation number, as it is the only one to which we shall make specific reference later.

1.7 Galilean invariance of Newton's laws

Consider Newton's second law for motion in one dimension. We want to show that this takes the same form in any two inertial frames. It will be sufficient to show this for the case of a standard configuration, as the proof can be generalized to any configuration. We begin by writing down N2 in the frame S and then consider how it transforms to the moving frame S'. Thus, in frame S:

$$F_x = m\frac{\mathrm{d}v_x}{\mathrm{d}t} = m\frac{\mathrm{d}^2x}{\mathrm{d}t^2}. \tag{1.55}$$

From (1.54), we have the transformation from S to S' as

$$x' = x - Vt.$$

Now let us differentiate both sides of this transformation equation, with respect to time, with the result:

$$\frac{\mathrm{d}x'}{\mathrm{d}t} = \frac{\mathrm{d}x}{\mathrm{d}t} - V. \tag{1.56}$$

This equation gives us the transformation of speeds between the two frames. We shall come back to this in Section 1.8.1. For the moment we carry on by differentiating both sides of (1.56) with respect to time, and remembering that V is constant, we find

$$\frac{\mathrm{d}^2x'}{\mathrm{d}t^2} = \frac{\mathrm{d}^2x}{\mathrm{d}t^2}, \tag{1.57}$$

and therefore

$$F_x = F'_x,$$

where the last step follows from N2. That is to say, the acceleration is the same in both frames, and this means that if N2 holds in one frame, it must hold in both. Therefore force is unaffected by Galilean transformation. Another way of stating this, is to say that Newton's second law takes the same form in both frames. In order to deduce N2 in S' we just have to take N2 in S and add primes to all the variables. It is convenient to refer to this property as **form invariance**. We can also say that Newton's second law is **Galilean invariant**.

The form invariance of Newton's first law follows from the above, as N1 is a special case of N2. The proof of the form invariance of the third law may also be seen as following trivially from the proof of the Galilean invariance of force. Alternatively, by considering two-body collisions, one can give a more formal proof, and this is deferred until the next chapter where we treat the case of a system of two particles.

1.8 The principle of Galilean relativity

Suppose we do a mechanical experiment in a railway carriage, which is: (a) at rest; and (b) moving at speed V. For instance, we might measure the period of a simple pendulum. In both cases we would get the same result: this follows from the above invariance of Newton's second law. If we use

N2 to work out the equation of motion of the pendulum, this will take exactly the same form in both cases; or, more specifically, in both frames of reference. It is intuitively obvious that this argument can be extended to any mechanical experiment and to any pair of inertial frames. In effect we are saying that the laws of mechanics are the same for all inertial frames. This is the principle of Galilean relativity and it may be stated more formally, as follows:

Principle: The laws of mechanics are Galilean invariant.

Corollary: No mechanical experiment can be used to tell whether an *inertial* frame is moving or at rest (with respect to any other frame).

In the second part of the book we shall see that the principle of special relativity may be obtained from the above principle of Galilean relativity, if we replace the word 'mechanics' by the word 'physics'.

1.8.1 *Relativity of speed*

Two properties of a body do depend on which inertial frame they are measured in. These are its apparent location in the frame, which follows trivially from the coordinate transformation given in (1.54), and its speed relative to the frame, which follows directly when we differentiate both sides of equation (1.54), and obtain (1.56) which connects velocities in the x-direction in both frames. However, the relativity of speed is such an important concept that we shall discuss it here at slightly greater length.

Consider a particle moving at speed c along the x-axis in S. What is its speed c' in S'? Assume that a time t has elapsed so that we now have the situation depicted in Figure 1.6. The particle has travelled a distance ct in S and a distance $c't'$ in S', while S' has travelled a distance Vt in S. From inspection of the figure, it follows therefore that the displacement of the particle in one frame is related to its displacement in the other by

$$ct = Vt + c't.$$

Then, cancelling t across and rearranging, we obtain:

$$c' = c - V.$$

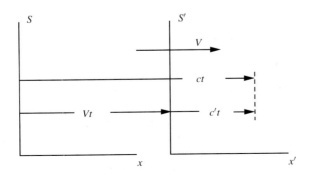

Fig. 1.6 A particle moving with speed c in S moves with speed c' in S'.

Thus, in the *classical* picture, the velocity of *any* process will depend on the frame in which it is measured. Again, as we shall see in the second part of this book, the specification of the principle of special relativity will require us to abandon this dependence of the velocity on the frame in which it is measured for the special and unique case of the speed of light *in vacuo*. Also note that the key element in going from equation (1.58) to (1.59) is the assumption that time is the same in both frames. This principle of a universal time will also have to be abandoned when we go to special relativity.

1.8.2 Example: *Motion of a projectile relative to S and S'*

A projectile in S is launched with speed u at an angle α to the horizontal (x-axis) at time $t = 0$. Its position at any subsequent time t is given in parametric form by the equations:

$$x = (u \cos \alpha)t, \quad y = (u \sin \alpha)t - \frac{1}{2}gt^2.$$

Find the value of t for which the height y is a maximum. Also find the value of t at which the projectile returns to the initial horizontal plane and the value of x at that time.

By using Galilean transformations, obtain the values of these quantities which would be observed by someone at rest in the frame S'.

In S: the condition for the height y to be a maximum is:

$$\frac{dy}{dt} = 0.$$

We are given

$$y = (u \sin \alpha)t - \frac{1}{2}gt^2;$$

and, differentiating with respect to time,

$$\frac{dy}{dt} = u \sin \alpha - gt = 0.$$

Solving this equation gives us the time for the projectile to reach its maximum height as:

$$t_{1/2} = \frac{u \sin \alpha}{g}.$$

The maximum value of x is when $y = 0$ (i.e. the projectile has come down again); thus:

$$t\left(u \sin \alpha - \frac{1}{2}gt\right) = 0,$$

and ignoring the trivial solution $t_0 = 0$, we have

$$t_1 = \frac{2u \sin \alpha}{g} = 2t_{1/2}.$$

Hence

$$x_{\text{max}} = (u \cos \alpha)t_1 = \frac{2u^2 \sin \alpha \cos \alpha}{g} = \frac{u^2 \sin 2\alpha}{g}.$$

In S': the Galilean transformations give us $x' = x - Vt$, $y' = y$, and $t' = t$. Hence:

$$t'_{1/2} = \frac{u \sin \alpha}{g} \quad \text{and} \quad t'_1 = \frac{2u \sin \alpha}{g} \quad \text{while} \quad x'_{\text{max}} = x_{\text{max}} - Vt_1.$$

The last of these can be written more explicitly as:

$$x'_{\text{max}} = \frac{2u^2 \sin \alpha \cos \alpha}{g} - \frac{2uV \sin \alpha}{g}.$$

Note that only the range in the x-direction is affected by the transformation from S to S'. The observer in S' measures the projectile as going a shorter distance.

1.8.3 Example: *Motion of a particle sliding down a slope in S' relative to S*

A particle slides down an inclined plane which is moving horizontally with constant speed V. Find the horizontal position of the particle as a function of time. You may take the initial displacement and velocity of the particle to be zero.

In S, the inclined plane moves with velocity V.

In S', the inclined plane is at rest.

Measure a coordinate ξ down the plane in S', as shown in Figure 1.7.

We begin in S': displacement at time t of particle with mass m is readily obtained from equation (1.10) by substituting $F = mg \sin \theta$, along with

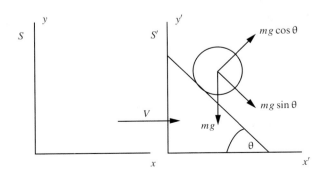

Fig. 1.7 Particle sliding down a moving inclined plane.

the initial conditions, thus:

$$\xi = \frac{1}{m} \int_0^t dt' \int_0^{t'} (mg \sin \theta) \, dt'' = g \sin \theta \int_0^t dt' [t'']_0^{t'}$$
$$= g \sin \theta \int_0^t t' \, dt',$$

and, performing the easy integral, we find for the distance down the slope at any time t,

$$\xi = g \frac{\sin \theta}{2} t^2.$$

Note that by using the dummy variables of integration—t' and t'', which are just intermediate times—we have built the initial conditions into the integrals, and there is therefore no need to evaluate separate constants of integration.

Still in S', we now resolve the displacement along the x-axis:

$$x' = \xi \times \cos \theta = \frac{gt^2}{2} \cos \theta \sin \theta.$$

In S, the Galilean transformation (1.54) gives us

$$x = x' + Vt,$$

and so

$$x = \frac{gt^2}{2} \cos \theta \sin \theta + Vt,$$

is the horizontal position of the particle relative to S at any time t.

1.9 The instantaneous rest frame of a moving body

As we shall see, an important concept in relativity is the *rest frame* of a particle or moving body. We have previously defined this in Section 1.5.1, and it is no more than the frame of reference in which the particle is stationary. When a particle moves at constant velocity then the rest frame is easily seen to belong to the class of inertial frames discussed earlier in Section 1.5.1. However, the question then arises: how do we cope with the case where the particle is accelerating? In order to see the answer to that, we shall consider a simple example.

1.9.1 Example: *A particle undergoing constant linear acceleration*

A particle moves along the x-axis with speed $u(t) = \lambda t$. For the case $\lambda = 1$, sketch the particle's comoving inertial frames S, S', S'' and S''' at times $t = 0, 1, 2$ and 3 seconds.

This is of course a very simple problem. We integrate the given expression for the particle's speed with respect to time and obtain an expression

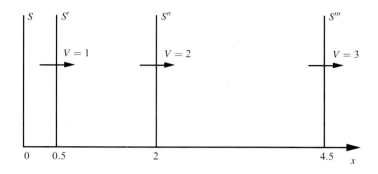

Fig. 1.8 The instantaneous rest frames of an accelerating particle.

for the distance travelled at time t, thus

$$x(t) = \int_0^t u(t')\,\mathrm{d}t' = \lambda \int_0^t t'\,\mathrm{d}t' = \lambda \left[\frac{t'^2}{2}\right]_0^t = 0.5\lambda t^2. \qquad (1.60)$$

For the case $\lambda = 1$ we make a table as follows:

$$
\begin{array}{lll}
t = 0 & u(0) = 0 & x(0) = 0 \\
t = 1 & u(1) = 1 & x(1) = 0.5 \\
t = 2 & u(2) = 2 & x(2) = 2 \\
t = 3 & u(3) = 3 & x(3) = 4.5
\end{array}
$$

The instantaneous rest frames may then be sketched, setting $V = u(1)$, $u(2)$ *etc.* in turn, as shown in Figure 1.8.

In this way we can think of a family of inertial frames $\{S, S', S'', S''', \ldots\}$, with the accelerating particle momentarily at rest in one after another. If we then extend this process by sketching our inertial frames at regular time intervals Δt, then in the limit as $\Delta t \to 0$, the set of frames $\{S, S', S'', S''', \ldots\}$ makes up the **instantaneous** rest frame of the particle at time t. This is also referred to as the **instantaneous comoving frame**, or **ICMF** of a particle.

1.10 Exercises

1.1 A charged particle moves with a constant speed u along the x-axis. If the motion of the particle is unaffected by the combined effects of an electrical field **E** in the direction of the y-axis and a magnetic induction **B** in the direction of the z-axis, show that the speed of the particle must satisfy $u = E/B$.

1.2 Consider an electrical dipole to consist of charges $\pm q$ separated by a distance a and show that equation (1.20) for the potential can be directly evaluated as $\phi = qa\cos\theta/x^2$, where θ is the angle between the vector **x**

and the line of the dipole, and terms of order $(a/x)^2$ have been neglected.

1.3 Consider the gravitational attraction of a uniform, circular ring of radius a and a mass density λ per unit length. Take the ring to lie in the yz-plane and to have its centre at the origin of coordinates. Assume that there is a unit test mass situated on the x-axis, a distance x from the ring and work out how the gravitational potential of the ring varies with x.

1.4 A spherical star of radius R is in static equilibrium, with internal pressure $p(r)$ and density $\rho(r)$. If the density is known to be of the form $\rho = \rho_0(1 - r^2/R^2)$, with ρ_0 constant, show that the pressure at the centre of the star is

$$\frac{13\pi G}{30}\rho_0^2/2R^2.$$

1.5 Assuming that the radius of the Earth is 6371 km, by what percentage would the measured value of the gravitational acceleration g differ between (a) the top of a mountain 5 km high; and (b) the bottom of a mine 5 km deep?

1.6 The length of a rod can be determined by placing it against a scale and aligning the two ends (as closely as possible) with marks on the scale. If the rod is about a metre long and the scale is graduated in millimetres then the probable error is about 0.1%. On the basis of plausible estimates of the time required to make such a comparison with the measuring scale, estimate how this error would be affected if the rod was moving parallel to the scale. Check this out for two speeds, say 1 cm per second and 1 metre per second. What would be the likely effect if one observer had to make both comparisons?

1.7 An inertial frame S' moves with velocity $\mathbf{V}$ with respect to another inertial frame S. Note that the axes of S' do not coincide with those of S, nor does the velocity vector $\mathbf{V}$ pass through the origin of S. With the aid of sketches, identify which spatial rotations and translations of time and space are required to bring the two frames into standard configuration.

1.8 In an inertial frame S, a particle is projected upwards with speed u. Show that it rises to a height of $u^2/2g$ and that it returns to the point of projection in a time $2u/g$.

What would be observed in S', which is an inertial frame in standard configuration with S and moving with speed V relative to S?

1.9 Verify that the Newtonian relationship between kinetic energy and momentum,

$$E = p^2/2m,$$

is Galilean invariant.

1.10 The one-dimensional equation of fluid motion may be written in the form

$$\frac{\partial u(x, t)}{\partial t} + u(x, t)\frac{\partial u(x, t)}{\partial x} = f(x, t),$$

where $u(x, t)$ is the velocity field of the fluid and $f(x, t)$ is the resultant of all forces (both internal and external) acting on unit mass of fluid. Show that this equation is invariant under Galilean transformation.
[Hint: remember that $u(x, t)$ is a continuous function of both variables. Also, you may assume that $f(x, t)$ transforms as an invariant.]

1.11 A boat is travelling upstream on a river which flows with speed V, and as it passes under a bridge it drops a marker buoy. After continuing upstream for a further 15 minutes, the boat turns (effectively, without loss of time) and travels downstream with the same speed, relative to the water, as before. When it reaches the marker buoy, this has floated downstream to a point 1 km from the bridge. What is the speed V of the water?
[Hint: Taking S to be the frame fixed to the river-bank, and with its origin at the bridge, and S' the frame moving with the river, and with its origin at the buoy, work in S' and transform the answer to S.]

Conservation laws

2

In the previous chapter, we pointed out that N1 was different from N2 in that it was qualitative in nature rather than quantitative. If we wished to regard it as being quantitative, then it could be seen as the null member of the set of quantitative statements about the momentum of a body: that is, the change in momentum is zero if no forces act.

Such a statement may be seen as really quite profound. It simply says: without a cause, there is no effect. Or, putting it in a rather colloquial way: you do not get something for nothing! Such a statement is likely to be true under all circumstances and for this reason we can expect N1 to be true even when we go over to Einstein's relativity in the second part of this book. There would surely be something wrong with any theory which allowed bodies to start into motion when there is no net force acting on them.

Where the motion of a body is controlled by Newton's laws, these can be re-expressed in terms of three *conservation laws*: viz., momentum, energy and angular momentum. Of course we all know that the main use of these laws is to provide an easy way of solving problems in mechanics! However, their profound importance is increasingly being recognized—particularly in the subject of quantum mechanics, where an early development was the relation between symmetry and conservation. Nowadays, this type of thinking is influencing other fields of physics, including classical mechanics, and is encapsulated in Noether's theorem, which may be stated as follows:

Noether's theorem: for each continuous symmetry which a system possesses there is a conserved quantity.

The term 'continuous symmetry' here implies that the operation which establishes the symmetry is continuous. For example, a continuous translation in space or time; or a continuous rotation: each of these would qualify. On the other hand, the symmetry of a cube under reflections in planes or rotations about axes would not qualify. The elements in such a symmetry group are denumerable, or countable, and hence the group is not continuous. To go any further into Noether's theorem would take us far beyond the scope of this book,[1] but we should note here the fundamental correspondences:

- Invariance under translation in space implies conservation of momentum.

[1] For the interested reader, a good account of Noether's theorem will be found in the book *Classical Mechanics* (2nd ed.) by H. Goldstein (Addison-Wesley, New York, 1980).

- Invariance under translation in time implies conservation of energy.
- Invariance under continuous rotations implies conservation of angular motion.

The main purpose of this chapter is to show how these principles are consistent with Newton's laws, and to illustrate some aspects of their use in applications.

2.1 Conservation of linear momentum

As we saw in the previous chapter, we can solve the equation of motion,

$$\mathbf{F} = \frac{d\mathbf{p}}{dt} = \frac{d}{dt}\left(m\frac{d\mathbf{x}}{dt}\right), \tag{2.1}$$

in principle, by integrating twice and using the initial conditions to fix the constants of integration. In this way, we get the trajectory of the particle:

$$\mathbf{x} = \mathbf{x}(t). \tag{2.2}$$

But in practice we are often only interested in initial and final states, and it can be easier to use conservation laws.

We take the first of these to be the conservation of linear momentum, as this is the one most readily derived from Newton's laws. If $\mathbf{F} = 0$, then N2 becomes

$$\frac{d\mathbf{p}}{dt} = 0.$$

Hence a trivial integration gives:

$$\mathbf{p} = m\mathbf{v} = \text{constant.} \tag{2.3}$$

That is, the momentum of a particle is constant in the absence of an external force. (Or, in other words, N1.)

2.2 Conservation of energy

Before beginning on this topic we should point out that the energy we are referring to here is *mechanical* energy. That is, either kinetic or potential energy; and, in most cases, the sum of the two. We are not concerned here with the general proposition that energy cannot be either created or destroyed. If we have to consider cases where energy is not conserved then we mean that it is lost from the mechanical system under consideration by the generation of heat or sound or by some other means.

In this section, we show that the sum of the kinetic energy T and the potential energy U is constant for a particle which is subject to only conservative forces (as defined and discussed in Section 1.2.1). We prove the relationship:

$$T + U = E \equiv \text{constant}. \qquad (2.4)$$

First, consider the work done by a force $\mathbf{F}$ in moving a particle from point 1 to point 2, using the expression given in equation (1.4):

$$W_{12} = \int_1^2 \mathbf{F} \cdot d\mathbf{x} = \int_1^2 \frac{d}{dt}(m\mathbf{v}) \cdot d\mathbf{x}, \qquad (2.5)$$

where the second step follows by N2. Hence

$$W_{12} = m \int_1^2 \frac{d\mathbf{v}}{dt} \cdot d\mathbf{x}, \qquad \text{(for } m = \text{constant)}$$

$$= m \int_1^2 \frac{d\mathbf{v}}{dt} \cdot \frac{d\mathbf{x}}{dt} \, dt, \qquad \text{(change of variable)}$$

$$= m \int_1^2 \left(\frac{d\mathbf{v}}{dt} \cdot \mathbf{v} \right) dt, \qquad \text{(i.e. using } d\mathbf{x}/dt = \mathbf{v}\text{)}$$

$$= m \int_1^2 \frac{1}{2} d(v^2), \qquad \text{(change of variable)}$$

and finally

$$W_{12} = \frac{1}{2} m(v_2^2 - v_1^2) = T_2 - T_1. \qquad (2.6)$$

Thus we have derived an expression for the change of kinetic energy of a particle due to the work done on it by an applied force.

Second, if $\mathbf{F}$ is *conservative*, it can be derived from a scalar potential U. For example, in one dimension, equation (1.14) would take the form:

$$F_x = -\frac{\partial U}{\partial x}.$$

Hence in three dimensions:

$$W_{12} = \int_1^2 \mathbf{F} \cdot d\mathbf{x} = -\int_1^2 \left(\frac{\partial U}{\partial x} dx + \frac{\partial U}{\partial y} dy + \frac{\partial U}{\partial z} dz \right),$$

and so

$$W_{12} = -\int_1^2 dU = U_1 - U_2. \qquad (2.7)$$

That is, we now also have an expression for the change in the potential energy of a particle due to the work done on it by an applied force.

Then conservation of energy follows when we equate (2.6) and (2.7) for the work done, thus

$$W_{12} = T_2 - T_1 = U_1 - U_2,$$

or

$$T_1 + U_1 = T_2 + U_2 = \text{constant} \equiv E. \tag{2.8}$$

In other words, we have the familiar result that the particle moves in such a way that the sum of its kinetic energy and potential energy is constant.

2.2.1 Graphical treatment of potentials

We can visualize the idea of a potential in a rather simple way by considering the example of a ball in a bowl. This constitutes a two-dimensional potential well and the steeper the sides (i.e. the greater the gradient) the greater the force acting on the ball.

Our sketches in Figure 2.1 are cross-sections. Another way of putting it is they show the motion of the ball projected on one dimension; or, alternatively, they may be thought of as one-dimensional potential wells.

Let us take the coordinate x to be measured in the horizontal direction, and consider this type of potential (i.e. symmetrical in x and concave upwards). Then we may usefully expand it in a Maclaurin series in powers of x about $x = 0$:

$$U(x) = U(0) + U'(0)x + \frac{1}{2} U''(0)x^2 + 0(x^3),$$

where the primes denote differentiation with respect to x. It is convenient to choose the origin of coordinates to be at the lowest point of the 'bowl', so that we have $U(0) = 0$. It also follows that $U'(0) = 0$, in view of the minimum value.

For small values of x, we can neglect x^3 (and higher-order terms) to write

$$U(x) \simeq \frac{1}{2} U''(0)x^2.$$

It is usual to write the constant $U''(0) = \kappa$ and the result

$$U(x) = \frac{1}{2} \kappa x^2, \tag{2.9}$$

is known as the harmonic potential. We shall discuss this form of potential further in Chapter 4.

Before turning to some examples, we should note that the motion of the ball in the bowl will be unaffected by the height of the bowl above some arbitrary reference level. We have explicitly recognized this fact

Fig. 2.1 Sketches to show a ball in a bowl as an example of a particle in a potential.

already in the above discussion when we chose our coordinate system to be such that $U(0) = 0$. In fact this arbitrariness is a general feature of this type of potential (i.e. scalar potentials) and in practice is eliminated by choosing some convenient reference level.

In the next chapter, when we consider central forces, we shall be dealing with a force $f(r)$ (say) which depends only on the radial distance of the particle from the origin, and acts along that radius. Under these circumstances it is convenient to define the potential for a conservative force $f(r)$ by

$$U(r) = - \int_{r_0}^{r} f(r') \, dr', \qquad (2.10)$$

where r_0 is the reference level. In the following examples, we see that the appropriate reference level can be zero at either $r = 0$ or $r = \infty$, depending on the shape of the potential.

2.2.2 Example: *Find the potentials for specified force fields*

Find the potentials corresponding to the central force fields defined by (a) $\mathbf{F} = -\kappa \mathbf{e}_r / r^3$; and (b) $\mathbf{F} = \kappa r \mathbf{e}_r$.

(a) $\mathbf{F} = -\kappa r^3 \mathbf{e}_r$.

$$U(r) = - \int_{r_0}^{r} \mathbf{F}(r') \cdot d\mathbf{r}' = - \int_{r_0}^{r} \left(\frac{-\kappa}{(r')^3} \right) dr' = \kappa \int_{r_0}^{r} (r')^{-3} \, dr'$$

$$= \kappa \left[\frac{(r')^{-2}}{-2} \right]_{r_0}^{r} = \frac{-\kappa}{2} \left[r^{-2} - r_0^{-2} \right].$$

Thus

$$U(r) = \frac{-\kappa}{2r^2},$$

for $r_0 \to \infty$.

(b) $\mathbf{F} = \kappa r \mathbf{e}_r$.

$$U(r) = - \int_{r_0}^{r} \mathbf{F} \cdot d\mathbf{r}' = -\kappa \int_{r_0}^{r} r' \, dr' = -\kappa \left[\frac{(r')^2}{2} \right]_{r_0}^{r}$$

$$= -\frac{1}{2} \kappa r^2 + \frac{1}{2} \kappa r_0^2.$$

Thus

$$U(r) = -\frac{1}{2} \kappa r^2,$$

as $r_0 \to 0$.

2.2.3 Example: *Time taken for a particle to move a given distance under the influence of a potential field*

A particle of mass m in S is constrained to move along the x-axis under the influence of a force which is due to a potential field $U(x)$. (a) Show that the time taken for the particle to go from x_1 to x_2 is

$$t_2 - t_1 = \int_{x_1}^{x_2} \frac{\mathrm{d}x}{\sqrt{[2E/m - 2U/m]}},$$

where E is the total energy of the particle. (b) What would be the time as recorded by an observer in S', if S and S' are the usual two inertial frames in standard configuration?

(a) For a particle of mass m and speed v, conservation of energy gives:

$$\frac{1}{2}mv^2 + U(x) = E \implies v = \frac{\mathrm{d}x}{\mathrm{d}t} = \sqrt{\frac{2}{m}[E - U(x)]}.$$

Separating the variables and rearranging gives

$$\frac{\mathrm{d}x}{\sqrt{(2/m)[E - U(x)]}} = \mathrm{d}t,$$

and integrating yields

$$t_2 - t_1 = \int_{x_1}^{x_2} \frac{\mathrm{d}x}{\sqrt{(2/m)[E - U(x)]}}.$$

(b) The same time is measured by an observer in S'.

2.2.4 Example: *Elastic collision of two particles*

A body of mass m_1 moves along the x-axis in S, with a constant velocity v_{1i}. It collides with a mass m_2, which is at rest in S. (a) Show that the velocities of the two masses after the collision are given by:

$$v_{1f} = \frac{m_1 - m_2}{m_1 + m_2}v_{1i} \quad \text{and} \quad v_{2f} = \frac{2m_1}{m_1 + m_2}v_{1i},$$

respectively.

(b) Transform these results into the frame S'. Do they agree with the results which you would obtain by performing the same experiment in S'? If not, why not?

(a) Invoke the principles of conservation of momentum and energy:

Conservation of momentum:

$$m_1 v_{1i} = m_1 u_{1f} + m_2 v_{2f}. \tag{2.11}$$

Conservation of energy:

$$\frac{1}{2}m_1 v_{1i}^2 = \frac{1}{2}m_1 v_{1f}^2 + \frac{1}{2}m_2 v_{2f}^2.$$ (2.12)

Given m_1, m_2 and v_{1i}, find v_{1f} and v_{2f}:

$$(2.11) \implies m_1(v_{1i} - v_{1f}) = m_2 v_{2f}$$ (2.13)

$$(2.12) \implies m_1(v_{1i} - v_{1f})(v_{1i} + v_{1f}) = m_2 v_{2f}^2.$$ (2.14)

Divide (2.13) into (2.14) to get

$$v_{1i} + v_{1f} = v_{2f}.$$ (2.15)

Now substitute from (2.15) for v_{1f} into (2.13) and solve for v_{2f}:

$$v_{2f} = \frac{2m_1}{m_1 + m_2}v_{1i} \qquad \text{and similarly} \qquad v_{1f} = \frac{m_1 - m_2}{m_1 + m_2}v_{1i}.$$

(b) Transform to S':

$$v_{2f}' = \frac{2m_1}{m_1 + m_2}(v_{1i}' + V) - V,$$

and similarly for v_{1f}'.

This is not the same as performing the same experiment in S', as in that case $v_{2i}' \neq 0$.

2.2.5 Example: *Simple pendulum oscillating with finite amplitude*

Show that the period τ of a simple pendulum of length L, oscillating with angular amplitude θ_{MAX} (not assumed small), is

$$\tau = \sqrt{\frac{L}{g}} \int_0^{\theta_{\text{MAX}}} \frac{d\theta}{[\cos\theta - \cos\theta_{\text{MAX}}]^{1/2}}$$

$$= \frac{\tau_0}{2\pi} \int_0^{\theta_{\text{MAX}}} \frac{d\theta}{[\cos\theta - \cos\theta_{\text{MAX}}]^{1/2}},$$

where τ_0 is the period of the same pendulum when the oscillations are restricted to small amplitudes. [Hint: Use energy conservation to express $\dot{\theta}$ in terms of θ.]

As suggested, we invoke the principle of conservation of energy. Referring to Figure 2.2 for the definitions of symbols, this takes the form:

$$\frac{1}{2}m(L\dot{\theta})^2 + mgL(1 - \cos\theta) = E,$$

where m is the mass of the bob and the first term on the left-hand side comes from the general result that the linear speed of the bob may be written in terms of its angular speed as $V = r\dot{\theta}$, where in this case $r = L$.

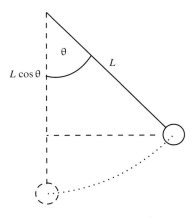

Fig. 2.2 Simple pendulum oscillating with finite amplitude.

Now divide across by Lm and rearrange:

$$\frac{1}{2}L\dot{\theta}^2 = C - g(1 - \cos\theta); \qquad \text{where} \quad C = E/Lm.$$

Hence

$$\dot{\theta}^2 = \frac{2}{L}[C - g(1 - \cos\theta)].$$

But $\dot{\theta} = 0$ at $\theta = \theta_{\text{MAX}}$ and therefore $C = g(1 - \cos\theta_{\text{MAX}})$. Substitute for C and take the square root of both sides:

$$\dot{\theta} = \sqrt{\frac{2g}{L}[\cos\theta - \cos\theta_{\text{MAX}}]} \equiv \frac{d\theta}{dt}.$$

Separate variables and integrate over one quarter-period:

$$\int_0^{\tau/4} dt = \int_0^{\theta_{\text{MAX}}} \frac{d\theta}{\sqrt{(2g/L)[\cos\theta - \cos\theta_{\text{MAX}}]}};$$

thus

$$\tau = 4\sqrt{\frac{L}{2g}} \int_0^{\theta_{\text{MAX}}} \frac{d\theta}{[\cos\theta - \cos\theta_{\text{MAX}}]^{1/2}}$$

$$= \sqrt{\frac{L}{g}} \int_0^{\theta_{\text{MAX}}} \frac{d\theta}{[\cos\theta - \cos\theta_{\text{MAX}}]^{1/2}}.$$

From Section 4.2.2 we have the result for the period of a simple pendulum undergoing SHM. The actual result given is for the natural frequency ω; but, recalling that this is related to the period by $\omega = 2\pi/\tau_0$, we have $\tau_0 = 2\pi\sqrt{L/g}$, and with the substitution of this in the above equation, the required result for τ follows immediately.

It is of interest to note that this expression shows that the period depends on the amplitude θ_{MAX}, in contrast to the case of small oscillations. It is easily shown that this result reduces to the correct form for small oscillations and this is left as an exercise. [Hint: take θ_{MAX} to be small enough for the cosine to be approximated by $\cos\theta_{\text{MAX}} = 1 - \theta_{\text{MAX}}^2/2$ and use the standard form: $\int_0^a dx/\sqrt{a^2 - x^2} = \pi/2$.]

2.3 Angular momentum

In many situations, ranging from the orbits of planets to the motion of spinning tops, we have to consider the motion of a particle relative to a fixed point. We term this *angular motion*, as distinct from *linear motion*, and accordingly we are interested in the moments of the relevant aspects of linear motion about the fixed point. Thus the moment of the linear momentum (i.e. the angular momentum) and the moment of the applied force (i.e. the applied torque) become the relevant quantities.

2.3.1 *Analogous form of N2 for angular motion*

Under these circumstances, we need a generalization of N2 to angular motion. We begin with the angular momentum $\mathbf{l}$ (of a particle) about the point O. This is defined as:

$$\mathbf{l} = \mathbf{x} \times \mathbf{p}, \tag{2.16}$$

where $\mathbf{x}$ is the vector giving the position of the particle at any time relative to O and $\mathbf{p}$ is its linear momentum.

We may define the torque $\boldsymbol{\Gamma}$ as the moment of the force $\mathbf{F}$ acting on the particle about the fixed point O, thus:

$$\boldsymbol{\Gamma} = \mathbf{x} \times \mathbf{F}. \tag{2.17}$$

Then the rate of change of $\mathbf{l}$ with time may be related to the torque. Differentiating both sides of equation (2.16) with respect to time, we obtain

$$\frac{d\mathbf{l}}{dt} = \dot{\mathbf{x}} \times \mathbf{p} + \mathbf{x} \times \dot{\mathbf{p}}. \tag{2.18}$$

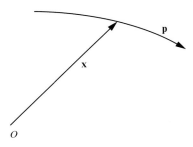

Fig. 2.3 Motion of a particle relative to a fixed point O.

Now the first term on the right-hand side of (2.18) may be shown to be zero, thus:

$$\dot{\mathbf{x}} \times \mathbf{p} = \mathbf{v} \times m\mathbf{v} = 0 \quad \text{(by vector properties)};$$

and, the second term becomes

$$\mathbf{x} \times \dot{\mathbf{p}} = \mathbf{x} \times \mathbf{F} \ \text{(by N2)} = \boldsymbol{\Gamma} \quad \text{(by definition)}.$$

Therefore equation (2.18) becomes:

$$\frac{d\mathbf{l}}{dt} = \boldsymbol{\Gamma}. \tag{2.19}$$

This result is the analogous form of N2 for angular motion.

2.3.2 *N1 for angular motion: conservation of angular momentum*

If the torque $\boldsymbol{\Gamma}$ is zero, then the angular momentum $\mathbf{l}$ of the particle is constant. That is, for zero applied torque, angular momentum is conserved (e.g. as in planetary motion). Thus for angular motion, conservation of angular momentum is the analogue of N1 for linear momentum.

It is perhaps worth emphasizing that this should not be interpreted as a proof or derivation of the principle of conservation of angular momentum from N2, or its generalization to angular motion, in the form of equation (2.19). Instead, as we have already pointed out in Chapter 1, the conservation principles should be regarded as more fundamental than N2 or its generalizations, and are a consequence of underlying symmetries. Accordingly, just as N1 should be seen as a **null principle**, so also should its angular analogue, which is the principle of conservation of angular momentum. As with linear momentum and energy, we are engaged principally in a demonstration that the conservation principle is consistent with N2. Later on, when we go over to special relativity, we shall have to abandon N2, and one consequence will be that we shall also have to reconsider the form taken by the principles of conservation of momentum and energy.

2.3.3 *Angular momentum of a particle relative to S and S'*

As always in this book (unless otherwise stated), we take S and S' to be inertial frames in standard configuration, as defined in Section 1.6.2, with S' moving at speed V relative to S.

In S we have $\mathbf{l} = \mathbf{x} \times \mathbf{p}$ where $\mathbf{x} \equiv (x_1, x_2, x_3)$ and $\mathbf{p} \equiv m(u_1, u_2, u_3)$. By definition

$$\mathbf{l} = \mathbf{x} \times \mathbf{p}$$
$$= \mathbf{i}m(x_2 u_3 - x_3 u_2) - \mathbf{j}m(x_1 u_3 - x_3 u_1) + \mathbf{k}m(x_1 u_2 - x_2 u_1).$$

In S', we have

$$\mathbf{l}' = \mathbf{x}' \times \mathbf{p}'$$
$$= (x_1 - Vt, x_2, x_3) \times m(u_1 - V, u_2, u_3),$$

where we have used the Galilean transformations of the components of $\mathbf{x}'$ and $\mathbf{p}'$ to re-express the angular momentum in S' in terms of the variables defined in S. Therefore

$$\mathbf{l}' = \mathbf{i}m(x_2 u_3 - x_3 u_2) - \mathbf{j}m([x_1 - Vt]u_3 - x_3[u_1 - V])$$
$$+ \mathbf{k}m([x_1 - Vt]u_2 - x_2[u_1 - V]),$$

and finally substituting in terms of the components of $\mathbf{l}$, we obtain

$$\mathbf{l}' = (l_1, l_2 + mV[u_3 t - x_3], l_3 - mV[u_2 t - x_2]),$$

as the Galilean transform of the angular momentum.

2.3.4 Example: *Angular momentum of a particle with impact parameter b incident from infinity on a centre of force in S and S'*

A particle of mass m moving parallel to the x_1-axis in S approaches the origin from $x_1 = -\infty$ with constant speed U_∞ and $x_2 = b$, where b and U_∞ are constants. Without loss of generality we can take $x_3 = 0$. What are the particle's initial kinetic energy and angular momentum? What values of these quantities would be measured in S'?

In S the initial kinetic energy is obvious:

$$T = \frac{1}{2}mU_\infty^2.$$

In order to obtain an expression for the initial angular momentum, we start with the general definition

$$\mathbf{l} = \mathbf{x} \times m\mathbf{u}.$$

Then we can reduce the individual components to their specific values, as follows.

For the velocity vector $\mathbf{u} \equiv (u_1, u_2, u_3)$ we have $u_1 = U_\infty$ and $u_2 = u_3 = 0$; while for the position vector $\mathbf{x} \equiv (x_1, x_2, x_3)$ we have $x_1 = -\infty$, $x_2 = b$, and $x_3 = 0$. Thus the general expression for $\mathbf{l}$ reduces to

$$l_3\mathbf{k} = \mathbf{k}m(0 - bU_\infty),$$

and so the initial angular momentum of the particle in S is

$$l_3 = -mbU_\infty.$$

In S'

$$T' = \frac{1}{2}mU'^2_\infty = \frac{1}{2}m(U_\infty - V)^2,$$

i.e. the particle has less kinetic energy relative to S'.

We can get $\mathbf{l}'$ from the results of the previous section:

$$l_1' = l_1 = 0$$
$$l_2' = l_2 + mV[u_3 t - x_3] = 0 - 0 = 0$$
$$l_3' = l_3 - mV[u_2 t - x_2] = mbU_\infty - mV[0 - b] = -mbU_\infty + mbV,$$

hence

$$l_3' = -mb[U_\infty - V].$$

These results will be useful when we discuss the subject of collisions in Chapter 9.

2.4 Two-particle systems

This topic is also known as the 'two-body problem'. For simplicity, we do a one-dimensional treatment. This allows us to introduce important concepts such as centroid and relative coordinates, and the reduced mass, which will be needed later on. Thus this section may be seen as a step on the way to: (a) systems of N particles, where N is usually large; and (b) a general treatment of orbital motion, including collisions and scattering.

2.4.1 *Momentum*

We consider two particles moving in the same straight line and examine the conservation of momentum. The force equation for each particle is as follows:

Particle 1:

$$\frac{d}{dt}(m_1 v_1) = F_1 + F_{12} \quad \text{(from N2),} \tag{2.20}$$

where F_1 is the external force acting on m_1 and F_{12} is the internal force acting on m_1 due to m_2.

Particle 2:

$$\frac{d}{dt}(m_2 v_2) = F_2 + F_{21} \quad \text{(from N2),} \tag{2.21}$$

where F_2 is the external force acting on m_2 and F_{21} is the internal force acting on m_2 due to m_1. **Note** the order of indices in the two cases!

Next, we add together the two equations of motion:

$$\frac{d}{dt}(m_1 v_1 + m_2 v_2) = F_1 + F_2 + F_{12} + F_{21}. \tag{2.22}$$

But

$$F_{12} = -F_{21} \quad \text{(by N3),} \tag{2.23}$$

and so equation (2.22) becomes

$$\frac{d}{dt}(m_1 v_1 + m_2 v_2) = F_1 + F_2. \tag{2.24}$$

Therefore the total momentum of the system changes due to the total external force: it is unaffected by mutual interaction.

2.4.2 *Centre of mass (CM) and relative coordinates*

We introduce the centre of mass of the two particles as follows. Let x_1 and x_2 be the positions of m_1 and m_2; and let R be the position of the

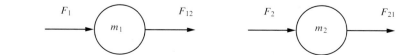

Fig. 2.4 Two particles moving in the same straight line.

centre of mass. Then, the definition of a centre of mass[2] takes the form

$$(m_1 + m_2)R = (m_1 x_1 + m_2 x_2). \tag{2.25}$$

We can extend the idea of centre of mass coordinates to the velocities as follows:

$$m_1 v_1 + m_2 v_2 = \frac{\mathrm{d}}{\mathrm{d}t}(m_1 x_1 + m_2 x_2) = (m_1 + m_2)\frac{\mathrm{d}R}{\mathrm{d}t}$$
$$= (m_1 + m_2)\bar{v}, \tag{2.26}$$

where the second equality follows from equation (2.25) and the third equality defines $\bar{v} = \mathrm{d}R/\mathrm{d}t$ as the velocity of the centre of mass. Therefore, assuming that the total mass of the system is constant, N2 and N3 can be written in terms of the centre of mass (or CM) coordinate as:

$$(m_1 + m_2)\frac{\mathrm{d}^2 R}{\mathrm{d}t^2} = (m_1 + m_2)\frac{\mathrm{d}\bar{v}}{\mathrm{d}t} = F_1 + F_2. \tag{2.27}$$

That is, the centre of mass moves as if the whole system mass were concentrated there and acted upon by the total external force.

We shall also find it helpful to work in terms of relative coordinates, which are the positions of the particles of the system when measured relative to the centre of mass, rather than the origin of coordinates. We begin by writing:

$$x_1 = R + r_1, \qquad x_2 = R + r_2, \tag{2.28}$$

where r_1 and r_2 are the respective positions of the two particles relative to the centre of mass. From the definition of the centre of mass:

$$m_1 r_1 + m_2 r_2 = 0. \tag{2.29}$$

We can express the velocities of the two particles relative to a CM velocity. Differentiate both equations making up (2.28) to get

$$\dot{x}_1 = v_1 = \dot{R} + \dot{r}_1, \quad \dot{x}_2 = v_2 = \dot{R} + \dot{r}_2. \tag{2.30}$$

Thus if we specify the velocity of the centre of mass relative to some frame S, we can regard the centre-of-mass frame as S', moving with $V = \dot{R}$. Then particle velocities can be specified in S' and subsequently transformed to S by Galilean transformation.

2.4.3 *Kinetic and potential energy*

By definition the kinetic energy of the two particles is

$$T = \frac{1}{2}m_1 v_1^2 + \frac{1}{2}m_2 v_2^2 = \frac{1}{2}m_1(\dot{R} + \dot{r}_1)^2 + \frac{1}{2}m_2(\dot{R} + \dot{r}_2)^2, \tag{2.31}$$

[2]We assume that the reader is already familiar with the idea of a centre of mass. Although we shall give the subject more attention in Chapter 5, for the present it should be noted that equation (2.25) expresses the notion that the sum of the moments of the individual masses about the origin is the same as the moment of the total mass concentrated at a fictitious point which we call the centre of mass.

where we have substituted from (2.30) for the particle velocities. Expanding out the brackets gives

$$T = \frac{1}{2}(m_1 + m_2)\dot{R}^2 + \frac{1}{2}m_1\dot{r}_1^2 + \frac{1}{2}m_2\dot{r}_2^2 + (m_1\dot{r}_1\dot{R} + m_2\dot{r}_2\dot{R}). \quad (2.32)$$

But the last term on the right-hand side of equation (2.32) reduces to:

$$\dot{R}(m_1\dot{r}_1 + m_2\dot{r}_2) = \dot{R}\frac{d}{dt}(m_1 r_1 + m_2 r_2) = 0, \quad (2.33)$$

where we have used equation (2.29). So the kinetic energy of two particles may be written as

$$T = \frac{1}{2}(m_1 + m_2)\dot{R}^2 + \frac{1}{2}m_1\dot{r}_1^2 + \frac{1}{2}m_2\dot{r}_2^2, \quad (2.34)$$

where $\frac{1}{2}(m_1 + m_2)\dot{R}^2$ is the kinetic energy of the total mass moving at the speed of the centre of mass and $\frac{1}{2}m_1\dot{r}_1^2 + \frac{1}{2}m_2\dot{r}_2^2$ is the sum of individual kinetic energies of motion relative to the centre of mass.

Alternatively, in S'—the centre-of-mass frame—the system has kinetic energy

$$T' = \frac{1}{2}m_1\dot{r}_1^2 + \frac{1}{2}m_2\dot{r}_2^2,$$

which is less than in S by an amount

$$\frac{1}{2}(m_1 + m_2)V^2 = \frac{1}{2}(m_1 + m_2)\dot{R}^2.$$

Turning now to the potential energy of the two particles, we restrict our attention to *conservative* external and internal forces and analyse as if for a single particle. The various forces acting may then be written as

$$F_1 = -\frac{d}{dx}U_1 \quad \text{and} \quad F_{12} = -\frac{d}{dx}U_{12} \quad (2.35)$$

$$F_2 = -\frac{d}{dx}U_2 \quad \text{and} \quad F_{21} = -\frac{d}{dx}U_{21}. \quad (2.36)$$

The total potential energy for the two particles is obtained quite simply by addition of the individual potential energies, viz.,

$$U = U_1 + U_2 + U_{12} \quad (2.37)$$

or

$$U = U_1 + U_2 + U_{21}, \quad (2.38)$$

or, in a more symmetric form,

$$U = U_1 + U_2 + \frac{1}{2}(U_{12} + U_{21}). \qquad (2.39)$$

Note that, whichever form we use, we do not count the mutual inter-action twice!

2.4.4 The reduced mass and equivalent single-body problem

In equation (2.27) we have an equation of motion for the CM coordinate of the system. Where there is relative motion, we are interested in the acceleration of the relative coordinate $r = r_1 - r_2$. We can obtain the equation for such an acceleration from (2.20) and (2.21). As, for the present, we are concentrating on the relative motion of the two particles, we can consider a problem without external forces and with only the interaction force between the two bodies. We rewrite equations (2.20) and (2.21) with the external forces F_1 and F_2 set to zero. We also divide across by the masses and express the velocities in terms of the coordinates relative to the centre of mass, thus:

$$\ddot{r}_1 = F_{12}/m_1 \qquad (2.40)$$

$$\ddot{r}_2 = F_{21}/m_2. \qquad (2.41)$$

Then we subtract the second of these from the first, and using $F_{12} = -F_{21}$, we obtain

$$\ddot{r} = \ddot{r}_1 - \ddot{r}_2 = \frac{F_{12}}{m_1} + \frac{F_{12}}{m_2} = \left(\frac{m_1 + m_2}{m_1 m_2}\right) F_{12}. \qquad (2.42)$$

Now we divide across by the term involving the masses, and renaming the new bracketed term as μ, we have

$$\mu\ddot{r} = F_{12}, \qquad (2.43)$$

where

$$\mu = \frac{m_1 m_2}{m_1 + m_2}, \qquad (2.44)$$

and is known as the **reduced mass**.

Two points should be noted about this.

- First, and most important, equation (2.43) is equivalent to N2 for a single body of mass μ moving under the action of a force F_{12}. This is known as the **equivalent single-body problem**.

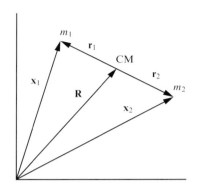

Fig. 2.5 Sketch showing the CM and relative coordinates for two particles of mass m_1 and m_2.

- Second, the reduced mass is less than either individual mass. This is easily shown and the reader should verify it.

2.4.5 Example: *Conservation of energy of two particles in terms of the reduced mass*

Two mass points m_1 and m_2 form a conservative system where the only forces are those due to an interaction potential U. If U is any function of the distance r between the two particles, show that the law of conservation of energy takes the form

$$U(r) + \frac{1}{2}\mu\dot{r}^2 = \text{constant},$$

where $\mu = m_1m_2/(m_1 + m_2)$ is the reduced mass.

Start by writing down the equation expressing conservation of energy, for the particular case of two particles. Referring to Figure 2.5, the relative position vector $\mathbf{r}$ is given by:

$$\mathbf{r} = \mathbf{r}_1 - \mathbf{r}_2 = \mathbf{x}_1 - \mathbf{x}_2.$$

For $\mathbf{r}_1$, $\mathbf{r}_2$ measured from the centre of mass, from (2.29) we have:

$$m_1\mathbf{r}_1 + m_2\mathbf{r}_2 = 0.$$

Substituting $\mathbf{r}_2 = \mathbf{r}_1 - \mathbf{r}$ into this, we get

$$m_1\mathbf{r}_1 + m_2\mathbf{r}_1 - m_2\mathbf{r} = 0$$

$$\implies \mathbf{r}_1 = \frac{m_2}{m_1 + m_2}\mathbf{r}. \tag{2.45}$$

Similarly,

$$\mathbf{r}_1 = \mathbf{r} + \mathbf{r}_2$$

$$\implies \mathbf{r}_2 = \frac{-m_1}{m_1 + m_2}\mathbf{r}. \tag{2.46}$$

Hence the kinetic energy of the system is

$$T = \frac{1}{2}M\dot{R}^2 + \frac{1}{2}m_1\dot{r}_1^2 + \frac{1}{2}m_2\dot{r}_2^2, \tag{2.47}$$

and substituting from (2.45) and (2.46) and cancelling gives:

$$T = \frac{1}{2}M\dot{R}^2 + \frac{1}{2}\left[\frac{m_1m_2^2}{(m_1 + m_2)^2} + \frac{m_2m_1^2}{(m_1 + m_2)^2}\right]\dot{r}^2$$

$$= \frac{1}{2}M\dot{R}^2 + \frac{1}{2}\mu\dot{r}^2. \tag{2.48}$$

in terms of the total mass M and the reduced mass μ.

Conservation of energy now takes the form:

$$\frac{1}{2}M\dot{R}^2 + U(r) + \frac{1}{2}\mu\dot{r}^2 = E = \text{constant}.$$

Hence

$$U(r) + \frac{1}{2}\mu\dot{r}^2 = E - \frac{1}{2}M\dot{R}^2 = \text{constant},$$

as $\dot{R} = $ constant if there are no external forces acting on the two-particle system.

2.4.6 *Verification of the Galilean invariance of N3*

In order to examine the transformation properties of N3, we consider two particles of mass m_1 and m_2 respectively, moving in S under the action of forces. We have already covered most of what we need in conjunction with equations (2.20)–(2.24); but in order to treat this important point formally, we shall repeat the necessary equations here. Thus we begin with equations (2.20) and (2.21), which are

$$\frac{\text{d}}{\text{d}t}(m_1 v_1) = F_1 + F_{12},$$

and

$$\frac{\text{d}}{\text{d}t}(m_2 v_2) = F_2 + F_{21},$$

where, as before, F_1 and F_2 are external forces, while F_{12} and F_{21} are forces of mutual interaction.

Add the two equations of motion to get:

$$\frac{\text{d}}{\text{d}t}(m_1 v_1 + m_2 v_2) = F_1 + F_2 + F_{12} + F_{21}.$$

Now N3 implies $F_{12} = -F_{21}$, in S. Therefore

$$\frac{\text{d}}{\text{d}t}(m_1 v_1 + m_2 v_2) = F_1 + F_2.$$

However, from our discussions in Section 1.7, we know that all these terms transform invariantly. Thus N3 in S implies N3 in S'. Hence the Galilean invariance of N3 has been established.

2.5 Exercises

These exercises generally correspond to the topics of the chapter, in the order presented. However, there is little of interest in applying conservation of linear momentum (and nothing else) to a single particle. Accordingly, we begin with a couple of problems where the momentum is not conserved and where the forces enter into the problem in interesting ways.

2.1 A hot-air balloon of effective mass m descends with acceleration a. What fraction of its mass should be jettisoned in order that it will have an upward acceleration of a? The air resistance may be neglected.

2.2 A body moves with constant speed down a slope which is inclined at an angle α to the horizontal. What will be the acceleration of the particle if the angle of inclination is increased to $\beta > \alpha$? Assume that the frictional forces have the same dependence on the normal reaction in both cases.

2.3 A force **F** acts on a particle, causing it to move with velocity **u**. Show that the kinetic energy of the particle changes at a rate given by $\mathbf{F} \cdot \mathbf{u}$.

2.4 A bullet of mass 25 g is discharged with a speed of 300 m s^{-1} from a gun. What is the kinetic energy of the bullet? Given that the length of the gun barrel is 1 m, find the rate at which the expanding gases do work on the bullet just as it leaves the muzzle. Assume that the pressure exerted on the bullet is constant during the discharge.

2.5 A repulsive force varies as $f(r) = \kappa/r^2$, where κ is a constant and r is the distance from the centre of force. Show that the corresponding potential is given by $U(r) = \kappa/r$.

2.6 A particle of unit mass is released from rest at a distance $x = a$ from the centre of a force, which varies as $f(x) = \lambda x^{-3}$. Obtain the corresponding potential and find the time taken for the particle to fall into the centre.

2.7 Two spheres of mass M_1 and M_2 move along the x-axis. Show that, if they undergo an elastic collision, the velocity of each sphere relative to the other is reversed in direction but is unaltered in magnitude. What happens in the limit where one mass tends to infinity?

2.8 A particle slides down an inclined plane which is itself moving horizontally with speed V in some frame S. Using the principle of conservation of energy, find the horizontal velocity of the particle in S. [Hint: a slightly different version of this problem is discussed in Section 1.8.3.]

2.9 A child's top, once set spinning, is observed to slow down. How do you reconcile this fact with the principle that angular momentum is conserved? What is the nature of any torque acting on the system?

2.10 The position of a particle of mass m relative to a fixed point O at any time is given by the vector **x**. Show that if the only forces acting on the particle are parallel to **x**, the angular momentum of the particle is constant.

2.11 If two particles of mass M_1 and M_2 are simultaneously projected under gravity show that their centre of mass will describe a parabola.

2.12 Two particles of mass M_1 and M_2 are joined by a string, which in turn is passed over a frictionless pulley. Find an expression for the acceleration of their centre of mass.

2.13 Two particles of mass M_1 and M_2 move along the x-axis, with their positions at any time denoted by x_1 and x_2. Show that in the limit $M_2/M_1 \to \infty$, the CM frame becomes identical to the target frame, in which the mass M_2 is at rest.

2.14 If two particles of mass M_1 and M_2 are subject only to their mutual gravitational attraction, show that their relative velocity u is given by

$$u^2 = 2G(M_1 + M_2)/r + C,$$

where G is the gravitational constant, r is the distance between the particles, and C is an arbitrary constant.

Central forces

3

This is a very important class of forces, with the gravitational and electrostatic forces being among the most important examples. The applications considered in this chapter range from planetary motions down to the scattering of charged particles by nuclei. Historically the subject developed from the problem of calculating the motions of the planets around the Sun and we shall treat this topic in some detail. Essentially the problem is to determine the trajectory $\mathbf{x}(t)$ of a particle, given the details of the force. This is just the classic initial value problem as discussed in Chapter 1, but sometimes we are interested in the inverse problem: given the motion of the particle, can we determine the nature of the force?

Yet another version of the problem arises when we consider the question of the orbit of a particle. For example, we know that the orbit of the Earth round the Sun is in the form of an ellipse. Thus by the term 'orbit', we mean the actual path traced out, irrespective of where the particle actually is on that path at a given time. In order to obtain the equation of the orbit, we eliminate the time as an independent variable from the problem and concentrate on finding the locus of points corresponding to a prescribed relationship between the radial distance from the centre of force and the angle through which the radial vector has turned.

Our general approach is to start by describing the motion of a particle in three dimensions under the action of a central force. We assume that the centre of the force is fixed and that one body moves under its influence. This picture can also be applied to the two-body problem, provided that we introduce the concept of effective mass. As we saw in Section 2.4.4, this allows us to introduce the **equivalent single-body problem.** It is, of course, easily shown that when one body is very much more massive than the other, the two pictures become the same.

Additionally, we can show that motion under a central force will be confined to a plane (this is not too surprising: we know, for instance, that the motions of the planets around the Sun are planar) which means that we need only consider a two-dimensional problem. And, lastly, we can use the fact that angular momentum is conserved, in the case of motion under a central force, in order to reduce the problem to the equivalent one-dimensional form. This may seem a little more surprising, but for the moment it can be visualized by picturing (say) the Earth's orbit round the Sun as viewed by someone in the plane of the ecliptic. To such an observer, the Earth would appear to move back and forward in a straight line, between limits given by an appropriate one-dimensional projection of the two-dimensional orbit.

3.1 Definition of a central force field

Our initial reference frame is a set of fixed coordinates:[1] the Cartesian basis $(\mathbf{i}, \mathbf{j}, \mathbf{k})$, with origin O, as illustrated in Figure 3.1. We take the position vector of a particle of mass m to be $\mathbf{x}(t)$. That is, the position of the particle is a function of time. Then the velocity of the particle is $\mathbf{v} = \dot{\mathbf{x}}$, although it should be noted that this is not necessarily in the same direction as $\mathbf{x}$.

Formally we may define a central force as follows:

Definition: $\mathbf{F}(\mathbf{x})$ is a central force field, if and only if:

- $\mathbf{F}(\mathbf{x})$ is always directed along $\hat{\mathbf{x}}$, where $\hat{\mathbf{x}}$ is the unit vector in the direction of $\mathbf{x}$.

- $|\mathbf{F}|$ depends only on $|\mathbf{x}|$.

When dealing with central forces, it is usual to work in plane polar coordinates and we shall introduce these shortly. However, at this point it is convenient to partly anticipate this step by introducing the radial coordinate. We define

$$r = |\mathbf{x}| \quad \text{where} \quad 0 \le r < \infty$$

and

$$\mathbf{e}_r = \mathbf{x}/r,$$

which is the unit radial vector and identical to $\hat{\mathbf{x}}$. Thus we can write the position vector equally well as

$$\mathbf{x} = |\mathbf{x}|\hat{\mathbf{x}} \quad \text{or} \quad \mathbf{x} = r\mathbf{e}_r.$$

Then the general form of a central force can be written as:

$$\mathbf{F}(\mathbf{x}) \equiv f(r)\mathbf{e}_r \equiv f(|\mathbf{x}|)\hat{\mathbf{x}}, \tag{3.1}$$

where

- $f(r) < 0$: attractive force;
- $f(r) > 0$: repulsive force.

Lastly, we should note that $f(r)$ is often known as '**the law of the force**'.

[1] In order to discuss planetary motions, we may, in principle, measure positions relative to the 'fixed stars'. In practice, it is usual to work in terms of a stationary reference frame with its origin at the centre of gravity of the solar system.

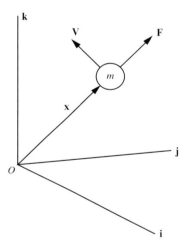

Fig. 3.1 Movement of a particle relative to a fixed coordinate system when subject to a central force.

3.2 The two-dimensional problem

Although we shall be considering problems where the motion is in principle in three space dimensions, it turns out (as we have mentioned previously) that, with a central force, the motion is actually only two-dimensional. We begin by establishing that this is a general characteristic of central force fields.

3.2.1 *Constant angular momentum*

We know from Section 2.3 that the angular momentum of a system is conserved in the absence of an applied torque. Now for central forces it follows from the properties of vector products that the torque is iden-tically zero. Thus it is really rather obvious that angular momentum must be conserved in motions where only central forces act. Nevertheless it is quite instructive to prove it generally and we do this as follows:

Theorem: In a central force field, angular momentum is conserved.

Proof In order to prove that the angular momentum is conserved, we show that its time derivative is zero. From (2.16) for the angular momen-tum and (2.3) for the linear momentum, we have:

$$
\begin{aligned}
\frac{d\mathbf{l}}{dt} = \frac{d}{dt}(m\mathbf{x} \times \dot{\mathbf{x}}) \\
= m(\dot{\mathbf{x}} \times \dot{\mathbf{x}}) + m(\mathbf{x} \times \ddot{\mathbf{x}}) \\
= 0 + m(\mathbf{x} \times \ddot{\mathbf{x}}) \quad \text{(vector products)} \\
= \mathbf{x} \times \mathbf{F}(\mathbf{x}) \quad \text{(by N2)} \\
= \mathbf{x} \times f(|\mathbf{x}|)\hat{\mathbf{x}} \quad \text{(central forces)} \\
= 0 \quad \text{(vector products).}
\end{aligned}
\tag{3.2}
$$

Hence $\mathbf{l}$ is a constant of the motion when the moving body is subject to purely central forces.

Corollary: The motion takes place *in a plane*.

First, consider the scalar product of the position vector of the particle with its angular momentum:

$$
\mathbf{x} \cdot \mathbf{l} = \mathbf{x} \cdot m(\mathbf{x} \times \dot{\mathbf{x}}) = 0.
\tag{3.3}
$$

This is because the vector product is itself a vector which is at right angles to both $\mathbf{x}$ and $\dot{\mathbf{x}}$. Hence the scalar product must be between two vectors at right-angles to each other and is therefore equal to zero.

Second, consider the scalar product of the velocity vector of the par-ticle with its angular momentum:

$$
\dot{\mathbf{x}} \cdot \mathbf{l} = \dot{\mathbf{x}} \cdot m(\mathbf{x} \times \dot{\mathbf{x}}) = 0,
\tag{3.4}
$$

and this vanishes for the same reasons as in the preceding case.

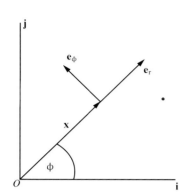

Fig. 3.2 Sketch defining the unit position vectors $(\mathbf{e}_r, \mathbf{e}_\phi)$

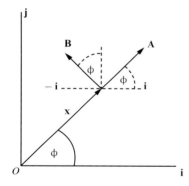

Fig. 3.3 Sketch showing the representation of the unit position vectors in terms of the fixed basis vectors.

Therefore $\dot{\mathbf{x}}$ and $\mathbf{x}$ are both perpendicular to $\mathbf{l}$ and the motion is in a plane with $\mathbf{l}$ normal to the plane.

3.2.2 *Motion in a plane*

We now introduce plane polar coordinates (r, ϕ), with their associated unit vectors $(\mathbf{e}_r, \mathbf{e}_\phi)$, and begin by examining their relationship to the fixed Cartesian basis $(\mathbf{i}, \mathbf{j}, \mathbf{k})$. The fact that the motion is in a plane means that we need only consider a two-dimensional position vector. Referring to Figure 3.2, we define the unit position vectors as:

$$\mathbf{e}_r = \frac{\mathbf{x}}{r} \equiv \text{unit radial vector} \tag{3.5}$$

$$\mathbf{e}_\phi \equiv \text{unit vector in direction of increasing } \phi. \tag{3.6}$$

Note that these unit vectors depend on the position and motion of the particle, whereas $(\mathbf{i}, \mathbf{j}, \mathbf{k})$ are fixed.

Next, we express $\mathbf{e}_r$ and $\mathbf{e}_\phi$ in terms of $\mathbf{i}$ and $\mathbf{j}$. As a first step, we consider arbitrary two-dimensional vectors $\mathbf{A}$ and $\mathbf{B}$. We take $\mathbf{A}$ along $\mathbf{x}$ and $\mathbf{B}$ at right angles to $\mathbf{x}$, as illustrated in Figure 3.3, but their magnitudes are quite arbitrary. Resolving $\mathbf{A}$ and $\mathbf{B}$ in the directions of $\mathbf{i}$ and $\mathbf{j}$, we have:

$$\mathbf{A} \equiv A\mathbf{e}_r = A \cos\phi \mathbf{i} + A \sin\phi \mathbf{j}$$
$$\mathbf{B} \equiv B\mathbf{e}_\phi = -B \sin\phi \mathbf{i} + B \cos\phi \mathbf{j}.$$

Therefore, since $\mathbf{e}_r$ and $\mathbf{e}_\phi$ are unit vectors,

$$\mathbf{e}_r = \cos\phi \mathbf{i} + \sin\phi \mathbf{j} \tag{3.7}$$

$$\mathbf{e}_\phi = -\sin\phi \mathbf{i} + \cos\phi \mathbf{j}. \tag{3.8}$$

Reminder: unlike $(\mathbf{i}, \mathbf{j})$, the unit vectors $(\mathbf{e}_r, \mathbf{e}_\phi)$ are **not** fixed. This follows from the fact that the angle ϕ changes with the position of the moving particle.

Later on we shall need the derivatives of these basis vectors and these are easily obtained from (3.7) and (3.8) as follows:

$$\frac{\partial}{\partial r}\mathbf{e}_r = 0, \qquad \frac{\partial}{\partial \phi}\mathbf{e}_r = -\sin\phi \mathbf{i} + \cos\phi \mathbf{j} = \mathbf{e}_\phi \tag{3.9}$$

$$\frac{\partial}{\partial r}\mathbf{e}_\phi = 0, \qquad \frac{\partial}{\partial \phi}\mathbf{e}_\phi = -\cos\phi \mathbf{i} - \sin\phi \mathbf{j} = -\mathbf{e}_r. \tag{3.10}$$

It is perhaps worth highlighting the fact that *neither* unit vector depends on the radial distance r but *both* unit vectors depend on the angular position ϕ. Also note that the angular derivative of $\mathbf{e}_r$ is just $\mathbf{e}_\phi$; whereas the angular derivative of $\mathbf{e}_\phi$ is $-\mathbf{e}_r$.

3.2.3 *Velocity of a particle in plane polar coordinates*

The particle velocity in plane polar coordinates can be obtained from its definition: $\mathbf{v} = \dot{\mathbf{x}}$. As the position vector of a particle takes the form

$\mathbf{x} = r\mathbf{e}_r$, it follows that

$$\mathbf{v} = \dot{\mathbf{x}} = \frac{d\mathbf{x}}{dt} = \frac{d(r\mathbf{e}_r)}{dt}$$

$$= \dot{r}\mathbf{e}_r + r\frac{d\mathbf{e}_r}{dt} \qquad \text{(differentiating a product)}$$

$$= \dot{r}\mathbf{e}_r + r\left(\frac{d\mathbf{e}_r}{d\phi}\frac{d\phi}{dt}\right) \quad \text{(chain rule)}$$

$$= \dot{r}\mathbf{e}_r + r\left(\mathbf{e}_\phi\frac{d\phi}{dt}\right)$$

$$= \dot{r}\mathbf{e}_r + r\dot{\phi}\mathbf{e}_\phi. \tag{3.11}$$

Hence we have the important result that the velocity vector of the particle expressed in plane polar coordinates takes the form:

$$\mathbf{v} = \dot{\mathbf{x}} = \dot{r}\mathbf{e}_r + r\dot{\phi}\mathbf{e}_\phi, \tag{3.12}$$

where

- $\dot{r}$ = radial component of velocity;
- $r\dot{\phi}$ = transverse component of velocity.

We now make use of this result to work out the angular momentum in plane polar coordinates and then we go on to examine the statement of conservation of energy, expressed in these same coordinates.

3.2.4 Angular momentum in plane polar coordinates

The particle motion takes place in a plane spanned by the unit vectors $\mathbf{i}$ and $\mathbf{j}$; and, from the results of Section 3.2.1, we know that the angular momentum vector is perpendicular to the plane and must lie along the vector $\mathbf{k}$. Thus we may write

$$\mathbf{l} = l\mathbf{k}. \tag{3.13}$$

From the definition of $\mathbf{l}$, we work out the magnitude of the particle's angular momentum about O, as follows. Re-writing the vector product in equation (2.16) in Cartesian tensor notation (and using the summation convention for repeated suffices), we have for the components of angular momentum

$$l_i = \epsilon_{ijk}x_j(m\dot{x}_k) = m\epsilon_{ijk}x_j\dot{x}_k,$$

where m is the mass of the particle and we have taken it outside the tensor expression for simplicity. Now, we only require the component of the angular momentum for which $i = 3$ (the other two are, of course, zero), and we may use the properties of the Levi-Civita density[2] ϵ_{ijk} to write

$$l_3 = m\epsilon_{312}x_1\dot{x}_2 + m\epsilon_{321}x_2\dot{x}_1 = mx_1\dot{x}_2 - mx_2\dot{x}_1.$$

[2]This is also known as the permutation tensor. It has the properties: $\epsilon_{ijk} = 1$, if the indices are a cyclic permutation of the numbers 1, 2 and 3; $\epsilon_{ijk} = -1$ if the indices are an anticyclic permutation of 1, 2 and 3. It is zero if any two indices are the same.

The position vector of the particle is $\mathbf{x} = r\mathbf{e_r}$, while the velocity vector is given by equation (3.12). In terms of the basis $(\mathbf{e}_r, \mathbf{e}_\phi, \mathbf{k})$, we can express these two quantities as:

$$\mathbf{x} = (r, 0, 0)$$

and

$$\dot{\mathbf{x}} = (\dot{r}, r\dot{\phi}, 0).$$

Substituting the appropriate components into the above equation for l_3 we have:

$$l_3 = m \cdot r \cdot r \cdot \dot{\phi} - m \cdot 0 \cdot \dot{r} = mr^2\dot{\phi}.$$

Or, making use of the fact that l_3 is in fact the total magnitude of the angular motion, we may write this result as:

$$l = mr^2\dot{\phi}.$$

As we have seen, this is a constant of the motion for a central force. It is useful to define a constant h such that:

$$h = \frac{l}{m} = r^2\dot{\phi}, \tag{3.14}$$

where h is the angular momentum per unit mass of the particle.

3.3 Conservation of energy for motion under a central force

We have already seen that the angular momentum of a particle moving under the influence of a central force is constant in both magnitude and direction. We have made use of this constancy of direction in reducing the problem from one involving three degrees of freedom to one involving only two degrees of freedom. In this section we shall show that problems in orbital motion can be solved using conservation of energy and in the next section that this approach can be further simplified by invoking the conservation of the magnitude of the angular motion.

We begin by expressing the principle of conservation of energy in a form appropriate to motion under a central force. Its general statement, as given in equation (2.4), is:

$$T + U = E.$$

For a particle of mass m and velocity $\mathbf{v}$, this becomes:

$$\frac{1}{2}(mv^2) + U(r) = E. \tag{3.15}$$

From (3.12), we put the velocity in term of polar coordinates, thus:

$$\mathbf{v} = \dot{\mathbf{x}} = \dot{r}\mathbf{e}_r + r\dot{\phi}\mathbf{e}_\phi,$$

and it follows that

$$v^2 = \mathbf{v} \cdot \mathbf{v} = \dot{r}^2 + r^2\dot{\phi}^2, \tag{3.16}$$

as $\mathbf{e}_r \cdot \mathbf{e}_r = \mathbf{e}_\phi \cdot \mathbf{e}_\phi = 1$ and $\mathbf{e}_r \cdot \mathbf{e}_\phi = 0$.

Our next step is to express the potential energy in terms of the central force. We know that for a conservative central force $\mathbf{F}(\mathbf{x})$ there exists a potential $U(r)$ such that

$$U(r) = -\int_{r_0}^{r} \mathbf{F}(\mathbf{x}') \cdot d\mathbf{x}' = -\int_{r_0}^{r} f(r')\, dr', \tag{3.17}$$

where r_0 is some reference point where the potential has a specified value: we shall discuss an example of how to handle this aspect of the problem in Section 3.4.3. The second equality has been seen before as equation (2.10). From (3.16) and (3.17), it follows that equation (3.15) for conservation of energy can be written as

$$\frac{1}{2}m(\dot{r}^2 + r^2\dot{\phi}^2) - \int_{r_0}^{r} f(r')dr' = E. \tag{3.18}$$

In the next section we shall see how conservation of angular momentum can be used to simplify the problem further.

3.4 Conservation of angular momentum and the equivalent one-dimensional problem

We have used the constancy of the *direction* of angular momentum to reduce the dimensionality of the problem from three space dimensions to two. This leaves us with a basic problem in which we specify r and ϕ for the particle's position at any time. However we can use conservation of the *magnitude* of angular momentum to eliminate the variable ϕ, thus leaving us with a one-dimensional picture of the motion, which only depends on the distance r of the particle from the centre of force, O.

3.4.1 *The effective potential*

We now eliminate the dependence on the angle ϕ in equation (3.18). From (3.14), we have $r^2\dot{\phi} = h$, therefore we substitute

$$r^2\dot{\phi}^2 = \frac{h^2}{r^2} = \frac{l^2}{m^2 r^2},$$

into the above energy equation to obtain:

$$\frac{1}{2}m\dot{r}^2 + U^*(r) = E,$$ (3.19)

where $U^*(r)$ is known as the effective potential and is defined as follows.

Definition: The **effective potential** U^* is given by:

$$U^*(r) = U(r) + \frac{l^2}{2mr^2} \quad \text{for } 0 \leq r < \infty,$$ (3.20)

where

- $U(r)$ = true potential;
- $l^2/2mr^2$ = centrifugal potential.

Note that (3.19) takes the conventional form of conservation of energy for motion in one dimension, depending only on the coordinate r. This dependence on a single variable justifies our description of this approach as 'one dimensional'. However, it should always be borne in mind that the motion is really two dimensional and that this is the *equivalent* one-dimensional problem. These aspects should become clearer as we go further with the formalism.

3.4.2 Classification of orbits

Motion under a central force often resolves itself into a question of orbits. For example:

- circular orbits of electrons in a cyclotron;
- elliptical orbits of planets;
- hyperbolic orbits of some comets, or alpha-particle scattering by nuclei.

We should remind ourselves that in the case of the orbital description, we are interested in finding out the path that the particle *must* travel along, once we have prescribed the law of the force and the initial conditions. In this approach, we are not concerned with the question of where the particle actually is at some given time. So what we have to do is eliminate the time as a variable from our description of the problem. We shall do this shortly by using conservation of angular momentum. First, however, we state the problem formally, as follows.

We wish to determine permitted values of r for the motion and hence deduce the nature of the orbit if we are given:

1. the initial values l and E;
2. the form of $f(r)$.

We start with the energy equation (3.19), and rearrange it as

$$\dot{r}^2 = \frac{2}{m}[E - U^*(r)].$$ (3.21)

Now $\dot{r}$ is the radial velocity of the particle and must be real. Therefore it follows that $\dot{r}^2$ must be positive. Hence equation (3.21) amounts to the requirement:

$$E - U^*(r) \geq 0,$$

and r can only take values which allow this relationship to be satisfied. We shall now examine the implications of this condition for an important class of laws of force (including the inverse-square law) and show how the general nature of the permitted orbit of a particle depends on its initial energy.

3.4.3 Example: *An attractive force vanishing at infinity*

We are given the information that this is an attractive force which vanishes as $r \to \infty$. As it is attractive, we know from Section 3.1 that this implies that we must have $f(r) < 0$. From these conditions on the law of the force, we may obtain conditions on the potential from which it is derived.

Equation (3.17) gives us:

$$U(r) = - \int_{r_0}^{r} \mathbf{F}(\mathbf{x}') \cdot d\mathbf{x}' = - \int_{r_0}^{r} f(r') \, dr',$$

where $U(r_0) = 0$. In this case, the reference level is at $r_0 \to \infty$, thus $U(\infty) = 0$, and hence the equation for the potential becomes:

$$U(r) = - \int_{\infty}^{r} f(r') \, dr' = \int_{r}^{\infty} f(r') \, dr'.$$

Therefore $U < 0$ for $f < 0$. That is to say, the potential U is negative for all values of r, and tends to zero as $r \to \infty$. A plausible shape for U is drawn as a dotted line below the axis in Figure 3.4. In Figure 3.4, we also plot the centrifugal potential, which appears in equation (3.20). We should note that this is always a repulsive potential, and has the same form for all particles, irrespective of the nature of the law of force; although clearly its magnitude depends on the initial angular momentum and on the mass of the particle. Provided that the actual potential $U(r)$ is something like a gravitational '$1/r$' potential (i.e. doesn't go to minus infinity as rapidly as the repulsive potential goes to plus infinity), then the centrifugal potential will win at short distances and the effective potential—which is the sum of the two, as defined by equation (3.20)—will look something like the form shown in Figure 3.4.

We shall go into more quantitative detail about the nature of orbits in the following sections. But for the moment, it is interesting to consider the qualitative aspects, by examining possible cases using only graphical methods. In Figure 3.4, we illustrate three significantly different forms of

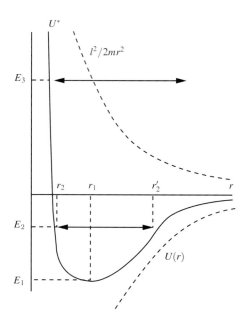

Fig. 3.4 The effective potential for a particle moving subject to a central force corresponding to a potential $U(r)$.

orbit, corresponding to particle energies denoted by E_1, E_2, or E_3. The possible types of orbit are as follows:

1. $E = E_1$: $U^* = E_1$; $\dot{r}^2 = 0$; $r = r_1$: circular orbit;
2. $E = E_2$: $\dot{r}^2 \geq 0$ for $r_2 \leq r \leq r_2'$: bounded orbit;
3. $E = E_3$: $\dot{r}^2 \geq 0$ for $r \geq r_3$: unbounded orbit.

It is worth emphasizing at this point that what we have shown is that, for a given shape of potential, the orbit of a body is determined solely by its initial energy. For the specific, and immensely important case, of the inverse-square law of force, we can go further and show that these orbits are actually conic sections, viz., ellipses, parabolas and hyperbolas. However, first we obtain a general equation for the orbit.

3.5 Equation for the orbit from conservation of energy

Later on, in Section 3.7, we shall obtain the equation for the orbit by simply solving N2 as an initial value problem. However, for our present purposes, we need to know the role played by the energy of the particle in determining the physical significance of the possible orbits. Accordingly, we begin by using the energy equation to find the orbital equation.

Rewrite the energy equation (3.21), with equation (3.20) substituted for the effective potential, as:

$$\frac{1}{2}m\dot{r}^2 = E - U(r) - \frac{1}{2}m\frac{h^2}{r^2}. \tag{3.22}$$

Then we solve for $\dot{r} = dr/dt$ and separate the variables:

$$dt = \frac{dr}{\sqrt{(2/m)(E - U - (mh^2/2r^2))}}. \tag{3.23}$$

We eliminate the time variable by using the conservation of angular momentum equation (3.14),

$$r^2\dot{\phi} = h \Longrightarrow \frac{r^2}{h}\left(\frac{d\phi}{dt}\right) = 1,$$

and so

$$dt = \frac{r^2}{h}d\phi. \tag{3.24}$$

We use this result to substitute for dt, thus:

$$dt = \frac{r^2}{h}d\phi = \frac{dr}{\sqrt{(2/m)(E - U - (mh^2/2r^2))}}, \tag{3.25}$$

and dividing across by r^2/h,

$$d\phi = \frac{h}{r^2}\frac{dr}{\sqrt{(2/m)(E - U - (mh^2/2r^2))}}$$

$$= \frac{dr}{r^2\sqrt{(2E/mh^2) - (2U/mh^2) - (1/r^2)}}. \tag{3.26}$$

Now we integrate from some earlier position (r_0, ϕ_0) up to the current position (r, ϕ) to obtain

$$\phi = \phi_0 + \int_{r_0}^{r} \frac{dr}{r^2\sqrt{(2E/mh^2) - (2U/mh^2) - (1/r^2)}}. \tag{3.27}$$

At this stage we make the important change of variable:

$$r = 1/u.$$

(Note: this is lower-case u: do not get confused with upper-case U for potential energy!) with the use of

$$dr = \frac{dr}{du}du = -\frac{1}{u^2}du, \tag{3.28}$$

equation (3.27) for ϕ becomes:

$$\phi = \phi_0 - \int_{u_0}^{u} \frac{du}{\sqrt{(2E/mh^2) - (2U/mh^2) - u^2}}. \tag{3.29}$$

This is the general equation for the orbit. Given the details of the force (or, equivalently, the potential U), we can use this equation to work out the form of the orbit.

3.5.1 *The inverse-square law: orbits as conic sections*

The inverse-square law of force takes the form

$$f = \frac{\kappa}{r^2}, \tag{3.30}$$

where

- $\kappa > 0$ for an attractive force;
- $\kappa < 0$ for a repulsive force;

with corresponding potential

$$U(r) = -\frac{\kappa}{r} = -\kappa u. \tag{3.31}$$

We begin by taking two steps:

(a) Rewrite equation (3.29) for ϕ in terms of an indefinite integral. This simply means dropping the limits on the integration and introducing an unknown constant of integration.

(b) Substitute for the specific inverse-square law potential. That is, we put $U(r) = -\kappa u$ with the result

$$\phi = \phi_1 - \int \frac{du}{\sqrt{(2E/mh^2) + (2\kappa u/mh^2) - u^2}}, \tag{3.32}$$

where ϕ_1 is the constant of integration.

For the integral, use the standard form:

$$\int \frac{dx}{\sqrt{A + Bx + Cx^2}} = \frac{1}{\sqrt{-C}} \arccos\left(\frac{-B + 2Cx}{D}\right), \tag{3.33}$$

where

$$D = \sqrt{B^2 - 4AC}. \tag{3.34}$$

Accordingly we set

$$A = \frac{2E}{mh^2}, \qquad B = \frac{2\kappa}{mh^2}, \qquad C = -1;$$

from which (3.34) becomes

$$D = \frac{2\kappa}{mh^2} \sqrt{1 + \frac{2mEh^2}{\kappa^2}},$$

so that the standard form of the integral now gives us:

$$\phi = \phi_1 - \arccos\left(\frac{(mh^2 u/\kappa) - 1}{\sqrt{1 + (2mEh^2/\kappa^2)}}\right). \tag{3.35}$$

Then, rearranging,

$$\arccos\left(\frac{(mh^2u/\kappa) - 1}{\sqrt{1 + (2mEh^2/\kappa^2)}}\right) = (\phi_1 - \phi), \qquad (3.36)$$

and inverting,

$$\left(\frac{(mh^2u/\kappa) - 1}{\sqrt{1 + (2mEh^2/\kappa^2)}}\right) = \cos(\phi_1 - \phi) = \cos(\phi - \phi_1), \qquad (3.37)$$

where the second equality relies on the cosine being an even function of its argument. Finally, we rearrange equation (3.37) as an expression for u:

$$u = \frac{\kappa}{mh^2}\left[\pm 1 + \sqrt{1 + \frac{2mEh^2}{\kappa^2}}\cos(\phi - \phi_1)\right], \qquad (3.38)$$

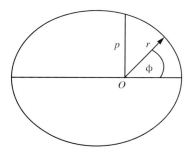

Fig. 3.5 Definition sketch for the parameters of a conic section.

where, from equations (3.30) and (3.31), $\kappa > 0$ gives $+1$, while $\kappa < 0$ gives -1 on the right-hand side. We shall continue this analysis with the '+' sign only, as this corresponds to an attractive force, and we are mainly interested in the force of gravity when considering orbital motion.

The solutions to this equation give rise to elliptic, parabolic, and hyperbolic orbits, depending on the values of the various parameters involved. That is, the orbits for the inverse-square law turn out to be conic sections. Referring to Figure 3.5, we may express the various conic sections in terms of a general equation:

$$r = \frac{p}{1 + e\cos\phi}, \qquad (3.39)$$

where e and p are constants. Loosely speaking, p determines the size of the orbit, while e determines its *eccentricity*. The conic sections may be listed as:

- ellipse: $e < 1$, $r = a(1 - e^2)/1 + e\cos\phi$;
- parabola: $e = 1$, $r = p/1 + \cos\phi$;
- hyperbola: $e > 1$, $r = a(e^2 - 1)/1 + e\cos\phi$.

Note that a is the semi-major axis and that a circle is a special case of an ellipse, when $e = 0$.

If we measure ϕ from an arbitrary angle ϕ_1, then the general equation of a conic becomes

$$r = \frac{p}{1 + e\cos(\phi - \phi_1)}. \qquad (3.40)$$

Changing back to $u = 1/r$, our equation for the orbit takes the form

$$\frac{1}{r} = \frac{\kappa}{mh^2}\left[1 + \sqrt{1 + \frac{2mEh^2}{\kappa^2}}\cos(\phi - \phi_1)\right]. \tag{3.41}$$

Comparison of these two equations gives

$$e = \sqrt{1 + \frac{2mEh^2}{\kappa^2}}, \tag{3.42}$$

as the eccentricity of the orbit, and hence the nature of the orbit depends on the energy E, as follows:

- $e > 1$ implies $E > 0$: the orbit is a hyperbola;
- $e = 1$ implies $E = 0$: the orbit is a parabola;
- $e < 1$ implies $E < 0$: the orbit is an ellipse;
- $e = 0$ implies $E = -\kappa^2/2mh^2$: the orbit is a circle.

3.5.2 Example: α-particle scattering

A particle of mass m is repelled from a fixed centre of force O by a force of magnitude $f = \kappa/r^2$, where $\kappa > 0$. Initially it is at a large distance from O and is projected with speed V and impact parameter b. If we choose the initial direction of motion to be parallel to the line $\phi = 0$, the orbital equation (3.38) may be written as

$$u(\phi) = -\kappa/mh^2 + A\cos(\phi - \phi_0).$$

Use the initial conditions to evaluate the constants A and ϕ_0.

Also find (a) the value of ϕ at closest approach; (b) the particle's distance of closest approach to O; (c) the value of ϕ when the particle is again at a large distance from O; and (d) the scattering angle.

The situation is illustrated in Figure 3.6. The initial conditions are:

- $\phi = 0$, $r = \infty$ (hence $u = 0$);
- $\dot{r}(0) = -V$ and hence $l = mbV$ (see Section 2.3.2).

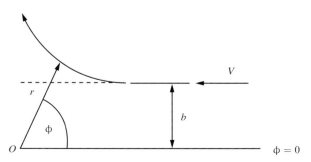

Fig. 3.6 A particle being scattered by a fixed centre of force at the point O.

We also need $du/d\phi \equiv u'(\phi)$ for $\phi = 0$. This is readily obtained as follows. It may be shown, using the definitions of u and h and the chain rule, that $u'(\phi) = -\dot{r}/h$, hence

$$u'(0) = +V/bV = 1/b.$$

Then from the given orbit equation, taken at $\phi = 0$, we have

$$u(0) = -\kappa/mh^2 + A\cos(-\phi_0) = 0.$$

Also, differentiating with respect to ϕ, and setting $\phi = 0$,

$$u'(0) = -A\sin(-\phi_0) = 1/b.$$

Hence, reacalling that cosine is even and sine is odd, this gives us $A\sin(\phi_0) = 1/b$ and $A\cos(\phi_0) = \kappa/mh^2$. It follows that the two constants in the orbit equation are given by:

$$\tan\phi_0 = mh^2/\kappa b = mbV^2/\kappa,$$

and, using the identity $\sin^2\theta + \cos^2\theta = 1$,

$$A = [1/b^2 + \kappa^2/m^2h^4]^{1/2}.$$

Now we can answer the individual questions as follows:

(a) At closest approach, $\dot{r} = 0$, thus

$$u'(\phi) = -\dot{r}/h = 0$$
$$\Longrightarrow 0 = -A\sin(\phi - \phi_0),$$

or $\phi = \phi_0$.

(b) When $\phi = \phi_0$,

$$u = -\kappa/mh^2 + A = -\kappa/mh^2 + [1/b^2 + \kappa^2/m^2h^4]^{1/2}$$
$$\Longrightarrow r_{min} = [-\kappa/mh^2 + (1/b^2 + \kappa^2/m^2h^4)^{1/2}]^{-1}.$$

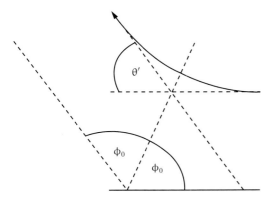

Fig. 3.7 Relationship of the scattering angle to the angular position coordinate ϕ_0.

(c) When $r = \infty$,

$$u = 0 = -\kappa/mh^2 + A\cos(\phi - \phi_0).$$

But, $\kappa/mh^2 = A\cos\phi_0$,

$$\implies 0 = -A\cos\phi_0 + A\cos(\phi - \phi_0).$$

This has solutions when $\phi = 0$ or $\phi = 2\phi_0$.

(d) Referring to Figure 3.7, we see that the scattering angle is given by:

$$\theta' = \pi - 2\phi_0$$

$$\implies \tan(\theta'/2) = \cot(\phi_0) = \kappa/mbV^2.$$

3.6 The inverse problem: obtain the force given the orbit

We start from the equation of conservation of energy in the form given by equation (3.18), viz.,

$$\frac{m}{2}(\dot{r}^2 + r^2\dot{\phi}^2) - \int f(r)\,\mathrm{d}r = E.$$

In order to eliminate the dependence on the time, we take two steps. First, we make use of the transformation

$$\dot{r} = \frac{\mathrm{d}r}{\mathrm{d}t} = \frac{\mathrm{d}r}{\mathrm{d}\phi}\cdot\frac{\mathrm{d}\phi}{\mathrm{d}t} = \frac{\mathrm{d}r}{\mathrm{d}\phi}\dot{\phi},$$

leading to

$$\frac{m}{2}\left[\left(\frac{\mathrm{d}r}{\mathrm{d}\phi}\right)^2 + r^2\right]\dot{\phi}^2 - \int f(r)\,\mathrm{d}r = E.$$

Our second step is to invoke conservation of angular momentum in the form $\dot{\phi} = h/r^2 = $ constant, in order to write the above equation as

$$\frac{mh^2}{2r^4}\left[\left(\frac{\mathrm{d}r}{\mathrm{d}\phi}\right)^2 + r^2\right] - \int f(r)\,\mathrm{d}r = E. \tag{3.43}$$

Now we rearrange this equation, and differentiate both sides with respect to the variable r to obtain

$$f(r) = \frac{mh^2}{r^4}\left[\frac{1}{2}\frac{\mathrm{d}}{\mathrm{d}r}\left(\frac{\mathrm{d}r}{\mathrm{d}\phi}\right)^2 - \frac{2}{r}\left(\frac{\mathrm{d}r}{\mathrm{d}\phi}\right)^2 - r\right].$$

The first term in the square bracket can be treated using the chain rule of differentiation, thus:

$$\frac{1}{2}\frac{d}{dr}\left(\frac{dr}{d\phi}\right)^2 = \frac{1}{2}\frac{d\phi}{dr}\frac{d}{d\phi}\left(\frac{dr}{d\phi}\right)^2 = \frac{d^2r}{d\phi^2},$$

and, in all,

$$f(r) = \frac{mh^2}{r^4}\left[\frac{d^2r}{d\phi^2} - \frac{2}{r}\left(\frac{dr}{d\phi}\right)^2 - r\right]. \tag{3.44}$$

Thus, given the orbit as an equation connecting r and ϕ, obtaining the law of the force involves little more than straightforward differentiation.

3.6.1 Examples: *Obtain the forces given the orbits*

Find the law of force corresponding to the following orbits: (a) $r = ae^{-\phi}$; and (b) a circular orbit of radius a where the particle passes through the centre of force O.

(a) Given $r = ae^{-\phi}$, we obtain

$$\frac{dr}{d\phi} = -ae^{-\phi} = -r,$$

$$\frac{d^2r}{d\phi^2} = ae^{-\phi} = r$$

and substitute into the right-hand side of equation (3.44), with the immediate result:

$$f(r) = -\frac{2mh^2}{r^3}.$$

(b) In this case we need the parametric equation of a circle passing through the origin and this takes the form $r = 2a\cos\phi$, where a is the radius. Substituting into the right-hand side of (3.44), we find

$$f(r) = \frac{mh^2}{r^4}\left[-2a\cos\phi - \frac{2}{2a\cos\phi}(-2a\sin\phi)^2 - 2a\cos\phi\right]$$

$$= \frac{mh^2}{r^4}\left[-\frac{4a}{\cos\phi}\right] = -\frac{4amh^2}{r^4\cos\phi}$$

$$= -\frac{8a^2mh^2}{r^5},$$

where we used the identity $\sin^2\phi = 1 - \cos^2\phi$.

3.7 Equation of the orbit using N2

The problem of determining orbits under central forces can be treated more straightforwardly as an initial value problem, using N2 and appropriate initial conditions. First we have to formulate N2 as an equation of motion in plane polar coordinates.

3.7.1 Acceleration in polar coordinates

In order to work out the acceleration of a particle in polar coordinates we extend the analysis already given in Section 3.2.3 for the velocity. Accordingly we write the acceleration $\mathbf{a}$ as the time derivative of the velocity, thus:

$$\mathbf{a} = \frac{d}{dt}\mathbf{v} = \frac{d}{dt}(\dot{r}\mathbf{e}_r) + \frac{d}{dt}(r\dot{\phi}\mathbf{e}_\phi), \tag{3.45}$$

where we have substituted for the velocity from equation (3.12). Now, noting that everything in each pair of brackets gets differentiated, and using the chain rule of differentiation, just as we did when working out the velocity earlier, we find

$$\mathbf{a} = \ddot{r}\mathbf{e}_r + \dot{r}\frac{d\mathbf{e}_r}{d\phi}\frac{d\phi}{dt} + \dot{r}\dot{\phi}\mathbf{e}_\phi + r\ddot{\phi}\mathbf{e}_\phi + r\dot{\phi}\frac{d\phi}{dt}\frac{d\mathbf{e}_\phi}{d\phi}. \tag{3.46}$$

Gathering terms, and substituting from (3.9) and (3.10) for the derivatives of the unit position vectors with respect to ϕ, we obtain for the separate radial and tangential components of the acceleration,

$$a_r = [\ddot{r} - r\dot{\phi}^2] \tag{3.47}$$

and

$$a_\phi = [2\dot{r}\dot{\phi} + r\ddot{\phi}], \tag{3.48}$$

where these components are defined by

$$\mathbf{a} = a_r\mathbf{e}_r + a_\phi\mathbf{e}_\phi. \tag{3.49}$$

It is easily verified (by working backwards!) that the result for a_ϕ may also be written in the form

$$a_\phi = \frac{1}{r}\frac{d(r^2\dot{\phi})}{dt}, \tag{3.50}$$

and this will turn out to be useful in the next section when we consider the equations of motion.

3.7.2 The equations of radial and circumferential motion

Now that we have expressions for the acceleration, the equations of motion may be written down directly using N2. For the radial direction we have from (3.47):

$$m[\ddot{r} - r\dot{\phi}^2] = f(r), \tag{3.51}$$

and, using equation (3.50) for the tangential acceleration, for the circumferential direction

$$m \frac{1}{r} \frac{\mathrm{d}}{\mathrm{d}t} (r^2 \dot{\phi}) = 0. \tag{3.52}$$

We should just remind ourselves that the zero on the right-hand side of the latter equation is purely a consequence of restricting our attention to central forces. Of course, we can integrate the left-hand side straight away to obtain

$$r^2 \dot{\phi} = \text{constant}, \tag{3.53}$$

which just recovers our earlier result (3.14) that the angular momentum is constant, thus:

$$r^2 \dot{\phi} = h = l/m.$$

There is also another first integral which yields conservation of energy, but we have already dealt with that aspect in Section 3.5.

3.7.3 Example: *Verify that an elliptic orbit corresponds to an inverse-square law of force*

Given that the orbit of a planet about the Sun is an ellipse, obtain an expression for the force per unit mass acting on it in the radial direction.

From equation (3.39) we have, with some rearrangement, the orbital equation as:

$$e \cos \phi = \frac{p}{r} - 1,$$

where p is as defined in Figure 3.5, and e is the eccentricity. The force per unit mass acting on the particle in the radial direction is just the radial component of acceleration as given by equation (3.47), viz.,

$$a_r = \ddot{r} - r \dot{\phi}^2.$$

Now differentiate both sides of the orbital equation with respect to time:

$$-e \sin \phi \cdot \dot{\phi} = -\frac{p}{r^2} \dot{r},$$

and substituting from (3.14) for $\dot{\phi}$, to get

$$e \sin \phi = \frac{p \dot{r}}{h}.$$

Again, differentiate both sides with respect to time, and substitute for $\dot{\phi}$, with the result

$$e \cos \phi = \frac{p r^2}{h^2} \ddot{r}.$$

Now substitute this result for $e \cos \phi$ back into the orbital equation to get

$$\frac{pr^2}{h^2} \ddot{r} = \frac{p}{r} - 1,$$

and hence:

$$\ddot{r} = \frac{h^2}{r^3} - \frac{h^2}{pr^2}.$$

Next, we substitute this expression into equation (3.47) for the radial acceleration, along once again with a substitution for $\dot{\phi}$ from (3.14), which leaves us with

$$a_r = -\frac{h^2}{pr^2}.$$

Thus the force attracting the planet (of mass m) towards the Sun is

$$f = \frac{m\gamma}{r^2},$$

where

$$\gamma = \frac{h^2}{p}. \tag{3.54}$$

Note that the force is of the expected inverse-square form. The coefficient γ is sometimes known as the 'acceleration at unit distance'.

3.7.4 The orbit equation

Equation (3.51) for the radial motion can be simplified by using the constancy of the angular momentum to eliminate $\dot{\phi}$ from it. Substituting $\dot{\phi} = l/mr^2$ from (3.14) we obtain

$$m\ddot{r} - \frac{l^2}{mr^3} = f(r). \tag{3.55}$$

As we saw earlier, we are often interested in the equation of the orbit, which can be written in terms of a relationship between r and ϕ, with the time variable t having been eliminated. As in Section 3.5, we find that the change of variable $r = 1/u(\phi)$ simplifies matters, as we then have

$$\dot{r} = \frac{\mathrm{d}}{\mathrm{d}t}(1/u) = -\frac{1}{u^2}\frac{\mathrm{d}u}{\mathrm{d}\phi}\frac{\mathrm{d}\phi}{\mathrm{d}t} = -r^2\frac{\mathrm{d}u}{\mathrm{d}\phi}\dot{\phi} = -h\frac{\mathrm{d}u}{\mathrm{d}\phi},$$

and hence

$$\ddot{r} = \frac{\mathrm{d}}{\mathrm{d}t}\left(-h\frac{\mathrm{d}u}{\mathrm{d}\phi}\right) = \frac{\mathrm{d}\phi}{\mathrm{d}t}\frac{\mathrm{d}}{\mathrm{d}\phi}\left(-h\frac{\mathrm{d}u}{\mathrm{d}\phi}\right) = -h\dot{\phi}\frac{\mathrm{d}^2u}{\mathrm{d}\phi^2}.$$

Substituting into (3.55) the above result for $\ddot{r}$, and from (3.14) for the angular momentum l,

$$-mh^2u^2\frac{d^2u}{d\phi^2} - \frac{m}{u}(h^2u^4) = f(r),$$

and, with some rearrangement,

$$\frac{d^2u}{d\phi^2} + u = \left(\frac{-1}{mh^2u^2}\right)f(1/u). \tag{3.56}$$

Evidently, once the law of force $f(r) \equiv f(1/u)$ is given, then this equation should be readily solved for the orbit. However, it is also quite possible to solve the inverse problem, and obtain the law of force corresponding to a given orbit. We illustrate this property of equation (3.56) in the next section, before going on to its more obvious applications.

3.7.5 Example: *Given the orbit, obtain the force*

Find the law of force corresponding to the orbit $r = \kappa/(1 + \epsilon \cos \phi)$.

Make the usual change of variables $u = 1/r$ so that the given orbital equation becomes

$$u = (1 + \epsilon \cos \phi)/\kappa.$$

Then denoting differentiation with respect to ϕ by a dash, we obtain

$$u' = -(\epsilon/\kappa)\sin \phi$$
$$u'' = -(\epsilon/\kappa)\cos \phi$$

and substituting into equation (3.56),

$$u'' + u = \frac{1}{\kappa} = -\frac{f(1/u)}{mh^2u^2}.$$

Thus, solving for $f(1/u)$, then changing back from u to r,

$$f(r) = \frac{mh^2}{r^2}.$$

3.7.6 Example: *Given the law of force, find the orbit*

A particle of mass m moves under the influence of a central force with $f(r) = -m\kappa(3r - 2a)/r^3$. If its orbit is a circle of radius a, show that its angular momentum is given by $m\sqrt{\kappa a}$. The particle is projected from the point at $r = a$, $\phi = 0$, with velocity V at right angles to the radius vector. If V is half the value required for a circular orbit, show that the orbit takes the form $r = 3a/(4 - \cos 3\phi)$.

With the substitution of the given law of force, the radial equation of motion (3.56) becomes

$$u'' + u = \frac{\kappa}{h^2}(3 - 2au).$$

For circular motion, we have $r = a$, $u = 1/a$, and accordingly $u' = u'' = 0$. Hence the radial equation of motion simplifies to:

$$\frac{1}{a} = \frac{\kappa}{h^2}(3 - 2) \implies h = \sqrt{\kappa a},$$

and so $h = \sqrt{\kappa a}$ with angular momentum

$$l = mh = m\sqrt{\kappa a},$$

as required. For the second part of the question, we also need the speed of the particle in circular motion. This is given by

$$l/ma = \sqrt{\kappa/a}.$$

In this part, we have initial conditions

$$r = a, \qquad r\dot{\phi} = V = \frac{1}{2}\sqrt{\kappa/a}.$$

From these it follows that $u(0) = 1/a$ and $u'(0) = -\dot{r}/h = 0$. The angular momentum per unit mass is given by

$$h = r^2\dot{\phi} = aV = \frac{1}{2}\sqrt{\kappa a}.$$

With this value substituted for h, the radial equation of motion now becomes

$$u'' + u = \frac{12}{a} - 8u,$$

or, with some rearrangement,

$$u'' + 9u = 12/a.$$

This differential equation is easily solved with:

- particular integral: $u = 4/3a$;
- complementary function: $u = A\cos 3\phi + B\sin 3\phi$;

giving a general solution

$$u(\phi) = \frac{4}{3a} + A\cos 3\phi + B\sin 3\phi.$$

Now use the initial conditions to fix the constants A and B. Setting $r = a$ and $\phi = 0$ in the general solution gives:

$$1/a = u(0) = 4/3a + A.$$

Then, differentiating the general solution with respect to ϕ and setting $\phi = 0$ and $u'(0) = 0$, we have:

$$0 = u'(0) = 3B.$$

Hence, solving these two equations simultaneously, we find the constants to be $A = -1/3a$ and $B = 0$. Substituting back, we then have

$$1/r = u(\phi) = \frac{1}{3a}(4 - \cos 3\phi)$$

and the required result follows upon inversion of both sides.

3.8 Kepler's laws of planetary motion

Kepler published his first two laws of planetary motion in 1609 and his third in 1619. They were not based on a theory, but represented an attempt to summarize the astronomical observations of his teacher and colleague Tycho Brahe. However, they were the basis upon which Newton, using his newly invented calculus, formulated his law of universal gravitation. Kepler's laws may be stated as follows:

K1 The planets describe ellipses with the Sun as one focus.

We have shown that this is the case when the law of force is the attractive inverse-square law: see Section 3.7.3.

K2 The radius vector drawn from the Sun to a planet sweeps out equal areas in equal times.

We have demonstrated this in Section 3.2, where we saw that the constancy of the angular momentum was a consequence of the restriction to a central force. In mathematical form, Kepler's second law may be restated as:

$$r^2\dot{\phi} = h,$$

where h is the angular momentum per unit mass and is, of course, constant. Note also that h is twice the areal velocity of the orbital motion.

K3 The squares of the periods of the different planets are proportional to the cubes of their respective mean distances from the Sun.

Note that the periodic time T is the time taken for the planet to complete one orbit, while the mean distance from the Sun is the semi-major axis a of the ellipse. This result may also be shown from our earlier theory of the inverse-square law, and its derivation will be given in the next section. For any two planets 1 and 2, it may be written as:

$$T_1^2/T_2^2 = a_1^3/a_2^3.$$

To sum up, we can show theoretically that K2 is a consequence of motion under the influence of a central force, whereas K1 and K3 are consequences of the central force taking the specific inverse-square form. Thus, although originally Kepler's laws were purely a summary of the experimental observations, their existence inspired Newton's general theory of gravitation and was influential in the development of classical mechanics, from which they can now all be derived as special cases of a much more general theory.

We shall conclude this chapter by giving an explicit derivation of K3, as this is the only one of Kepler's laws which we have not yet derived, and then we shall consider how this law should be modified slightly in order to take account of the two-body nature of the problem.

3.8.1 *Kepler's third law*

From K2, we have the rate of change of the area A swept out by the radius vector during the orbit as

$$\frac{\mathrm{d}A}{\mathrm{d}t} = \frac{1}{2}r^2\dot{\phi} = \frac{h}{2}.$$

From the theory of conic sections, we shall need the results

$$p = b^2/a,$$

where p is defined in Figure 3.5, a and b are the semi-major and semi-minor axes of an ellipse, and the area of an ellipse is given by

$$A = \pi ab.$$

The areal velocity is constant, so the area of the ellipse is also given by

$$A = \left(\frac{\mathrm{d}A}{\mathrm{d}t}\right)T = \frac{h}{2}T,$$

and equating these two expressions for the area swept out by the position vector of the planet during one complete orbit, we obtain for the orbital period

$$T = \frac{2\pi ab}{h}.$$

Now we can obtain an expression for h from equation (3.54). Making use of the result for conic sections given above, we have

$$h = \sqrt{\gamma p} = \sqrt{\gamma b^2/a},$$

and substituting into the above equation for T yields

$$T = \frac{2\pi}{\sqrt{\gamma}} a^{3/2}.$$

Or, alternatively, squaring both sides,

$$T^2 = \frac{4\pi^2}{\gamma} a^3, \tag{3.57}$$

which is just Kepler's third law.

We recall that γ is the acceleration at unit distance, and note that if Kepler's law holds, in the form quoted in the previous section for any pair of planets, then γ should be the same for all planets. In the next section, we shall see that this is only an approximation, albeit a good one.

3.8.2 The two-body problem and the Galilean-relativistic corrections to Kepler's third law

In practice, Kepler's laws do not give a completely accurate description of the motion of the planets round the Sun. Their apparent theoretical verification given here has been based on the premise that the planets are attracted by a fixed Sun and do not in turn move this Sun nor indeed affect each other.

We can give a more correct description by treating the motion of a planet round the Sun as a two-body problem. A simple version of this is to include in the analysis the acceleration induced by a planet of mass m_p in the Sun of mass m_s, and simply add this on to our previous calculation of the absolute acceleration of the planet, in order to obtain the relative acceleration. However, it is more in keeping with the spirit of this book to examine this more formally in a relativistic way.

Our previous derivation of K3 was carried out under the assumption that the Sun did not move. Accordingly we now denote the rest frame of the Sun by S and consider matters in a coordinate frame in which the centre of mass of the planet–Sun system is at rest. We denote this 'centre of mass' frame by S'. We ignore all rotational effects and regard S and S' as inertial frames. Now we repeat our derivation of K3—more correctly—in frame S'.

In S', the distance of the planet from the centre of mass is (see equation (2.45)) given by

$$r_p = \frac{m_s}{m_s + m_p} r \equiv \lambda_m r, \tag{3.58}$$

where r is the distance between the planet and the Sun, which is of course the same in both frames. It will be convenient to introduce a temporary symbol λ_m for the mass ratio in equation (3.58), and this is defined by the second equality.

Now we repeat our derivation of K3, but this time in S'. The acceleration of the planet towards the centre of mass is just

$$\ddot{r}_p = \frac{\gamma'}{r_p^2}.$$

However, acceleration is invariant under Galilean transformation, so this must be the same as the result obtained in S, viz.,

$$\ddot{r}_p = \frac{\gamma}{r^2}.$$

It is readily shown, following substitution from equation (3.58) for r_p, that these two results will be the same provided that:

$$\gamma' = \lambda_m^2 \gamma.$$

Also, in S' the planet still moves in an elliptical orbit, but now with semi-major axis given by

$$a_p = \lambda_m a,$$

in accordance with equation (3.58).

The arguments which led to equation (3.57) in S, now lead to the same form in S', thus:

$$T^2 = \frac{4\pi^2}{\gamma'} a_p^3,$$

and transforming back into S we have:

$$T^2 = \frac{4\pi^2}{\gamma} a^3 \lambda_m.$$

So for any two planets 1 and 2 say, with masses m_1 and m_2 the ratio of the squares of periodic times satisfies the relation

$$\frac{a_1^3}{a_2^3} = \left(\frac{m_s + m_1}{m_s + m_2} \right) \frac{T_1^2}{T_2^2}. \tag{3.59}$$

We can show that this is a small correction by taking out a common factor of m_s on the right-hand side and cancelling above and below to obtain:

$$\frac{a_1^3}{a_2^3} = \left(\frac{1 + m_1/m_s}{1 + m_2/m_s} \right) \frac{T_1^2}{T_2^2}.$$

Thus, in view of the small ratios of the various planetary masses to the Sun's mass, these corrections to Kepler's third law are actually very small.

3.9 Exercises

3.1. Prove that angular momentum is conserved for motion under a central force because the associated torque is zero.

3.2. A particle of mass m moves in a central force field which is directed towards a fixed point O. Show that the angular momentum of the particle about the point O may be written as

$$l = mr^2\dot{\phi}\mathbf{k},$$

where all the symbols have their usual meanings.

If dA is the area swept out by the radius vector in time dt, show that the areal velocity dA/dt is constant. [Hint: in vector notation, $dA = \frac{1}{2}|\mathbf{x} \times d\mathbf{x}|$.]

3.3. A particle of mass m moves about a point O with angular momentum l, and is subject to a central force of magnitude $f(r)$, which is directed towards O. Show that the introduction of an effective one-dimensional potential $U^*(r)$ is equivalent to the assumption that the particle is subject to a fictitious central force f^*, which is given by

$$f^* = f(r) + l^2/mr^3.$$

3.4. In clouds of charged particles or in fluid suspensions of particles, the many-body problem can be solved to good approximation for dilute systems by replacing the Coulomb potential by a screened potential,

$$U(r) = -\frac{\kappa e^{-\lambda r}}{r},$$

where κ and λ are positive constants. Discuss in qualitative terms the motion of a particle under the influence of such a potential, by making use of the equivalent one-dimensional potential.

3.5. Sketch the actual and effective potentials for a particle moving in a central force field, where the force is: (a) repulsive, and vanishes as $r \to \infty$; and (b) attractive, and vanishes as $r \to 0$. Show that the possible orbits are unbounded in the first case, but bounded in the second. In both cases, illustrate how bounds on the possible values of r are related to the total energy of the particle.

3.6. Sketch the equivalent one-dimensional potential for an attractive inverse-square law of force and illustrate the condition for: (a) circular orbits; and (b) bounded orbits.

3.7. Obtain the law of force acting on a particle of mass m which would result in the following orbits:
(a) $r = a\exp(-b\phi)$;
(b) $r = a + b\cos\phi$;
(c) $r = a/(1 + b\cos\phi)$,
where a and b are constants.

3.8. A particle moves in a circular orbit under the influence of an attractive force directed towards a point on the circumference of the circle. Show that the central force varies as the inverse fifth power of the radial distance.

3.9. A particle of mass m is attracted to a fixed point O by a central force with law given by

$$f(r) = -(a/r^2 + b/r^3),$$

where a and b are constants. By making the usual transformation $u = 1/r$, show that the general equation for the orbit is

$$u = \frac{a}{mh^2\lambda^2} + A\cos(\lambda\phi - B),$$

where A and B are arbitrary constants and $\lambda = (1 - b/mh^2)^{1/2}$. If the particle is projected from a point P in a direction perpendicular to the line OP, show that its radial velocity is zero when $\phi = n\pi/\lambda$, for integer n.

3.10. Two identical particles of mass m are connected by a light string of length s. The string passes through a small hole in a smooth horizontal plane so that one particle is free to move on the surface of the plane while the other particle can only move up or down in the vertical direction beneath it. Show that the radial coordinate of the particle on the plane satisfies the equation of motion:

$$2m\ddot{r} - \frac{l^2}{mr^3} + mg = 0,$$

where l is its angular momentum. Obtain the corresponding energy equation and show that the radial motion is the same as that of a single particle of mass $2m$ moving under the influence of the equivalent one-dimensional potential

$$U^*(r) = \frac{l^2}{2mr^2} + mg(r - s).$$

For what value of the angular momentum will the particle on the plane move in a circle of radius $s/2$?

3.11. A particle of mass m moves under the influence of an attractive central force. Find the speed V, for which the particle can move in a circular orbit of radius R, in terms of the law of force $f(r)$. Show that if the particle is given a small radial impulse, the quantity $d = r - R$ satisfies approximately the equation

$$m\ddot{d} + \left[\frac{3f(R)}{R} + f'(R)\right]d = 0,$$

where the prime denotes differentiation with respect to r. Deduce the condition for the orbit to be stable.

3.12. The motion of each of the planets around the Sun is governed, to a first approximation, by the inverse-square law of force $f(r) = \kappa/r^2$. Write down an explicit form for the constant κ. Also, state and prove Kepler's laws of planetary motion.

Mechanical vibrations and waves

4

We study this topic for two reasons. Firstly, it is a very important application of Newton's laws of motion. Secondly, it raises certain topics (e.g. the Doppler effect) which will be of interest once we begin to discuss special relativity. We begin by considering small displacements from stable equilibrium and derive the generic equation of simple harmonic motion (SHM) from N2. We treat various topics of interest in this area, but the main thrust of this chapter is to go from systems with one degree of freedom, to those with two, three and more degrees of freedom, until, in the limit, we consider continuous waves in systems with an infinite number of degrees of freedom. However, before beginning, we should emphasize that our treatment here of vibrations and waves will be the merest outline. It is a specialist subject in its own right, with a wide choice of monographs at all levels.

4.1 Stable and unstable equilibrium

We start with the idea of equilibrium. When a particle is at rest it is in equilibrium. We can express this idea quite generally by saying that the forces acting on a particle in equilibrium must sum to zero. Or, if the particle is subject to a net potential $U(x)$, then we have the equivalent condition

$$F = -\frac{\partial U}{\partial x} = 0.$$

This may be illustrated by the example (as discussed in Chapter 2) of a ball in a bowl. If the ball is at rest at the bottom of the bowl, then it is in equilibrium: see Figure 4.1.

If we give the ball a small displacement, then it will roll back under the influence of the restoring force $F = -\partial U/\partial x$, and will continue to oscillate until (in the real world!) friction damps out the motion. We call this a position of **stable equilibrium,** as it is stable with respect to small

Fig. 4.1 An illustration of stable equilibrium.

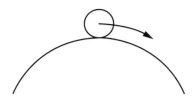

Fig. 4.2 An illustration of unstable equilibrium.

perturbations. What happens if we invert the bowl and replace the ball on top of it? As shown in Figure 4.2, we still have our position of equilibrium. However, if we now give the ball a small displacement, we see that it will continue rolling away from its equilibrium position. This is known as a position of **unstable equilibrium**.

In Section 2.2.1, we showed that an arbitrary potential could be expanded in powers of x. For the case of a symmetrical potential, as considered here, with a restriction to small displacements of the particle from the equilibrium position, we can write the force as

$$F = -\frac{\partial}{\partial x}\left(\frac{1}{2}\kappa x^2\right) = -\kappa x. \tag{4.1}$$

Note that κ is just $U''(0)$ and thus both the sign and the magnitude of κ depend on the shape of the potential function, as this is what determines the values of its derivatives. If we take the mass of the particle to be m, then application of N2 gives

$$m\ddot{x} = F = -\kappa x;$$

or

$$\ddot{x} + \omega^2 x = 0, \tag{4.2}$$

where

$$\omega^2 = \kappa/m.$$

With the restriction to small oscillations in our present problem, only the sign of κ depends on whether the 'bowl' is the right way up or inverted. It is easily verified that if the bowl is the right way up, then κ is positive. Under these circumstances, $F = -\kappa x$ is directed towards the origin (i.e. the direction of the force is negative for positive x and positive for negative x), and this corresponds to stable equilibrium. Clearly the reverse is true, in that if κ is negative, the negative signs cancel, and $F = \kappa x$ is directed away from the origin (i.e. F is positive for positive x and *vice versa*) and this corresponds to unstable equilibrium.

4.1.1 *Small oscillations*

Equation (4.2), being a second-order ODE with constant coefficients, may be solved by trying exponential functions in the usual way. Substitution allows one to verify that the general solution for the displacement $x = X(t)$ of the particle from equilibrium at any time t takes the form

$$X(t) = C_1 e^{-i\omega t} + C_2 e^{i\omega t}, \tag{4.3}$$

where $i = \sqrt{-1}$, the constant κ is positive and ω is real, being the angular frequency of oscillation. As always, integration of a second-order differential equation requires two arbitrary constants C_1, C_2 which must be fixed by reference to the initial conditions.

Conversely, if κ is negative, then ω^2 is real but ω is not, and so the solution involves real exponentials, viz;

$$X(t) = D_1 e^{-\lambda t} + D_2 e^{\lambda t},$$

where $\lambda^2 = \omega^2$ and D_1 and D_2 are the arbitrary constants.

In the latter case, the solution admits the possibility of the amplitude growing exponentially with time, so that the motion of the particle can become unbounded. Ultimately, this must result in the basic assumption which underpins the analysis being invalidated, viz; that the motion should be restricted to small amplitudes.

However, we shall not consider this matter further here. Instead, we concentrate on the case of small oscillations about stable equilibrium. In this case the motion is always bounded and the underlying conditions are thus always satisfied. This accounts for the universal occurrence of simple harmonic motion, both in nature and in applications, and it is to this that we now turn.

4.2 Systems with one degree of freedom

For the sake of a simple exposition, we begin by considering systems which involve a single mass moving in a single direction. Such systems are referred to as having one degree of freedom. Later we shall discuss systems with two or more degrees of freedom and in the process we shall make it clear what this terminology means. For the present we shall concentrate on the simplest generic system and then discuss some physical realizations of the idea.

4.2.1 The simple harmonic oscillator

The simple harmonic oscillator (or SHO) is a concept of extraordinary applicability in physics, varying from the simple pendulum through vibrating atoms in crystal lattices to concepts in high energy particle physics. Essentially a SHO is any physical system which can be described by an equation of motion of the form (4.2), with real ω such that its general solution is given by (4.3). In a real system, the displacement is always a real quantity. Thus we have to take the real part of (4.3) and it is customary to write this in the form:

$$X(t) = A \cos(\omega t + \epsilon), \tag{4.4}$$

and it is readily verified by direct substitution that this solution satisfies (4.2) for arbitrary constants A and ϵ.

This solution shows that the particle undergoing simple harmonic motion (SHM) has an oscillatory displacement which varies between the limits $X = \pm A$. We may list the salient features of this expression as follows:

- A is the amplitude of the motion;

- $\omega t + \epsilon$ is the phase of the motion;
- ϵ is the initial phase of the motion.

Before we proceed to specific examples, where we shall use initial conditions to fix values of the constants, it is useful to consider another general form. Invoking a compound-angle formula of trigonometry, we may expand

$$X = A \cos \omega t \cos \epsilon + A \sin \omega t \sin \epsilon = B \cos \omega t + C \sin \omega t, \quad (4.5)$$

where $B = A \cos \epsilon$ and $C = A \sin \epsilon$.

Differentiating this form with respect to time, we obtain the velocity as:

$$\dot{X} = -\omega B \sin \omega t + \omega C \cos \omega t. \quad (4.6)$$

Then, assuming initial conditions $X(0) = x_0$ and $\dot{X}(0) = U_0$, we set $t = 0$ in (4.5) and (4.6), to obtain $B = x_0$ and $\omega C = U_0$, and hence

$$X(t) = x_0 \cos \omega t + (U_0/\omega) \sin \omega t,$$

so that the displacement at any time t can be expressed in terms of the initial displacement x_0 and the initial velocity U_0.

Before turning to some specific physical systems, we shall briefly consider a useful interpretation of SHM as 'circular motion projected on a diameter'. To do this, we introduce the 'circle of reference' which is a circle of radius A, with a given radius vector rotating with angular velocity ω, as shown in Figure 4.3. At $t = 0$ the radius vector has already turned through an angle ϵ and at time t it has turned through an angle $(\omega t + \epsilon)$. Take a diameter of the circle to lie on the x-axis. Then the circular motion of a particle projected on this diameter is just

$$X(t) = A \cos(\omega t + \epsilon),$$

which is identical to equation (4.4). The rotating vector is called a **phasor** and the projection of its circular motion on a diameter takes the form of simple harmonic motion (or SHM) about $x = 0$.

In general this is a useful representation of SHM as it allows us to work out the effect of adding up many different SHMs, with different amplitude or frequency or phase, by using the normal vector rules for addition and subtraction.

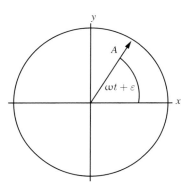

Fig. 4.3 Vector representation of SHM.

4.2.2 *The simple pendulum*

This is perhaps the most familiar example of SHM, having been the basic principle of time-keeping for hundreds of years. We take the pendulum to consist of a mass m suspended by a string of length l from some pivot. In this case, gravity provides the restoring force and (referring to Figure 4.4) this can be resolved both along the string and at right angles

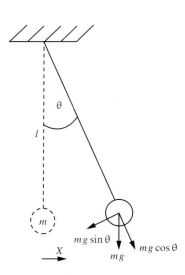

Fig. 4.4 The simple pendulum.

to it, leading to the resulting force balances:

- In the radial direction we have

$$T = -mg\cos\theta,$$

where T is the tension in the string.

- In the tangential direction, N2 gives

$$m\ddot{X} = -mg\sin\theta \simeq -mgX/l,$$

where $X(t)$ is the displacement of the particle from equilibrium at time t. **Note:** The second step in the above equation is an approximation which is only valid for small values of the displacement X.

Thus for small displacements the equation of motion is

$$\ddot{X} + (g/l)X = 0,$$

where comparison with (4.2) shows that the angular frequency of oscillation is

$$\omega = \sqrt{g/l}.$$

Note that in this system the restoring force is proportional to the mass which then cancels across, leaving us with a result for the natural frequency which does not depend on the mass of the particle.

Also note that in Section 2.2.5 we used the principle of conservation of energy to get a more general result.

4.2.3 *The spring–mass system*

The vertical oscillations of a mass m hanging from a fixed spring are another example of SHM, provided the extension of the spring is small enough for Hooke's law to apply. That is, if the mass is displaced a distance X below the equilibrium position, then the force acting upwards is kX where k is a constant. The physical system is illustrated in Figure 4.5.

Again, we apply N2 and write it as

$$m\ddot{X} = -kX,$$

and as before we compare this with equation (4.2) and deduce that the frequency of oscillation is

$$\omega = \sqrt{k/m}. \tag{4.7}$$

Unlike the case of the pendulum, this result depends on the particle's mass. However, this dependence may be eliminated as follows.

Initially we take the spring to have an unstretched length l (i.e. we assume that the effect of its own weight has already been taken into

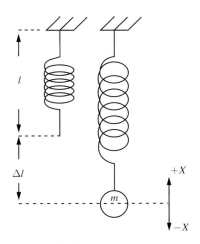

Fig. 4.5 Oscillation of a spring–mass system.

account). Then if we add a particle of mass m the spring stretches to $l + \Delta l$ and we have the equilibrium condition

$$mg = k\Delta l \implies m = k\Delta l/g.$$

Substituting this expression for the mass in (4.7), we find

$$\omega = \sqrt{g/\Delta l},$$

which is, in effect, just the same result as that for the pendulum.

4.2.4 A mass on a stretched string

In this case a particle of mass m is attached to the middle point of a string which is stretched between fixed points with a tension T. If the mass is displaced a small distance $x = X$ laterally, then the restoring force arises because a component of the tension can be resolved along the direction of the displacement without the magnitude of the tension being affected. We also take the inertia (or mass) of the string to be negligible. Referring to Figure 4.6, we see that the restoring force in the direction of decreasing x is given by

$$2T\cos\theta = 2T\cos\left(\frac{\pi}{2} - \phi\right) = 2T\sin\phi.$$

Note that the factor of 2 comes from the fact that both halves of the string contribute to the restoring force. N2 then yields

$$m\ddot{X} + 2T\sin\phi = 0.$$

For small X, we make the further approximations

$$\sin\phi \simeq \tan\phi \simeq \frac{X}{l/2} = \frac{2X}{l},$$

and hence the equation of motion becomes

$$\ddot{X} + \frac{4T}{ml}X = 0,$$

and again we have SHM with angular frequency

$$\omega = 2\sqrt{T/ml}.$$

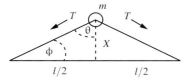

Fig. 4.6 Lateral displacement of a mass attached to a stretched string.

4.2.5 The effect of damping

All real mechanical systems are dissipative in character and so we need to know how to take into account the effects of friction on their oscillatory motion. However, one complicating factor is that the precise form of the resistive force due to friction is not always known. In practice, it is often assumed that the resistive force is simply proportional to the velocity of the particle (and its sign, of course, is such that it acts in the opposite direction to the velocity), and this is usually referred to as the **linear resistance law**. Other resistance laws can be assumed if the linear law does not prove adequate and a quadratic law is often tried.

However, there are two systems for which a linear law of resistance is known to be exact. One is the low-speed flow of a Newtonian fluid and the other is the flow of electricity in a passive circuit (i.e. one which does not contain any element other than inductance, capacitance and resistance). As the first of these problems is inherently non-linear, we shall illustrate the subject of damped harmonic motion by considering the flow of electric charge q round a closed circuit. We begin with the undamped oscillations of the charge in a circuit containing a capacitor C and an inductance L. Then we extend the analysis to the inclusion of a resistance R in the circuit. The resulting analysis can be generalized by analogy to any oscillatory motion which involves linear damping.

(a) LC **circuit**

Let the charge on the capacitor C at any time t be q, and assume that initially the capacitor has been charged up to $q = Q$. At time $t = 0$, we close the switch and the capacitor starts to discharge. The resulting current[1] $i = \mathrm{d}q/\mathrm{d}t$ flows through the inductance L and generates a magnetic field. This process carries on until the capacitor is charged up to $q = Q$ again, but with the opposite polarity. Then the process reverses, with i in the opposite direction, and so we have an oscillation in which electric charge moves back and forward.

We may obtain an 'equation of motion' for the electric charge stored in the capacitor as follows. At any time t, the energy stored in the capacitor is

$$U_E = \frac{1}{2}\left(\frac{q^2}{C}\right). \tag{4.8}$$

Also, the energy stored in the inductance is

$$U_B = \frac{1}{2}(Li^2) = \frac{1}{2}L\left(\frac{\mathrm{d}q}{\mathrm{d}t}\right)^2. \tag{4.9}$$

Hence in this system, the principle of conservation of energy takes the form:

Total energy $U = U_B + U_E = $ constant.

[1]The symbol for the current should not be confused with $i = \sqrt{-1}$, which occurs in equation (4.3).

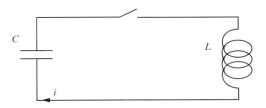

Fig. 4.7 An electrical circuit with undamped charge oscillations.

Now, differentiate both sides of this equation with respect to time, thus:

$$\frac{dU}{dt} = \frac{dU_B}{dt} + \frac{dU_E}{dt} = 0, \tag{4.10}$$

and substituting from (4.8) and (4.9) for U_E and U_B,

$$Li\frac{di}{dt} + \left(\frac{q}{C}\right)\frac{dq}{dt} = 0. \tag{4.11}$$

Then cancel $i = dq/dt$ across and divide by L to obtain:

$$\frac{di}{dt} + \frac{q}{LC} = 0,$$

and, finally, substituting $i = dq/dt$ we have

$$\frac{d^2q}{dt^2} + \frac{q}{LC} = 0. \tag{4.12}$$

This is an equation of SHM, and we note that this is a rare instance of us deriving this equation by means other than using N2! With an appropriate choice of initial conditions, the solution takes the form

$$q = Q\cos(\omega_0 t),$$

with angular natural frequency given by

$$\omega_0 = (1/LC)^{1/2}. \tag{4.13}$$

(b) *LCR* **circuit**

This time electrical energy is lost due to resistive heating and so the total electrical energy in the system U is no longer constant. Hence we now have that the total rate of change of electrical energy is given by

$$\frac{dU}{dt} = -i^2 R,$$

and (4.10) becomes

$$\frac{dU}{dt} = \frac{dU_B}{dt} + \frac{dU_E}{dt} = -i^2 R.$$

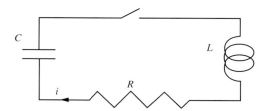

Fig. 4.8 An electrical circuit with damped charge oscillations.

Equations (4.8) and (4.9) are unaffected by the addition of thermal dissipation, hence the energy equation now takes the form:

$$Li\frac{di}{dt} + \left(\frac{q}{C}\right)\frac{dq}{dt} = -i^2 R.$$

Next cancel across $i = dq/dt$, divide by L and rearrange:

$$\frac{d^2 q}{dt^2} + \left(\frac{R}{L}\right)\frac{dq}{dt} + \frac{q}{LC} = 0. \tag{4.14}$$

This equation differs from equation (4.12) by the presence of the term involving the time derivative of the charge and is known as the equation of damped SHM. If we put $1/LC = \omega_0^2$ and set $R/L = 2\beta$ (known as the damping coefficient), it takes the general form:

$$\frac{d^2 q}{dt^2} + 2\beta\frac{dq}{dt} + \omega_0^2 q = 0. \tag{4.15}$$

Note that q is the analogue of displacement while dq/dt corresponds to the velocity in a mechanical system.

It is easily shown by direct substitution that the solution takes the form:

$$q = Q\,e^{-\beta t}\cos(\omega_0' t),$$

where the real exponential is the damping factor, and the damped natural frequency is given by

$$\omega_0' = (\omega_0^2 - \beta^2)^{1/2}. \tag{4.16}$$

Clearly, for vanishingly small damping, when $\beta \to 0$, this reduces to the undamped natural frequency ω_0, as given by equation (4.13).

4.2.6 Forced vibrations and resonance

Let us now consider a mechanical system which is subject to a resistive damping force given by $-b(dx/dt)$, where x is the displacement of the particle from its equilibrium position. We shall also assume that the system is driven by an oscillatory force $F = F_{max}\cos(\omega t)$. The resulting equation of motion is readily deduced from N2 by the addition of a damping force, and is a generalization of equation (4.2), thus:

$$m\frac{d^2 x}{dt^2} + b\frac{dx}{dt} + kx = F_{max}\cos\omega t.$$

Its resemblance to equation (4.15) should be noted. This resemblance may be increased, if we further write it as

$$\frac{d^2 x}{dt^2} + 2\beta\frac{dx}{dt} + \omega_0^2 x = \frac{F_{max}}{m}\cos\omega t;$$

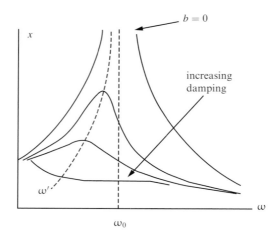

Fig. 4.9 Displacement versus driving frequency for a system undergoing forced oscillation.

where we have divided across by m and set $\omega_0^2 = k/m$, the undamped natural frequency, and $\beta = b/2m$, the damping coefficient.

Now, if we switch the driving force on at $t = 0$, the resulting motion is made up of a transient part plus a continuous part. The transient solution is just the result for unforced motion, viz.,

$$x_t = x_{max}e^{-\beta t}\cos\omega_0' t,$$

which oscillates at the damped natural frequency ω_0' and dies away in time $t = 1/\beta$.

The continuous solution may be found as the particular integral of the equation of motion, but it is readily verified by direct substitution that the correct form is:

$$x_c = x_{max}\cos\omega t.$$

This oscillates at the applied frequency ω until we switch the driving force off.

Typical behaviour of the continuous part of the solution is sketched out in a qualitative way in Figure 4.9, where it may be seen that the amplitude of the motion tends to increase when forcing frequencies approach the natural frequency of the system. This phenonenon is known as **resonance** and it occurs when $\omega = \omega_0'$. If the damping coefficient β is zero, then the amplitude at resonance is infinite, and $\omega = \omega_0' = \omega_0$.

4.3 Systems with several degrees of freedom

In the previous section, we have looked at some systems which have a single degree of freedom. By this we mean that only a single independent variable—the angular displacement or the linear displacement—was needed to specify completely the state of the system. If we now go on to the problem of two masses attached to a stretched string, then two displacements are needed to specify the state of the system; for three masses, clearly we should need three displacements. We can refer to these as examples of systems with, respectively, two and three degrees of freedom.

A general statement is possible, as follows:

A mechanical system is said to have n degrees of freedom if n independent variables are necessary and sufficient to specify the state of the system.

4.3.1 Two masses on a stretched string

Let us now consider a stretched string with two particles of mass m attached to it. We impose the same conditions as in Section 4.2.4: that is, the displacements are small so that one can treat the tension T as being unchanged from its equilibrium value, and also we can make the standard small-angle approximations.

The system is shown in Figure 4.10. The length of the stretched string is $2(a + b)$ and each particle is situated distance a from an end. The displacements from equilibrium are represented by X_1 and X_2.

We can extend the arguments of Section 4.2.4 to obtain an equation of motion for each mass. However, note that as drawn, one portion of string exerts a restoring force on the left-hand mass, whereas the other string pulls it away from equilibrium. With this in mind, the two equations may be written as:

$$m\ddot{X}_1 = -T\frac{X_1}{a} + T\frac{X_2 - X_1}{2b}$$

and

$$m\ddot{X}_2 = -T\frac{X_2}{a} - T\frac{X_2 - X_1}{2b}.$$

Now assume a trial solution:

$$X_1 = A\cos(\omega t + \phi)$$
$$X_2 = B\cos(\omega t + \phi),$$

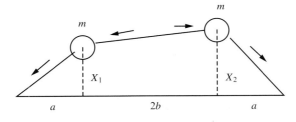

Fig. 4.10 Transverse displacements of two identical masses attached to a stretched string.

and substitute into both sides of the above equations of motion, with the result:

$$\left(\omega^2 - \frac{T}{m}\frac{a+2b}{2ab}\right)A = -\frac{T}{2mb}B \qquad (4.17)$$

$$\frac{T}{2mb}A = -\left(\omega^2 - \frac{T}{m}\frac{a+2b}{2ab}\right)B. \qquad (4.18)$$

If we cross-multiply these two equations and cancel the common factor of $-AB$, we end up with one equation, viz.,

$$\left(\omega^2 - \frac{T}{m}\frac{a+2b}{2ab}\right)^2 = \left(\frac{T}{2mb}\right)^2.$$

Taking square roots and rearranging gives us for ω^2:

$$\omega^2 - \frac{Ta + 2Tb}{2mab} \pm \frac{Ta}{2mab} = 0.$$

The permissible values of ω are obtained from this equation. Denoting these by ω_1, ω_2 we have

$$\omega_1^2 = \frac{T}{ma}, \qquad (4.19)$$

corresponding to a choice of the positive sign in the last term, and

$$\omega_2^2 = \frac{T(a+b)}{mab}, \qquad (4.20)$$

corresponding to the negative sign.

The constants A and B are related by equation (4.16) or (4.17). Choosing the first of these and substituting (4.18) for ω^2, we have

$$\left[\frac{T}{ma} - \frac{T}{m}\frac{(a+2b)}{2ab}\right]A = \frac{-T}{2mb}B;$$

and, after some cancellation and rearrangements,

$$A = B.$$

Similarly, substituting (4.19) for ω^2 leads to

$$A = -B.$$

The corresponding normal modes of vibration are as shown in Figure 4.11.

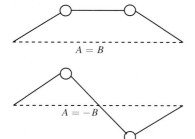

Fig. 4.11 Normal modes of transverse vibration for two masses attached to a stretched string.

4.3.2 Normal modes of vibration

A full discussion of the idea of normal modes of vibration, along with technical considerations like 'normal coordinates', would take us far

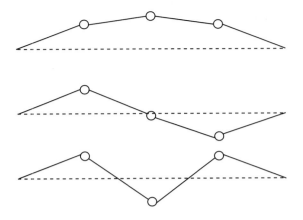

Fig. 4.12 Normal modes of transverse vibration for a system with three degrees of freedom.

beyond the scope and purpose of the present work. Therefore, we shall confine ourselves here to a few remarks which should help to establish the general idea.

We are already familiar with the idea that a system with one degree of freedom will, when started into motion, oscillate at its natural (or resonant) frequency ω_0. We simply extend this idea in an obvious way to systems with more than one degree of freedom.

For instance, to take the next step in order of complexity, a system with two degrees of freedom will have two natural frequencies. When started into motion, it will oscillate at one of these frequencies, depending on the initial conditions. However, the essential feature is that both the constituent masses oscillate at the **same** frequency, which will be one of the **two** natural frequencies for a system with two degrees of freedom.

We have already considered the normal modes of two masses on a stretched string. In the next section, we shall solve for the normal frequencies of the double pendulum. Here, the form of the normal modes is readily deduced to be as shown in Figure 4.15 below.

If we go on to systems with three degrees of freedom—and the problem of three masses attached to a stretched string will be a convenient example—then we may expect there to be three normal frequencies, and the modes of vibration will look like the illustration in Figure 4.12.

We can carry on in this way, and in the limit we can think of a string with its own mass as being a continuum with an infinite number of degrees of freedom. The normal modes of a stretched string are just the so-called standing waves and may be pictured as shown in Figure 4.13, for the first three, and so on. In Section 4.4.5 we shall see that a standing wave can be considered to be the superposition of two travelling waves.

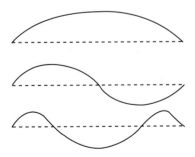

Fig. 4.13 Standing waves in a stretched string, showing the first three modes.

4.3.3 Example: *The double pendulum*

A mass m_1 hangs from a fixed point O by a string of length l_1, while a second mass m_2 hangs from m_1 by a string of length l_2. Obtain the

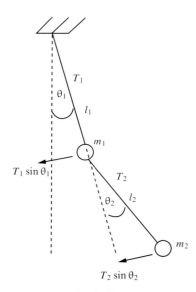

Fig. 4.14 The double pendulum.

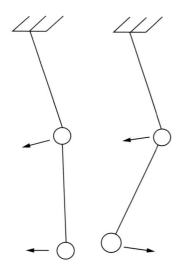

Fig. 4.15 Normal modes of vibration of a double pendulum.

equation for the frequencies of the normal modes. Solve this equation for the case $l_1 = l_2$, and verify that the two frequencies are approximately equal when $m_2 \ll m_1$.

To obtain the equations of motion, we assume that the motion is in a vertical plane through the point O and that the angles are small. We also assume that the static tensions are unaffected by the angular displacements. Thus, referring to Figure 4.14 for definitions of symbols, we have for the tension:

$$T_1 = (m_1 + m_2)g, \qquad T_2 = m_2 g, \qquad (4.21)$$

which requires $\cos\theta_1 \simeq \cos\theta_2 \simeq 1$.

Generalizing our earlier procedure for the simple pendulum, we denote the displacements of m_1 and m_2 from the vertical by X_1 and X_2, and apply N2 to each mass, thus:

$$m_1\ddot{X}_1 = -T_1 \sin\theta_1 + T_2 \sin\theta_2$$
$$m_2\ddot{X}_2 = -T_2 \sin\theta_2,$$

and taking the static tensions to be given by (4.20),

$$m_1\ddot{X}_1 = -(m_1 + m_2)g\frac{X_1}{l_1} + m_2 g\frac{(X_2 - X_1)}{l_2} \qquad (4.22)$$

$$m_2\ddot{X}_2 = -m_2 g\frac{(X_2 - X_1)}{l_2}. \qquad (4.23)$$

Note the presence of the coupling terms in both these equations. For $X_2 > X_1$, this force acts opposite to the 'simple pendulum' restoring force, and, of course for $X_2 < X_1$, it changes sign. It is instructive to compare these equations with the corresponding equations in the previous problem of two masses on a stretched string and this is left as an exercise for the reader. Again, as before, we assume the trial solutions

$$X_1 = A\cos(\omega t + \epsilon), \qquad X_2 = B\cos(\omega t + \epsilon),$$

and substituting these into (4.21) and (4.22) yields:

$$\left[\omega^2 - \frac{(1+\rho)g}{l_1} - \frac{\rho g}{l_2}\right]A + \frac{\rho g}{l_2}B = 0$$

$$\frac{g}{l_2}A + \left(n^2 - \frac{g}{l_2}\right)B = 0,$$

where $\rho = m_2/m_1$. Eliminating AB, we then obtain:

$$\omega^4 - (1+\rho)g\left(\frac{1}{l_1} + \frac{1}{l_2}\right)\omega^2 + (1+\rho)\frac{g^2}{l_1 l_2} = 0, \qquad (4.24)$$

which is a quadratic in ω^2.

In general it can be shown that real roots ω_1^2 and ω_2^2 exist, but we shall concentrate here on the simpler special case where $l_1 = l_2$. Thus the general solution equation (4.23) reduces to

$$\omega^2 = (1 + \rho)\frac{g}{l} \pm \frac{1}{2}\sqrt{(1 + \rho)^2\frac{g^2}{l^2} - (1 + \rho)\frac{g^2}{l^2}}$$

$$= (1 + \rho)\frac{g}{l}\left[1 \pm \left(\frac{\rho}{1 + \rho}\right)^{1/2}\right].$$

Evidently if $m_2 \ll m_1$, then $\rho \to 0$, $\rho/(1 + \rho)^{1/2} \to 0$ and $\omega^2 \simeq (1 + \rho)g/l$, as required.

4.4 Waves

So far we have used the concept of SHM to study periodic motion which is localized at a point in space. Now we wish to extend the concept of SHM to periodic motion which travels through space.

We can envisage a helpful model of wave motion in continuous media by considering a chain of oscillators coupled together. For example, we can model waves on stretched strings by coupling together a lot of spring–mass systems along a line in the x-direction. If we displace the first mass in the y-direction, then it will execute SHM, which will start the second mass oscillating, and so on.

We note that the second mass will lag behind the first, the third will lag behind the second, and so on. Thus the phase will vary along the chain of oscillators and we can think of the wave as being a fixed value of the phase which is travelling along the chain in the x-direction. We should also note that, although the physical system does not move through space, the wave motion on it does transfer both energy and momentum in the x-direction. These ideas can be taken over to the case of a continuous string, stretched under tension T between two supports. The individual masses are then replaced by the mass per unit length, and the string tension replaces the individual spring constants.

4.4.1 SHM in a moving reference frame

As a way of introducing some of the mathematical features of wave motion, let us consider a particle (which could be an infinitesimal element of mass in some continuous medium) which is executing one-dimensional simple harmonic motion in the inertial frame of reference S'. We take y' as its displacement at any time t and its angular frequency to be given by

$$\omega = 2\pi/T.$$

The SHM is an oscillation in time at any particular value of x' and, for convenience, we shall choose the value $x' = 0$.

Now, to an observer in S, which is another inertial frame in standard configuration with S', and moving with speed u in the direction of negative x, the motion of the particle will appear to be extended in the

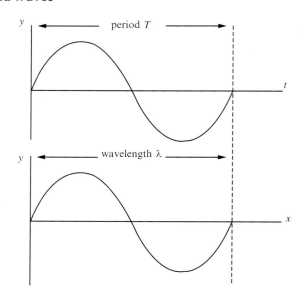

Fig. 4.16 Correspondence of the period of SHM in its rest frame to its wavelength in a moving frame.

positive x-direction and to take the form of a wave with spatial frequency given by

$$k = 2\pi/\lambda.$$

Both these interpretations are illustrated in Figure 4.16. The wavelength in the frame S is given by the Galilean transformation, where we take the inverse of (1.54), and write

$$x = x' + ut,$$

as, in this case, the speed of the moving frame S is u. The wavelength λ then corresponds to the distance x from the origin in S at time $t = T$ (recall that the oscillator is situated at $x' = 0$ and that the two frames are in standard configuration), so that we find

$$\lambda = uT. \tag{4.25}$$

Of course, the wave in this case is no more than a trace or snapshot of a wave (see Exercise 4.1, at the end of this chapter), but the relationship just derived also holds for a real travelling wave. We shall consider the properties of such waves in the next section.

4.4.2 The mathematical representation of a travelling wave

Now we restrict our attention to one frame of reference and examine the characteristics of a wave travelling through that frame. At each value of x there is SHM with period T. Also, the wave travels a distance λ during the period T. Hence, with simple rearrangements of equation (4.24) above, we can write expressions for the wave speed in the forms

$$u = \lambda/T \quad \text{or} \quad u = \omega/k. \tag{4.26}$$

Similarly, at each position x, the SHM is given by:

$$y = A\sin(\omega t + \phi). \tag{4.27}$$

We should note that the **phase constant** ϕ depends on x. You can see this from the picture in Figure 4.16, which is a 'snapshot' of a wave varying with x at a given time t. Thus the 'phase constant' is $\phi = \phi(x)$.

Now we will consider a fixed point on a wave and follow it through space in the x-direction. This means taking a fixed value of the phase and for convenience we take the phase to be zero. This corresponds to one point on the wave where the displacement y is also zero. So we consider a point on the wave travelling from left to right, such that:

$$(\omega t + \phi(x)) = 0.$$

The general equation for the displacement at any x is just

$$y = A\sin(\omega t + \phi(x)),$$

and we shall start our wave off with $\phi = 0$ at $x = 0$ and $t = 0$, and hence (from the above equation) $y = 0$.

In order to move with the reference frame of the wave, we have to keep the total phase constant (and equal to zero) as we move from left to right. But ωt increases with increasing time, so $\phi(x)$ must decrease as x increases, in order to compensate and keep the total phase equal to zero, and so we must have

$$\phi(x) = -\omega t.$$

However, the time taken for point to go a distance x is $t = x/u$, so it follows that

$$\phi(x) = -\omega t = -\omega x/u = -kx,$$

and hence the result:

$$\text{The phase of a travelling wave} = \omega t - kx.$$

Finally, we may now write the equation for the displacement of a travelling wave as

$$y = A\sin(\omega t - kx), \tag{4.28}$$

in the case where the wave is moving in the direction of *increasing* x.

It is readily shown, by means of arguments similar to those just given, that the the equation for the displacement of a wave travelling in the direction of *decreasing* x is

$$y = A\sin(\omega t + kx), \tag{4.29}$$

and this is left as an exercise for the reader.

(a)

direction of SHM

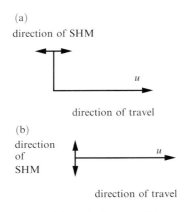

direction of travel

(b)

direction
of
SHM

u

direction of travel

Fig. 4.17 (a) Longitudinal and (b) transverse waves.

We close this section with some general remarks on types of waves. The most general of these is that waves can be divided into two kinds, as follows:

- **Longitudinal waves** are waves in which the displacement is in the direction of travel of the wave.

- **Transverse waves** are waves in which the displacement is perpendicular to the direction of travel.

The two types are illustrated in Figure 4.17. Compressional waves in solids and fluids (e.g. sound waves) are examples of longitudinal waves while electromagnetic waves (e.g. light, radio waves), and the vibrations of stretched strings in musical instruments are examples of transverse waves. We should also remark that wave motion can be associated with vector fields in two or more dimensions. However, in the interests of brevity, we shall restrict our attention to one-dimensional displacements only. All the results which we obtain can be quite easily generalized to higher numbers of dimensions.

4.4.3 *Wave speeds in elastic media as obtained by Galilean transformation*

We shall give a conventional derivation of an expression for the wave speed in an elastic medium, along with a derivation of the wave equation in the next section. Here we show how a Galilean transformation to the comoving frame of the wave may be used to obtain an expression for the speed of a wave in a stretched string.

As usual, we take the string to be stretched between supports. If we displace the string slightly, the tension T acts as a restoring force and the displaced part of the string oscillates. This oscillation travels along the string as a wave. The wave speed will depend on the mass of the string (strictly, on the mass per unit length of the string or the linear density ρ) and on how tightly the string is stretched (that is, the tension force T). This oscillation is, of course, an example of a transverse wave.

Our problem is this: there is a small transverse disturbance moving from left to right: what is its speed u? Our basic trick is to transform to the reference frame of the moving wave: in this frame of reference, the string then appears to move from right to left with speed $-u$. This situation is illustrated in Figure 4.18.

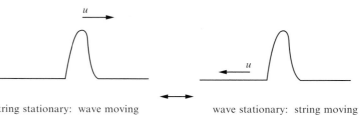

Fig. 4.18 Transformation to a frame of reference moving with the wave.

string stationary: wave moving

wave stationary: string moving

Consider a small part of the crest of a wave, of length l, as shown in Figure 4.19. The 'dotted box' means that we are moving with the wave. As the string has mass per unit length ρ, the mass of the element of length l is ρl.

Particles of string move round the arc of a circle of radius R. Due to their circular motion, they experience a centrifugal force which acts radially outward. Thus by the usual elementary result from circular motion, we have the radial outward force acting on the element of mass $m = \rho l$ as:

$$\frac{mu^2}{R} = \frac{\rho l u^2}{R}.$$

The tension force in the string keeps the particles from moving out radially. We can see this by resolving the tension force into its radial and circumferential components, as shown in Figure 4.20.

We note that the circumferential components oppose each other and therefore cancel. However, there is a net radial inward force, given by:

$$2T\sin\alpha = \frac{2Tl}{2R} = \frac{Tl}{R},$$

where for small angles α, we approximate $\sin\alpha = (l/2)R = l/2R$.

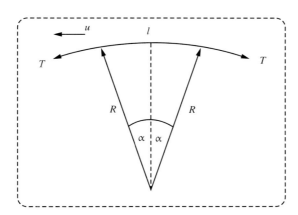

Fig. 4.19 The motion of the string in the rest frame of the wave.

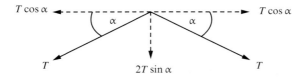

Fig. 4.20 Tension in the string resolved into radial and circumferential components.

For equilibrium, we equate the outward and inward forces to get:

$$\frac{\rho l u^2}{R} = \frac{Tl}{R}.$$

Then, cancelling across the common factor l/R, and taking the square root of both sides, we have for the wave speed of transverse waves:

$$u = (T/\rho)^{1/2}. \tag{4.30}$$

In fact this result can be generalized to other forms of waves in elastic media, but we shall leave that until we have given the more general treatment of the wave equation in the next section.

4.4.4 The general form of the wave equation and its solutions

We shall continue to use the example of a stretched string as a convenient specific system, and again we consider a string, with mass density ρ per unit length, which is stretched between fixed supports under a tension T. As always, we restrict our attention to a wave motion where the tension is unaltered from its equilibrium value by this displacement, and thus displacements have to be small.

Referring to Figure 4.21, initially the string lies along the x-axis and we concentrate on an element lying between x and $x + \mathrm{d}x$. At some later time the string has a displacement $y(x)$ in the y-direction. We apply N2 to the mass element $\rho\mathrm{d}x$ to write its equation of motion as

$$\rho\mathrm{d}x\frac{\mathrm{d}^2 y}{\mathrm{d}t^2} = F(x + \mathrm{d}x) - F(x),$$

where the right-hand side gives the net force due to the string tension in the y-direction. Expanding out $F(x + \mathrm{d}x)$ to first-order in **Taylor series** and cancelling the terms $F(x)$, we have:

$$\rho\mathrm{d}x\frac{\mathrm{d}^2 y}{\mathrm{d}t^2} = \mathrm{d}x\frac{\partial F}{\partial x}.$$

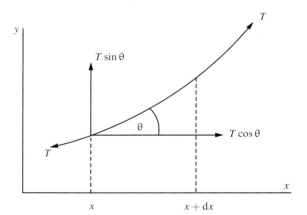

Fig. 4.21 An infinitesimal portion of a transverse wave in a stretched string.

Then, further cancelling across factors of dx, and substituting $F = T \sin \theta$, which is resolved in the y-direction, we obtain:

$$\rho \frac{d^2 y}{dt^2} = \frac{d}{dx}(T \sin \theta).$$

Now, we make the approximations based on small displacements. That is, we treat T as constant and invoke the usual small-angle relationships

$$\sin \theta \simeq \tan \theta = dy/dx.$$

Then, having done that, we divide across by T and rearranging, we get the standard result for the one-dimensional wave equation:

$$\frac{d^2 y}{dx^2} = \frac{1}{u^2} \frac{d^2 y}{dt^2}, \tag{4.31}$$

where the wave speed u is given by

$$u^2 = T/\rho. \tag{4.32}$$

This result can be generalized quite easily to disturbances moving in three space dimensions. If we let ψ stand for any displacement—it could, for example, be an electrical field or a hydrostatic pressure or even a simple linear displacement, as in the preceding discussion—then the general equation of wave motion can be written as

$$\nabla^2 \psi(\mathbf{x}, t) = \frac{1}{u^2} \frac{\partial^2 \psi}{\partial t^2}(\mathbf{x}, t),$$

where the expression analogous to equation (4.31) for the wave speed will depend on the specific physical system under consideration. We conclude this section with a brief discussion of some possible extensions of these results to compression and torsional waves in elastic media.

The transverse waves which we have been discussing rely on tension to provide the restoring force. However, the result embodied in equation (4.31) for transverse waves can be expressed in a general way as:

$$(\text{wave speed})^2 = (\text{restoring force})/(\text{mass per unit length}).$$

Now, in elastic media, the restoring force will depend on the relevant elastic modulus. For instance, in compression waves, the restoring force will be due to the bulk modulus B, which we shall take as an example, as follows.

For the case of compression waves in a liquid or gas, we consider a volume element $V = A \times l$, where A is the area and l is the length of the element. We note that B has units of stress = force/area and so the restoring force is proportional to $B \times A$. Substitute this into our wave speed formula:

$$(\text{wave speed})^2 = (B \times A)/(\text{mass per unit length}) = B/\rho,$$

where $\rho = $ mass per unit volume is the density of the fluid. Finally, therefore, we have for the wave speed of compression waves in a fluid:

$$u = (B/\rho)^{1/2}. \tag{4.33}$$

We can extend this result to other cases by making an appropriate change of elastic constant. For example:

- Compression waves in a metal bar: $u = (Y/\rho)^{1/2}$, where Y is Young's modulus;
- torsional waves in a metal rod: $u = (\sigma/\rho)^{1/2}$, where σ is the shear modulus.

4.4.5 Travelling waves and standing waves

In Section 4.4.2, we discussed the idea of waves moving through a medium in the sense of a surface of constant phase moving in (say) the x-direction. For obvious reasons, such waves are known as **travelling waves**. However, if we superimpose two identical travelling waves going in opposite directions along (say) a stretched string the result can be a standing wave. We may see this as follows.

Consider two identical waves going in the direction of positive x and the direction of negative x, respectively, as given by equations (4.27) and (4.28); thus:

$$\longrightarrow Y_1 = A\sin(\omega t - kx) \quad \text{and} \quad Y_2 = -A\sin(\omega t + kx)\longleftarrow.$$

The resulting motion of the stretched string is found by adding the two waves together. The resultant is:

$$Y = Y_1 + Y_2 = A(\sin(\omega t - kx) + \sin(\omega t + kx)) = 2A\sin(kx)\cos(\omega t), \tag{4.34}$$

where we have used the trigonometric identity:

$$\sin B - \sin C = 2\sin[(B - C)/2]\cos[(B + C)/2].$$

Equation (4.33) for y is the equation of a **standing wave**. We can interpret it as meaning that each particle in the string is executing SHM in the following way, according to the particular value of its x coordinate:

$$y = \text{'Amplitude'} \times \cos\omega t,$$

where

$$\text{'Amplitude'} = 2A\sin kx.$$

That is, the amplitude of the motion depends on where the particle is, and some particles will even have zero amplitude and therefore be stationary. It is usual to categorize standing waves by noting where they are permanently stationary and also where they always have maximum amplitude of oscillation. These points are so important that they are given special names which may be formally defined as follows:

- **Antinodes** These are points of maximum amplitude of oscillation. They occur where $\sin kx = 1$; or $kx = \pi/2, 3\pi/2, 5\pi/2, \ldots$; or: $x = \lambda/4, 3\lambda/4, 5\lambda/4, \ldots$
- **Nodes** These are points of zero amplitude of oscillation. They occur where $\sin kx = 0$; or: $kx = \pi, 2\pi, 3\pi, \ldots$; or: $x = \lambda/2, \lambda, 3\lambda/2, \ldots$

The positions of the nodes and antinodes in a system determine the **modes of vibration**. As an example, we shall discuss the modes of a stretched string of length l. As the string is fixed at both ends, any vibrations must have nodes at the ends of the string. This limits the possible vibrations to the series illustrated (for the first three only) in Figure 4.22. It should be obvious that, although we have only shown three modes, there is in fact an infinite series of such modes with an increasing number of nodes and antinodes in each.

We can also work out the temporal frequency of a standing wave, as we already know its spatial frequency from the pattern made by the given mode of vibration. From the results given in Section 4.4.1, we also know that angular time and spatial frequencies are related by $\omega = ku$; or (equivalently) by $f = u/\lambda$, for the frequency f. As a specific example, we may work out an expression for the fundamental note of a stretched string. In this case, we have $\lambda = 2l$ and $u = (T/\rho)^{1/2}$, where T is the tension and ρ is the linear density of the string in units of mass per unit length. Thus, from the above expression for the frequency we obtain

$$f = \frac{(T/\rho)^{1/2}}{2l}.$$

This procedure applies to any vibrating system: we just need to know the permitted wavelengths, as determined by the boundary conditions, and the speed of wave propagation in the system.

4.4.6 Acoustics in moving reference frames: the classical Doppler effect

The study of the Doppler effect on sound waves is especially worthy of our attention on two grounds. First, it is experienced in everyday life. When, for instance, an ambulance passes us, the siren appears to change its note. As it approaches, the pitch is high. As it passes us, the pitch falls. The effect occurs with any type of wave but we shall discuss sound waves here. Second, the fact that the speed of sound is very small compared to the speed of light means that the acoustic Doppler effect is inherently a matter of Galilean relativity. As we shall see later, corrections to classical results in special relativity are expressible in powers of the characteristic speed divided by the speed of light *in vacuo*. For any typical sonic speed, this would lead to immeasurably small corrections. However, for the optical Doppler effect, the converse is true. As we shall see in Section 12.2.7, this is inherently a matter of special relativity.

For the purposes of the present section, we again consider sound waves of velocity u and wavelength λ. There are two general cases of interest and we examine these in turn.

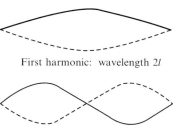

First harmonic: wavelength $2l$

Second harmonic: wavelength l

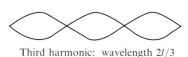

Third harmonic: wavelength $2l/3$

Fig. 4.22 The first three modes of transverse vibration of a stretched string.

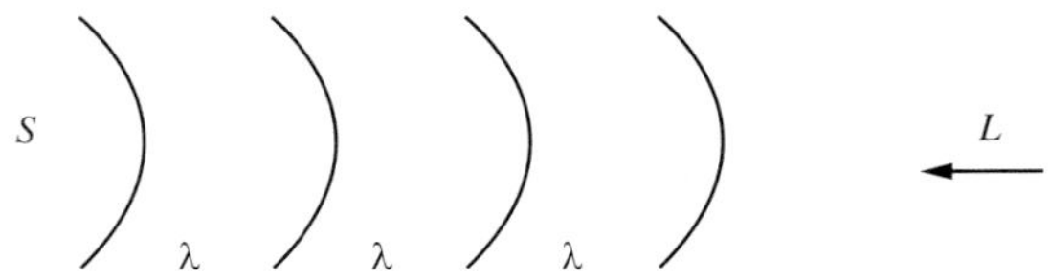

Fig. 4.23 Stationary source, moving listener.

Case 1: Source S at rest, listener L moving with speed v_L. This case is illustrated in Figure 4.23. An experimentalist measures the frequency of the sound waves f by counting the number of complete cycles in time t, and dividing by t. We shall consider in turn the result of carrying out this operation in the two different inertial frames.

1. The rest frame of the source S:

$$\text{number of cycles in time interval } t = \frac{\text{length of wave train}}{\text{length of one wave}} = \frac{ut}{\lambda}.$$

Hence it follows that the frequency is given by

$$\text{the number of cycles per second} = ut/\lambda t = u/\lambda = f.$$

2. The rest frame of the listener L. Now L is moving at speed v_L relative to S. So, by Galilean transformation, waves approach the experimentalist at speed $u + v_L$. Hence:

$$\text{number of cycles in time interval } t = \frac{\text{length of wave train}}{\text{length of one wave}}$$
$$= \frac{(u + v_L)t}{\lambda},$$

and so now the frequency is given by:

$$\text{the number of cycles per second} = \frac{(u + v_L)}{\lambda} = \left(\frac{u}{\lambda}\right)\frac{(u + v_L)}{u}$$
$$= \frac{(u + v_L)f}{u} = f_L.$$

It is easily seen that the general result for this case is given by

$$f_L = \frac{(u \pm v_L)f}{u} = \left(1 \pm \frac{v_L}{u}\right)f,$$

where the positive sign means that the listener approaches the source, while the minus sign means that the listener recedes from the source.

Case 2: Source S moving with speed v_s, listener L at rest. This situation is shown in Figure 4.24. As the source moves towards the listener, the wavelength received by L is $\lambda' < \lambda$. We can see this as follows: Referring to Figure 4.25, we again examine the situation in the two different frames.

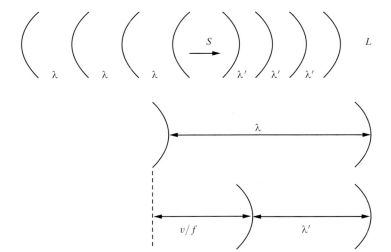

Fig. 4.24 Source moving, listener at rest.

Fig. 4.25 Shortening of the perceived wavelength due to the approach of the source.

1. In its rest frame, S emits two waves, time $T = 1/f$ apart.
2. In the frame of L, the moving S emits two waves, time $T = 1/f$ apart[2] and during that time S moves a distance $v_s T = v_s/f$.

[2] As we shall see in Chapter 11, this assumption of the universality of time would not be valid in special relativity.

It follows that the experimentalist at rest in L's frame measures a wavelength given by

$$\lambda' = \lambda - v_s/f.$$

Therefore L hears an apparent frequency of sound which is given by

$$f' = \frac{u}{\lambda'} = \frac{u}{(\lambda - v_s/f)} = f\left[\frac{u}{(u - v_s)}\right],$$

where we have multiplied top and bottom by f.

It is easily shown that the general result is just

$$f' = f[u/(u \mp v_s)],$$

where the positive sign means that the source recedes from the listener, and the minus sign means that the source approaches the listener.

4.5 Exercises

4.1 Work out the period of a simple pendulum of length 0.49 m. A small ink jet is attached to the bob of such a pendulum. A strip of paper under the bob is pulled along horizontally and at right angles to the plane of oscillation of the pendulum. Write down the equation for the wave drawn on the paper, assuming that the oscillations are undamped.

4.2 A right circular cylinder floats with its axis vertical in a liquid of density ρ. If the cylinder is given a small displacement in the vertical direction, show that the period of its subsequent vertical oscillations is given by the expression

$$\tau = 2\pi\sqrt{\frac{h}{g}},$$

where h is the mean depth of the cylinder below the surface of the water, and g is the gravitational acceleration.

4.3 If a straight smooth tunnel is bored through the Earth (from London to Paris, say) and a particle is dropped into one end, show that the particle will execute simple harmonic motion with period given by:

$$\tau = 2\pi \sqrt{\frac{a}{g}},$$

where a is the radius of the Earth.

4.4 In Section 2.2.5, we found that the period of a simple pendulum, executing oscillations of finite amplitude, was given by:

$$\tau = \sqrt{\frac{L}{g}} \int_0^{\theta_{\max}} \frac{d\theta}{[\cos\theta - \cos\theta_{\max}]^{1/2}},$$

where L is the length of the pendulum and θ is the angle made with the vertical at any time. Show that this form reduces to the result for small oscillations provided that the maximum amplitude $\theta_{\max}$ is small enough for the cosine to be approximated by

$$\cos\theta_{\max} = 1 - \frac{\theta_{\max}^2}{2}.$$

4.5 A light string of length $3a$ is stretched between two fixed points on a smooth horizontal plane. Two particles of mass m are fixed to the string such that they divide it into three equal lengths. Show that, if the particles are given small displacements from equilibrium, they describe transverse oscillations on the plane with equations of motion given by:

$$\ddot{x}_1 + n^2(2x_1 - x_2) = 0$$
$$\ddot{x}_2 + n^2(-x_1 + 2x_2) = 0,$$

where $n^2 = T/ma$. Find the frequencies of the normal modes of vibration and sketch their form.

4.6 Repeat the previous question, on the assumption that the system has been set into longitudinal oscillation. Sketch the form of the normal modes of vibration.

4.7 A light string of length $3a$ is stretched between two fixed points on a smooth horizontal plane. Two particles of mass m are fixed to the string such that they divide it into three equal lengths. The plane is rough and a particle moving over it experiences a frictional force of magnitude $2mk$ times its velocity. Show that, if the particles are given small displacements from equilibrium, they describe transverse oscillations on the plane with equations of motion given by:

$$\ddot{x}_1 + 2k\dot{x}_1 + n^2(2x_1 - x_2) = 0$$
$$\ddot{x}_2 + \dot{x}_2 + n^2(-x_1 + 2x_2) = 0,$$

where $n^2 = T/ma$. If $k < n$, find the normal frequencies by considering trial solutions of the form

$$x = A \exp[i(\omega t + \phi)],$$

and sketch the normal modes of oscillation.

4.8 A particle moves under the influence of a force directed towards a fixed point and of magnitude proportional to the distance of the particle from the point. Show that the motion of the particle can be resolved into simple harmonic motion along the x and y directions. What is the form of the orbit? [Hint: the orbit is an example of a Lissajous figure.]

4.9 In the classical Doppler effect, as discussed in Section 4.4.6, the magnitude of the effect of a moving source on a stationary listener is not the same as when the source is stationary and the listener is moving. Show that the two effects become identical when the speeds of source and listener are small compared to the speed of sound.

Why are the two effects different when apparently one situation can be transformed into the other on the basis of a Galilean transformation? [Hint: evaluate the frequency shifts for the two cases when the speed of source or listener is the same as the speed of sound.]

Systems of N particles

5

In this chapter, we are going to extend our two-particle analysis, as given in Section 2.4, to a general system, consisting of N particles. It should be noted that we restrict our attention to interaction forces between pairs of particles that act along the lines joining the pairs of particles. The particles making up the system are to be thought of as being labelled with an index s, which is an integer and can take any value from unity to N. The coordinates that specify the positions of the individual particles are referred to a fixed Cartesian basis. As coordinates, we use the three-dimensional position vector $\mathbf{x}_1$ to denote the position of particle 1, which is taken as having mass m_1; and similarly $\mathbf{x}_2$ for m_2, $\mathbf{x}_s$ for m_s, and so on, for all N particles.

The general system is illustrated schematically in Figure 5.1, where the dotted lines are intended to hint at the presence of other particles, too numerous to show explicitly. One must be very careful to remember that the index on the vector refers only to the identity of the particle and has nothing to do with the components of the vector. In any of the various coordinates systems or conventions, the position vector of each particle has the components appropriate to that system. For instance, to take two specific examples, in Cartesian coordinates and spherical polar coordinates, the position vectors have components like:

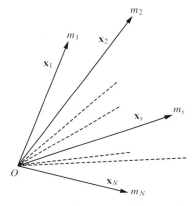

Fig. 5.1 Coordinate system for a set of N particles.

- $\mathbf{x}_1 \equiv (x_1, y_1, z_1) \equiv (r_1, \theta_1, \phi_1)$;
- $\mathbf{x}_2 \equiv (x_2, y_2, z_2) \equiv (r_2, \theta_2, \phi_2)$;
- and so on.

Note also that the familiar Cartesian convention $\mathbf{x} \equiv (x_1, x_2, x_3)$ is not quite so convenient in many-body problems, in that one has to resort to a superscript, as well as the component subscript, in order to identify the particle.

5.1 The centre of mass of a system

In certain circumstances, the behaviour of a system of particles may be equivalent to the behaviour of a single fictitious particle, with a mass M, situated at a point called the **centre of mass of the system**, where

$M = \sum_{s=1}^{N} m_s$ is the total mass of the system. We may establish the position of the centre of mass by taking moments about the origin of coordinates, as follows.

Let the centre of mass (CM) be defined as the point whose position vector **R** is given by:

$$MR = \left(\sum_{s=1}^{N} m_s \right) \mathbf{R} = \sum_{s=1}^{N} m_s \mathbf{x}_s,$$

where the left-hand side is the moment of the total mass M, supposed concentrated at a point located by the vector **R**, about the origin; and the right-hand side is the sum of the moments of the individual particles about the origin. This expression can be rearranged to give the conventional equation for the position of the centre of mass, thus:

$$\mathbf{R} = \frac{1}{M} \sum_{s=1}^{N} m_s \mathbf{x}_s. \tag{5.1}$$

This is, of course, just the generalization of equation (2.25) for the two-particle case, to the case of many particles.

5.1.1 Example: *Centre of mass of three given masses at specified positions*

A system consists of three particles of mass 3, 5 and 2 units, which are situated respectively at the points $(1, 0, -1)$, $(-2, 1, 3)$ and $(3, -1, 1)$. Obtain the coordinates of the centre of mass of the system. Verify your result by checking that the moments of the particles, in coordinates relative to the centre of mass, sum to zero.

We may set out the details of the system as follows:

$$m_1 = 3 \text{ at the point } (1, 0, -1)$$
$$m_2 = 5 \text{ at the point } (-2, 1, 3)$$
$$m_3 = 2 \text{ at the point } (3, -1, 1).$$

First we work out the total mass of the three-particle system. This is just:

$$M = \sum_{s=1}^{3} m_s = 10.$$

Now we use equation (5.1) to work out the position of the centre of mass. The contributions of the individual particles to the summation are:

$$m_1 \mathbf{x}_1 = (3, 0, -3)$$
$$m_2 \mathbf{x}_2 = (-10, 5, 15)$$
$$m_3 \mathbf{x}_3 = (6, -2, 2)$$

and hence the sum of the three is

$$\sum_{s=1}^{3} m_s \mathbf{x}_s = (-1, 3, 14).$$

Then, from equation (5.1), we have

$$\mathbf{R} = \frac{1}{M} \sum_{s=1}^{3} m_s \mathbf{x}_s = \left(-\frac{1}{10}, \frac{3}{10}, \frac{14}{10}\right),$$

as the position of the centre of mass. Next we consider the relative coordinate, as defined by equation (2.28). This is given by

$$\mathbf{r}_s = \mathbf{x}_s - \mathbf{R}.$$

The property that the moments in coordinates relative to the centre of mass sum to zero must hold separately for each Cartesian component $r_x = x - R_x$, $r_y = y - R_y$, $r_z = z - R_z$, where $\mathbf{x} \equiv (x, y, z)$. We shall verify it here for the x-component only. For each of the three particles in turn, we have

1. $m_1 = 3$, $x - R_x = \frac{11}{10}$, $m(x - R_x) = \frac{33}{10}$;
2. $m_2 = 5$, $x - R_x = -\frac{19}{10}$, $m(x - R_x) = -\frac{95}{10}$;
3. $m_3 = 2$, $x - R_x = \frac{31}{10}$, $m(x - R_x) = \frac{62}{10}$;

and adding up the three final terms, we have $\frac{33}{10} - \frac{95}{10} + \frac{62}{10} = 0$, as required, with similar results for the other two coordinate directions.

5.2 Linear momentum of a system

In this section we shall introduce the total linear momentum, P, of the system and show that N2 for the system takes the form:

$$\dot{\mathbf{P}} = M\ddot{\mathbf{R}} = \mathbf{F}_{\text{tot}},$$

where the dots denote differentiation with respect to time. The procedure by which we do this could hardly be simpler. We write down N2 for a single particle and then add up the individual terms for all N particles of the system.

The equation of motion for particle s is just N2 applied to that particle, thus:

$$\dot{\mathbf{p}}_s = \mathbf{F}_s + \sum_{k \neq s} \mathbf{F}_{sk}, \tag{5.2}$$

where $\mathbf{p}_s = m\mathbf{v}_s$ is the momentum of particle s, and the forces on the right-hand side are:

- $\mathbf{F}_s$ external force on particle s;
- $\mathbf{F}_{sk}$ internal force acting on particle s due to particle k.

It should be noted that the sum on the right-hand side is over all the other $(N-1)$ particles to give the total *internal* force acting on $\mathbf{m}_s$.

Next we add up all the equations of motion to get the total rate of change of momentum for the system:

$$\dot{\mathbf{P}} = \sum_{s=1}^{N} \dot{\mathbf{p}}_s = \sum_s \mathbf{F}_s + \sum_s \sum_k \mathbf{F}_{sk} \quad \text{(where } k \neq s\text{).} \qquad (5.3)$$

Just as in the two-particle case, the double sum on the right-hand side gives zero. This is because all the terms cancel in pairs, as a consequence of N3, viz., $F_{12} + F_{21} = 0$, $F_{13} + F_{31} = 0$, and so on. Hence, for the system as a whole, N2 takes the form:

$$\dot{\mathbf{P}} = \sum_{s=1}^{N} \mathbf{F}_s = \mathbf{F}_{\text{tot}}, \qquad (5.4)$$

where $\mathbf{F}_{\text{tot}}$ is the total externally applied force. As in the two-particle case, we can express the results in terms of the motion of the CM using equation (5.1). Thus,

$$\dot{\mathbf{P}} = M\ddot{\mathbf{R}} = \mathbf{F}_{\text{tot}}. \qquad (5.5)$$

Thus the whole system behaves like a particle whose mass equals the total mass of the system, which is located at the centre of mass and which is acted upon by the total external force $\mathbf{F}_{\text{tot}}$. If $\mathbf{F}_{\text{tot}}$ is zero, then the total *linear* momentum of the system is conserved.

5.2.1 Example: *Velocity of the centre of mass of three particles*

Three particles of mass 2, 1 and 5 units have the following position vectors:

$$x_1 = 5t\mathbf{i} - 2t^2\mathbf{j} + (3t - 2)\mathbf{k}$$
$$x_2 = (2t - 3)\mathbf{i} + (12 - 5t^2)\mathbf{j} + (4 + 6t + 3t^2)\mathbf{k}$$
$$x_3 = (2t - 1)\mathbf{i} + (t^2 + 2)\mathbf{j} - t^3\mathbf{k},$$

where t is time. Find the velocity of the centre of mass of the system at $t = 1$, in arbitrary units. Check your answer by adding up the individual momenta of the three particles.

The details of the three-particle system are as follows:

$$m_1 = 2 \quad \text{at} \quad 5t\mathbf{i} - 2t^2\mathbf{j} + (3t - 2)\mathbf{k}$$
$$m_2 = 1 \quad \text{at} \quad (2t - 3)\mathbf{i} + (12 - 5t^2)\mathbf{j} + (4 + 6t + 3t^2)\mathbf{k}$$
$$m_3 = 5 \quad \text{at} \quad (2t - 1)\mathbf{i} + (2 + t^2)\mathbf{j} - t^3\mathbf{k},$$

and the total mass is

$$M = \sum_{s=1}^{3} m_s = 8.$$

As before, we work out the individual contributions to the sum $\sum_s m_s \mathbf{x}_s$ as:

$$m_1 \mathbf{x}_1 = 10t\mathbf{i} - 4t^2\mathbf{j} + (6t - 4)\mathbf{k}$$
$$m_2 \mathbf{x}_2 = (2t - 3)\mathbf{i} + (12 - 5t^2)\mathbf{j} + (4 + 6t + 3t^2)\mathbf{k}$$
$$m_3 \mathbf{x}_3 = (10t - 5)\mathbf{i} + (10 + 5t^2)\mathbf{j} - 5t^3\mathbf{k},$$

and hence, from equation (5.1), the position vector of the centre of mass is

$$\mathbf{R}(t) = \frac{1}{8}[(22t - 8)\mathbf{i} + (22 - 4t^2)\mathbf{j} + (12t + 3t^2 - 5t^3)\mathbf{k}].$$

The velocity of the centre of mass is obtained by differentiating $\mathbf{R}$ with respect to time, thus:

$$\dot{\mathbf{R}}(t) = \frac{1}{8}[22\mathbf{i} - 8t\mathbf{j} + (12 + 6t - 15t^2)\mathbf{k}],$$

and for $t = 1$,

$$\dot{\mathbf{R}}(1) = \frac{1}{8}[22\mathbf{i} - 8\mathbf{j} + 3\mathbf{k}],$$

so the total linear momentum is:

$$\mathbf{P}_{\text{tot}}(1) = M\dot{\mathbf{R}}(1) = 22\mathbf{i} - 8\mathbf{j} + 3\mathbf{k}.$$

We check this result by first working out the linear momenta of the individual particles, thus:

$$m_1 \dot{\mathbf{x}}_1 = 10\mathbf{i} - 8\mathbf{j} + 6\mathbf{k}$$
$$m_2 \dot{\mathbf{x}}_2 = 2\mathbf{i} - 10\mathbf{j} + 12\mathbf{k}$$
$$m_3 \dot{\mathbf{x}}_3 = 10\mathbf{i} + 10\mathbf{j} - 15\mathbf{k}.$$

Then, adding up the individual momenta for each of the i, j and k directions, we obtain

$$\mathbf{P}_{\text{tot}} = \sum_{s=1}^{3} m_s \dot{\mathbf{x}}_s = 22\mathbf{i} - 8\mathbf{j} + 3\mathbf{k},$$

in agreement with the previous result.

5.3 Energy of a system

Expressions for the kinetic energy and potential energy of the N-particle system can be found by generalizing our earlier results for two particles. Changing from one dimension to three dimensions merely requires the change of position coordinate from scalar to vector form (that is, $x_1 \rightarrow \mathbf{x}_1$) and so on. In this way, we take over the results previously obtained in Section 2.4.3.

5.3.1 Kinetic energy

We simply add up the kinetic energy of all the particles, following the steps which led to equation (2.34) for two particles, to find:

$$T = \sum_{s=1}^{N} \frac{1}{2} m_s \dot{R}^2 + \sum_{s=1}^{N} \frac{1}{2} m_s \dot{r}_s^2. \tag{5.6}$$

The sum consists of the kinetic energy of a particle with mass M and velocity $\dot{R}$, situated at the centre of mass of the system, and the sum of the kinetic energies of the individual particles relative to the centre of mass.

Aside: Remember $\dot{R} \cdot \dot{R} = \dot{R}^2$ and $\dot{r}_s \cdot \dot{r}_s = \dot{r}_s^2$.

5.3.2 Potential energy

Again we generalize the two-particle result. Recall that there will be an external potential acting on each particle, as well as one due to the mutual interaction of pairs of particles.

Adding up the total potential energy of the system (this time for N particles), we obtain the generalization of (2.39) as

$$U = \sum_{s} U_s + \frac{1}{2} \sum_{s,k} U_{sk}, \tag{5.7}$$

where:

- $U_s \equiv$ the potential acting on particle s due to the external field;
- $U_{sk} \equiv$ the potential acting on particle s due to particle k.

An alternative method of writing this result is:

$$U = \sum_{s} U_s + \sum_{s<k} U_{sk}, \tag{5.8}$$

and this is sometimes helpful in applications. It is left as a simple exercise for readers to convince themselves that the two forms are equivalent.

5.3.3 Example: Energy of N particles in CM and relative coordinates

A system of particles consists of the set of masses $\{m_1, m_2, \ldots, m_s, \ldots, m_N\}$, situated respectively at the points $\{x_1, x_2, \ldots, x_s, \ldots, x_N\}$. Show that the total kinetic energy of the system may be written in the form

$$T = \frac{1}{2} M \dot{R}^2 + \frac{1}{2} \sum_{s=1}^{N} m_s \dot{r}_s^2,$$

where M is the total mass of the system, $\mathbf{R}$ is the position of the centre of mass and $\mathbf{r}_s$ is the position of m_s, relative to the centre of mass.

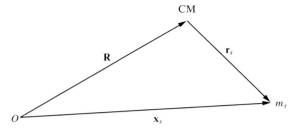

Fig. 5.2 CM and relative coordinates for a particle.

The kinetic energy of the system is found by adding up the individual kinetic energies of the constituent particles, thus:

$$T = \frac{1}{2} \sum_s m_s \dot{x}_s^2 = \frac{1}{2} \sum_s m_s \dot{\mathbf{x}}_s \cdot \dot{\mathbf{x}}_s.$$

From Figure 5.2, we see that the position vector, relative to the origin of coordinates, may be written as the sum of the position vector of the centre of mass plus the vector giving the position of the particle relative to the centre of mass, thus

$$\mathbf{x}_s = \mathbf{R} + \mathbf{r}_s.$$

Now we differentiate each term in this equation with respect to time, to give the corresponding relation for velocities, thus

$$\dot{\mathbf{x}}_s = \dot{\mathbf{R}} + \dot{\mathbf{r}}_s,$$

and it follows immediately that we can write

$$\dot{x}_s^2 = \dot{\mathbf{x}}_s \cdot \dot{\mathbf{x}}_s = (\dot{\mathbf{R}} + \dot{\mathbf{r}}_s) \cdot (\dot{\mathbf{R}} + \dot{\mathbf{r}}_s)$$
$$= \dot{R}^2 + \dot{r}_s^2 + 2\dot{\mathbf{R}} \cdot \dot{\mathbf{r}}_s.$$

Thus the expression for the total kinetic energy of the system of particles now takes the form

$$T = \frac{1}{2} \sum_s m_s \dot{\mathbf{x}}_s \cdot \dot{\mathbf{x}}_s = \frac{1}{2} \sum_s m_s \dot{R}^2 + \frac{1}{2} \sum_s m_s \dot{r}_s^2 + \sum_s m_s \dot{\mathbf{R}} \cdot \dot{\mathbf{r}}_s$$
$$= \frac{1}{2} M \dot{R}^2 + \frac{1}{2} \sum_s m_s \dot{r}_s^2 + \sum_s m_s \dot{\mathbf{R}} \cdot \dot{\mathbf{r}}_s.$$

The last term can be shown to be zero, thus:

$$\dot{\mathbf{R}} \cdot \sum_s m_s \dot{\mathbf{r}}_s = \dot{\mathbf{R}} \cdot \frac{\mathrm{d}}{\mathrm{d}t} \sum_s m_s \mathbf{r}_s = 0,$$

as

$$\sum_s m_s \mathbf{r}_s = 0,$$

and the required result follows as:

$$T = \frac{1}{2} M \dot{R}^2 + \frac{1}{2} \sum_{s=1}^{N} m_s \dot{r}_s^2.$$

This is an important result, as it shows that the total kinetic energy of the system can be written as the kinetic energy of the whole mass of the system moving like a single particle with the velocity of the centre of mass plus the sum of the kinetic energies of the individual particles moving relative to the system's centre of mass.

5.3.4 Example: *Kinetic energy of two particles in terms of the reduced mass*

Two mass points m_1 and m_2 form a conservative system where the only forces are those due to an interaction potential U. If U is any function of the distance r between the two particles, show that the law of conservation of energy takes the form

$$U(r) + \frac{1}{2} \mu \dot{r}^2 = \text{constant},$$

where $\mu = m_1 m_2 / (m_1 + m_2)$ is the reduced mass.
[Hint: start by writing down the equation expressing conservation of energy from the previous example, for the particular case of two particles.]

Referring to Figure 5.3, we may write the relative position vector $\mathbf{r}$ in the form

$$\mathbf{r} = \mathbf{r}_1 - \mathbf{r}_2 = \mathbf{x}_1 - \mathbf{x}_2.$$

As $\mathbf{r}_1$, $\mathbf{r}_2$ are measured from the centre of mass,

$$m_1 \mathbf{r}_1 + m_2 \mathbf{r}_2 = 0.$$

Into this relationship, we substitute

$$\mathbf{r}_2 = \mathbf{r}_1 - \mathbf{r}$$

from the previous relationship to get

$$m_1 \mathbf{r}_1 + m_2 \mathbf{r}_1 - m_2 \mathbf{r} = 0 \implies \mathbf{r}_1 = \frac{m_2}{m_1 + m_2} \mathbf{r}.$$

Similarly,

$$\mathbf{r}_1 = \mathbf{r} + \mathbf{r}_2 \implies \mathbf{r}_2 = \frac{-m_1}{m_1 + m_2} \mathbf{r}.$$

Now from the previous section, the kinetic energy of the system is just

$$T = \frac{1}{2} M \dot{R}^2 + \frac{1}{2} m_1 \dot{r}_1^2 + \frac{1}{2} m_2 \dot{r}_2^2.$$

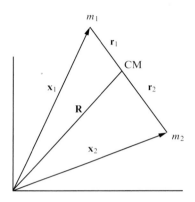

Fig. 5.3 CM and relative coordinates for a two-particle system.

Substituting from above for $\dot{\mathbf{r}}_1, \dot{\mathbf{r}}_2$ and differentiating with respect to time, we obtain

$$T = \frac{1}{2} M \dot{R}^2 + \frac{1}{2} \left[\frac{m_1 m_2^2}{(m_1 + m_2)^2} + \frac{m_2 m_1^2}{(m_1 + m_2)^2} \right] \dot{r}^2$$
$$= \frac{1}{2} M \dot{R}^2 + \frac{1}{2} \mu \dot{r}^2.$$

The principle of conservation of energy takes the form:

$$\frac{1}{2} M \dot{R}^2 + U(r) + \frac{1}{2} \mu \dot{r}^2 = E = \text{constant},$$

and so

$$U(r) + \frac{1}{2} \mu \dot{r}^2 = E - \frac{1}{2} M \dot{R}^2 = \text{constant},$$

as $\dot{R} = \text{constant}$, if there are no external forces.

5.4 The angular momentum of a system

We shall show that the total angular momentum of an N-particle system may be written as:

$$\mathbf{L} = \mathbf{R} \times \mathbf{P} + \mathbf{L}',$$

where the first term is the angular momentum of the total mass, at the centre of mass, *about the origin of coordinates* and $\mathbf{L}'$ is the angular momentum of the system *about its centre of mass*.

As before, we need a coordinate system which refers the positions of the individual particles to the centre of mass, and referring to Figure 5.2, we use $\mathbf{r}_s$ as the distance of particle s from the centre of mass. We refer to these coordinates as:

- $\mathbf{R}$ is the centroid coordinate;
- $\mathbf{r}$ is the relative coordinate.

Also, as before, we can write

$$\mathbf{x}_s = \mathbf{R} + \mathbf{r}_s$$
$$\dot{\mathbf{x}}_s = \dot{\mathbf{R}} + \dot{\mathbf{r}}_s.$$

The total angular momentum is obtained by adding up the individual angular momenta of the particles which make up the system. We begin by specializing equation (2.16) for the angular momentum of a single particle to particle s, thus the angular momentum of particle s is given by

$$\mathbf{l}_s = \mathbf{x}_s \times \mathbf{p}_s,$$

and hence the total angular momentum is obtained by summing over all the particles, thus:

$$L = \sum_s \mathbf{x}_s \times \mathbf{p}_s. \tag{5.9}$$

This result can be broken down into its constituent parts, in terms of centroid and relative coordinates, thus:

$$L = \sum_s \mathbf{x}_s \times m_s \dot{\mathbf{x}}_s = \sum_s (\mathbf{R} + \mathbf{r}_s) \times m_s (\dot{\mathbf{R}} + \dot{\mathbf{r}}_s)$$
$$= \sum_s m_s \mathbf{R} \times \dot{\mathbf{R}} + \sum_s \mathbf{R} \times m_s \dot{\mathbf{r}}_s + \sum_s m_s \mathbf{r}_s \times \dot{\mathbf{R}} + \sum_s m_s \mathbf{r}_s \times \dot{\mathbf{r}}_s. \tag{5.10}$$

The second and third terms vanish because

$$\sum_s m_s \mathbf{r}_s = 0,$$

and

$$\frac{\mathrm{d}}{\mathrm{d}t} \sum_s m_s \mathbf{r}_s = 0.$$

Thus

$$L = \sum_s m_s \mathbf{R} \times \dot{\mathbf{R}} + \sum_s m_s \mathbf{r}_s \times \dot{\mathbf{r}}_s$$
$$= \mathbf{R} \times M\dot{\mathbf{R}} + \sum_s \mathbf{r}_s \times \mathbf{p}_s$$
$$= \mathbf{R} \times \mathbf{P} + \mathbf{L}', \tag{5.11}$$

where

$$\mathbf{L}' = \sum_s \mathbf{r}_s \times \mathbf{p}_s. \tag{5.12}$$

Therefore the total angular momentum of the system about the point O is

$$\mathbf{L} = \mathbf{R} \times \mathbf{P} + \sum_{s=1}^{N} \mathbf{r}_s \times \mathbf{p}_s, \tag{5.13}$$

which is the sum of

1. the angular momentum of the total mass (concentrated at the centre of mass) about the point O; and,
2. the angular momentum of the system about its centre of mass.

5.5 Torque

For the case of a single particle, the concept of the torque Γ acting on a particle, relative to some fixed point O, is introduced in Section 2.3 by equation (2.17), thus:

$$\Gamma = \mathbf{x} \times \mathbf{F},$$

where $\mathbf{F}$ is the force acting on the particle and $\mathbf{x}$ is its position vector relative to the fixed point O.

5.5.1 *Form of N2 for angular motion*

The generalization of N2 which is appropriate for angular motion is given in Section 2.3, in the form of equation (2.19). If we consider the whole system as one big particle, then N2 for angular momentum immediately gives:

$$\frac{d\mathbf{L}}{dt} = \Gamma_{\text{tot}}, \tag{5.14}$$

where Γ_{tot} is the total external torque applied to the system, and $\mathbf{L}$ is the total angular momentum of the system.

We shall now demonstrate that this equation results from summing over the constituent particles of the system, each of which obeys N2 in both linear and angular forms. Let us begin by considering each of the relevant system parameters in turn; first $\mathbf{L}$, then Γ_{tot}.

The total angular momentum of the system is just given by equation (5.9) and we repeat it here for the sake of convenience, thus:

$$\mathbf{L} = \sum_s \mathbf{x}_s \times \mathbf{p}_s. \tag{5.15}$$

Using the linear form of N2, $\mathbf{F}_s = \dot{\mathbf{p}}_s$, for particle s, the torque is given by

$$\Gamma_s = \mathbf{x}_s \times \mathbf{F}_s = \mathbf{x}_s \times \dot{\mathbf{p}}_s,$$

and so the total torque exerted on the system by external forces is

$$\Gamma_{\text{tot}} = \sum_s \Gamma_s = \sum_s \mathbf{x}_s \times \dot{\mathbf{p}}_s. \tag{5.16}$$

Now we verify the 'angular' form of N2, as given by (5.14), for the system. If we substitute from (5.15) for the total angular momentum, equation (5.14) becomes

$$\Gamma_{\text{tot}} = \frac{d\mathbf{L}}{dt} = \frac{d}{dt} \sum_s \mathbf{x}_s \times \mathbf{p}_s,$$

hence

$$\Gamma_{\text{tot}} = \sum_s \dot{\mathbf{x}}_s \times \mathbf{p}_s + \sum_s \mathbf{x}_s \times \dot{\mathbf{p}}_s,$$

by the chain rule, and

$$\Gamma_{\text{tot}} = \sum_s \mathbf{x}_s \times \dot{\mathbf{p}}_s,$$

as, by the properties of vector products,

$$\dot{\mathbf{x}}_s \times \mathbf{p}_s = \dot{\mathbf{x}}_s \times m\dot{\mathbf{x}}_s = 0,$$

and so

$$\Gamma_{\text{tot}} = \sum_s \mathbf{x}_s \times \dot{\mathbf{p}}_s. \qquad (5.17)$$

Next we substitute for $\dot{\mathbf{p}}_s$ from equation (5.2) to get,

$$\Gamma_{\text{tot}} = \sum_s \mathbf{x}_s \times \left(\mathbf{F}_s + \sum_{k \neq s} \mathbf{F}_{sk} \right), \qquad (5.18)$$

and so

$$\Gamma_{\text{tot}} = \sum_s \mathbf{x}_s \times \mathbf{F}_s + \sum_s \sum_k \mathbf{x}_s \times \mathbf{F}_{sk} \quad (k \neq s), \qquad (5.19)$$

therefore

$$\Gamma_{\text{tot}} = \sum_s \mathbf{x}_s \times \mathbf{F}_s, \qquad (5.20)$$

provided the double sum vanishes (we shall prove this next), and therefore (5.14) holds. Thus Γ_{tot} is the sum of the individual torques due to the external forces on the individual particles.

Lastly, we shall prove that the double sum in (5.18) is zero and hence the interaction force doesn't contribute. Considering the interaction force on its own, we may rewrite it as

$$\sum_{s \neq k} \mathbf{x}_s \times \mathbf{F}_{sk} = \frac{1}{2} \sum_s \sum_k (\mathbf{x}_s \times \mathbf{F}_{sk} + \mathbf{x}_k \times \mathbf{F}_{ks}),$$

because

$$\sum_s \sum_k (\mathbf{x}_s \times \mathbf{F}_{sk}) = \sum_s \sum_k (\mathbf{x}_k \times \mathbf{F}_{ks}).$$

Now, we have

$$(\mathbf{F}_{sk} = -\mathbf{F}_{ks}) \implies \frac{1}{2} \sum \sum (\mathbf{x}_s - \mathbf{x}_k) \times \mathbf{F}_{sk} = 0,$$

by the rules of vector products, as $\mathbf{F}_{sk}$ is constrained to lie along $(\mathbf{x}_s - \mathbf{x}_k)$.

Equation (5.14) takes on a particularly important form if the external forces or their moment is zero. Then it states that the total angular momentum of the system is conserved. As we saw in Chapter 1, this is actually a more fundamental result than N2, and must hold under all circumstances provided only that the total system is invariant under rotations.

5.6 Continuous distributions of mass

We can extend all the results of this chapter to situations which are a limiting case of systems of discrete particles, and can be represented by distributions of mass which are continuous functions of the space coordinates. We characterize such a system by a mass density $\rho(\mathbf{x})$ (mass/volume) and the total volume V occupied by the system. In these circumstances, we replace the concept of a discrete particle, by an infinitesimal element of mass

$$\mathrm{d}m = \rho(\mathbf{x})\,\mathrm{d}V,$$

where $\mathrm{d}V$ is an element of volume belonging to V. Then, instead of summing over N particles, we use an integration to add up all the mass elements and find that the total mass of the system is given by

$$M = \int_V \rho(\mathbf{x})\,\mathrm{d}V.$$

Now, just as in the discrete particle case, we determine the centre of mass by taking moments about the point O. Taking the element $\mathrm{d}m$ to be located at $\mathbf{x}$, the moment of the element of mass $\mathrm{d}m$ about O is just

$$\mathbf{x}\,\mathrm{d}m = \mathbf{x}\rho(\mathbf{x})\,\mathrm{d}V.$$

We obtain the 'total' moment by integrating over the volume V. Thus equation (5.1) for the discrete case can be extended to give the centre of a continuous distribution of mass,

$$\mathbf{R} = \frac{1}{M} \int_V \mathbf{x}\rho(\mathbf{x})\,\mathrm{d}V. \tag{5.21}$$

If the material being considered is of uniform density, then

$$\rho(\mathbf{x}) = \rho = \text{constant},$$

and the equation for the centre of mass becomes

$$\mathbf{R} = \frac{\rho}{M} \int_V \mathbf{x}\,\mathrm{d}V = \frac{1}{V} \int_V \mathbf{x}\,\mathrm{d}V. \tag{5.22}$$

5.6.1 Example: *Centre of mass of a solid hemisphere*
Find the coordinates of the centre of mass of a solid hemisphere of radius a and constant density ρ.

Take the base of the hemisphere to be in the xy-plane, with its centre at the origin of coordinates. Also take the axis of symmetry to coincide with the z-axis, with $z = 0$ at the base of the hemisphere and $z = a$ at its top.

The elementary masses are obtained by imagining the hemisphere sliced into parallel circular discs, each of thickness dz, and each with its centre on the z-axis. If we take the position vector of the centre of mass to have the components $\mathbf{R} \equiv (X, Y, Z)$, then from considerations of symmetry, it follows that the centre of mass of each disc, and hence the whole hemisphere, must lie on the z-axis, and so:

$$X = 0 \quad \text{and} \quad Y = 0.$$

Now, to find the position of Z, we make use of equation (5.22). Our elementary mass at a height z above the base is a disc of radius $r = \sqrt{a^2 - z^2}$, and thus has volume

$$dV = \pi(a^2 - z^2)\,dz.$$

We simply multiply this expression by the density to obtain the elementary mass dm and, invoking equation (5.22), we have for the centre of mass

$$Z = \frac{\rho}{M} \int_0^a \pi z(a^2 - z^2)\,dz = \frac{\pi a^4}{4V} = \frac{3a}{8},$$

where we have used the fact that the volume of the hemisphere is $V = \frac{2}{3}\pi a^3$.

5.7 Exercises

5.1 A particular system of particles is an aggregate of several subsystems of particles. Show that the centroid of the total system is given by $MR = \sum_s M_s R_s$, where M_s is the total mass of the sth subsystem and R_s is its centroid.

5.2 A solid body is constructed by joining a right circular cone, of height and base radius a, to the plane surface of a hemisphere of the same radius. Assuming that the same material of constant density is used throughout, find the position of the centre of mass of the body.

5.3 In general, the total angular momentum of a system will depend on the choice of the fixed reference point O. Show that, if the centre of gravity of the system is at rest, the angular momentum is independent of the choice of O.

5.4 In a frame of reference S, a particle A of mass m_A moves in a circle of radius a about the origin O. A second particle B, of mass m_B moves in a circle of radius b about A, both rotations being in the xy plane. Show that the coordinates of the centre of mass of the system at any time are given by:

$$X = a\cos\alpha + \lambda b\cos(\alpha + \beta)$$
$$Y = a\sin\alpha + \lambda b\sin(\alpha + \beta),$$

where, at the given time, α is the angle between the line OA and the x-axis, and β is the angle between the line AB and the line OA and $\lambda = m_B/(m_A + m_B)$.

5.5 By transforming to relative and absolute coordinates, show that the total kinetic energy of the system in Exercise 5.4 is given by:

$$2T = F\dot{\alpha}^2 + G\dot{\alpha}\dot{\beta} + H\dot{\beta}^2,$$

where

$$F = m_A a^2 + m_B(a^2 + 2ab\cos\beta + b^2),$$
$$G = 2m_B(a\cos\beta + b),$$

and $H = m_B b^2$.

Solid-body motion

6

A solid body can be viewed as the limiting case of a system of particles. That is to say, the N-particle system can be used to represent a solid body if we prescribe the relative distances between mass points. For the simplest case, which is a *rigid* body, the intervals $\mathbf{x}_s - \mathbf{x}_k$ are *constant* for all s and k.

We have seen that the general motion of a system of N particles can be decomposed into:

- The motion of a 'point mass' (its mass given by $M = \sum_s m_s$) concentrated at the centre of mass of the system; and,
- the motion of the system relative to the centre of mass.

This is also true of the special case which corresponds to a solid body. However, as the 'single particle' behaviour can be treated by the methods which we used earlier for point particles, we shall concentrate here on the second aspect: the motion relative to the centre of mass.

For a rigid body, such motion relative to the centre of mass must be made up of *rotations*. We shall consider two general cases:

- rotation about a fixed axis;
- rotation about a fixed point.

6.1 Motion about a fixed axis

In this case, the position of the body is specified by the angular displacement θ of a plane passing through the fixed axis and part of the solid body. The general situation is shown in Figure 6.1, where we have taken a right circular cylinder as an example of a rigid body. In this situation, the angular velocity vector lies along the axis and the magnitude of the angular velocity is denoted by ω.

If the angle of rotation changes by an amount $\delta\theta$ in time δt, the angular velocity at instant t is given by:

$$\omega = \frac{\mathrm{d}\theta}{\mathrm{d}t} = \lim_{\delta t \to 0} \frac{\delta\theta}{\delta t}. \tag{6.1}$$

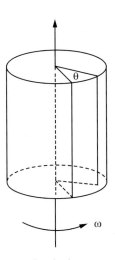

Fig. 6.1 Angular displacement of a rigid body about a fixed axis.

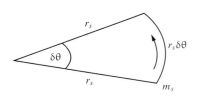

Fig. 6.2 Elementary displacement of a material particle in time δt.

Similarly, the angular acceleration is obtained by differentiating the expression for the angular velocity with respect to the time, thus:

$$\frac{d\omega}{dt} = \frac{d^2\theta}{dt^2}. \tag{6.2}$$

Or, if ω is a function of θ,

$$\frac{d\omega}{dt} = \frac{d\omega}{d\theta}\frac{d\theta}{dt} = \omega\frac{d\omega}{d\theta}. \tag{6.3}$$

This latter result can be useful in applications.

With the velocity defined, we are now in a position to work out the angular momentum about the fixed axis. Referring to Figure 6.2, we consider one of the constituent particles of the body. This specific particle, labelled by s, is of mass m_s and is a distance r_s from the fixed axis. In time δt, m_s at r_s is displaced a distance $r_s\delta\theta$. Therefore the velocity of particle s is

$$v_s = r_s\frac{d\theta}{dt} = \omega r_s. \tag{6.4}$$

The linear momentum of the particle is

$$p_s = m_s v_s = m_s\omega r_s,$$

which is perpendicular to both r_s and the fixed axis, and so the magnitude of the angular momentum of the particle is

$$l_s = r_s p_s = m_s\omega r_s^2.$$

Note that this is a special case of the general expression for the angular momentum of a particle, as given by equation (2.16), and reduces to this simple form because the three vectors involved are mutually perpendicular, which simplifies the vector product.

The total angular momentum of the body is obtained by adding up the contributions from all the individual particles which make up the body, thus:

$$L = \left(\sum_{s=1}^{N} m_s r_s^2\right)\omega. \tag{6.5}$$

In the next section, we shall use this expression in order to introduce the concept of the moment of inertia of a body, when the reason for the bracket will become clear.

6.2 Moments of inertia

The idea of a moment of inertia can present some problems to the student and so it is worth introducing the basic concept with some care.

However, even when explained carefully, there are some counter-intuitive aspects to the idea and we shall highlight these at the appropriate point. We begin by developing the analogy between linear and angular motion.

In equation (6.5), we have the definition of 'momentum' for the case of the angular motion of an extended solid body. At this juncture it will prove helpful to go back and remind ourselves of the equivalent statement for linear motion, as revised in Section 1.1.1. For linear motion, we define the momentum p (for motion in one dimension) by the relationship

$$p = mv,$$

for a particle of mass m and speed v. Now, what we commonly call the 'mass of a body' is really a measure of the amount of matter in it, and it is this everyday usage which forms our intuition about the idea of mass. On the other hand, the quantity which appears in N2 is actually the inertial mass; or, perhaps for our present purposes, simply just the inertia of the particle. So, purely from the point of view of Newtonian mechanics, we could write the definition of the linear momentum in words as:

$$\text{linear momentum} = \text{inertia} \times \text{linear velocity},$$

and abandon for the moment the misleading idea of the 'mass'.[1]

Now, in order to pursue our analogy between angular and linear motion, we can rewrite equation (6.5) in words as:

$$\text{angular momentum} = \text{'something'} \times \text{angular velocity}.$$

In arriving at the 'something', in equation (6.5), we took moments of the particle mass—or its inertia—about the fixed axis; it would seem appropriate to complete the analogy with the linear case and now write (6.5) as:

$$\text{angular momentum} = \text{moment of inertia} \times \text{angular velocity},$$

and this step defines what we mean by **moment of inertia**.

6.2.1 *The moment of inertia I of a body*

Formally we define the moment of inertia I of a body, which is made up of N particles, each of mass m_s, by rewriting equation (6.5) as:

$$L = I\omega, \tag{6.6}$$

where comparison of the two equations shows that the moment of inertia I is

$$I = \sum_{s=1}^{N} m_s r_s^2, \tag{6.7}$$

[1]Actually, from our earliest childhood, we encounter masses (including our own!) which are subject to the Earth's gravitational pull. As a result, the concept with which we are most familiar is the *weight* of a body. This is not entirely helpful when we need to consider the inertial mass of a body. Needless to say, it is even less helpful when we come to consider moments of inertia.

about the fixed axis. It is perhaps worth highlighting another counter-intuitive aspect of the moment of inertia which is that this is actually the *second moment* of the mass element about the designated axis.

We can extend this result to the more commonly encountered case, where the matter of the body is distributed continuously, with some density distribution $\rho(r)$. Then the summation over discrete masses becomes an integral. We have, of course, already met this idea of a transition from a discrete summation to a continuous integration in Section 5.6.

Let us consider an element of the body, of volume dV, situated a distance r from the fixed axis. The corresponding element of mass dm is just $dm = \rho(r)dV$ and the moment-of-inertia sum now can be written as the integral

$$I = \int_V \rho(r)r^2 \, dV. \tag{6.8}$$

Very often the density of the body in question is a constant. This permits us to take the density outside the integral and then the equation for the moment of inertia takes its simplest form:

$$I = \rho \int_V r^2 \, dV, \tag{6.9}$$

where the constant density is $\rho(r) = \rho$.

These results are readily extended to calculate the moment of inertia of thin sheets of material, where we specify the *surface* mass density $\sigma(r)$, which has dimensions of mass per unit area. The analogue of equation (6.8) then takes the form:

$$I = \int_S \sigma(r)r^2 \, dS. \tag{6.10}$$

Similarly, for a linear body, we have

$$I = \int_0^l \mu(x)x^2 \, dx, \tag{6.11}$$

where l is the length of the body and $\mu(x)$ is the mass per unit length.

6.2.2 The radius of gyration of a body

Another useful concept is that of the radius of gyration of a body rotating about a fixed axis. This is denoted by k, and for a body made up of discrete particles is given by:

$$k^2 = \frac{\sum_s (m_s r_s^2)}{\sum_s m_s}. \tag{6.12}$$

For a continuous body, the obvious generalization is

$$k^2 = \frac{\int_V \rho(r) r^2 \, dV}{\int_V \rho(r) \, dV}. \tag{6.13}$$

If the body has constant density $\rho(r) = \rho$, then the density cancels, so that the radius of gyration is then given by:

$$k^2 = \frac{\int_V r^2 \, dV}{\int_V \, dV}. \tag{6.14}$$

The physical significance of the radius of gyration is that, if the whole mass of the body were concentrated at a point a distance k from the axis, then the resulting moment of inertia would be the same as that of the original body. This can be seen by rewriting equation (6.12) as

$$I = Mk^2. \tag{6.15}$$

6.2.3 Example: *Moment of inertia of a slender rod about an axis perpendicular to its length*

Consider a uniform slender rod, of length l, mass M and free to rotate about a fixed axis perpendicular to its length, as shown in Figure 6.3. In order to work out the moment of inertia about the perpendicular axis, we measure the coordinate x from that axis. Then, assuming that both the cross-sectional area and the density of the rod are uniform, the element of mass is given by:

$$\rho dV = \rho A dx = \frac{\rho A l}{l} dx = \frac{M}{l} dx.$$

Then, substituting into equation (6.9), and integrating, we have for the moment of inertia

$$I = \frac{M}{l} \int_{-h}^{l-h} x^2 dx = \left[\frac{M}{l} \frac{x^3}{3} \right]_{-h}^{l-h} = \frac{1}{3} M(l^2 - 3lh + 3h^2). \tag{6.16}$$

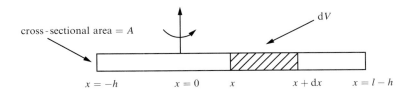

cross-sectional area $= A$

dV

$x = -h$ $x = 0$ x $x + dx$ $x = l - h$

Fig. 6.3 A rod free to rotate about an axis perpendicular to its length.

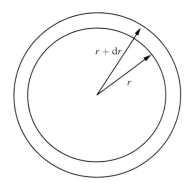

Fig. 6.4 Cross-section of an annular element of volume.

6.2.4 Example: *Moment of inertia of a solid cylinder about its axis of symmetry*

Let us consider a solid body which is a right circular cylinder and which is free to rotate about its axis of symmetry. We take its length to be l, its mass M, and its radius a.

Consider an annular element between r and $r + dr$, as in Figure 6.4. The corresponding elementary mass is

$$\rho dV = \rho(2\pi r\, dr)l.$$

Substituting this into equation (6.9) for the moment of inertia, we obtain:

$$I = 2\pi\rho l \int_0^a r^2(r\, dr) = 2\pi\rho l \int_0^a r^3\, dr$$
$$= \frac{\pi\rho l}{2}a^4 = \frac{Ma^2}{2}. \qquad (6.17)$$

6.2.5 Parallel axes theorem

The parallel axes theorem is stated for a body which is a lamina (or thin sheet), which may be taken to be of zero thickness. It can be stated as follows:

$$I_p = I_{cm} + Md^2, \qquad (6.18)$$

where

- I_{cm} = moment of inertia about an axis through the centre of mass;
- I_p = moment of inertia about a second axis parallel to the first;
- d = distance between the two axes;
- M = mass of the lamina.

Alternatively, we can express it in terms of the radius of gyration as

$$I_p = M(k^2 + d^2), \qquad (6.19)$$

where k is the radius of gyration, as given by equation (6.12).

Proof Referring to Figure 6.5, we take the origin of coordinates O at the centre of mass and take two axes in the z-direction (i.e. normal to the plane of the figure) through the points O and P. From (6.15), we have

$$I_{cm} = \sum_s m_s(x_s^2 + y_s^2), \qquad (6.20)$$

and also

$$I_p = \sum_s m_s(x_s - a)^2 + (y_s - b)^2.$$

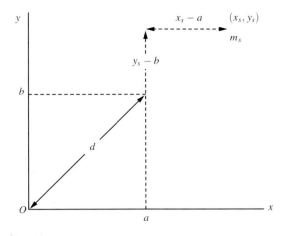

Fig. 6.5 Coordinates for the proof of the parallel axis theorem.

Expanding out the right-hand side then gives

$$I_\mathrm{p} = \sum_s m_s(x_s^2 + y_s^2) - 2a \sum_s m_s x_s - 2b \sum_s m_s y_s$$
$$+ (a^2 + b^2) \sum_s m_s. \qquad (6.21)$$

Now, on the right-hand side, the first term equals I_{cm}; the second and third terms are zero, as they are the coordinates of the centre of mass and we have chosen a coordinate system such that the centre of mass is at $(0, 0)$. Finally, the fourth term is

$$(a^2 + b^2) \sum_s m_s = Md^2,$$

and equation (6.18) follows.

6.2.6 Perpendicular axes theorem

If we know the moment of inertia of a plane area about each of two perpendicular axes in the plane (say I_x and I_y), then the moment of inertia of the area about a third axis perpendicular to the area, and through the point of intersection of the other two axes, is given by the sum:

$$I_z = I_x + I_y. \qquad (6.22)$$

Proof A rectangular lamina of mass M lies in the xy-plane and is bounded by $-a \leq x \leq a$ and $-b \leq y \leq b$, respectively. As a first step, we obtain its moments of inertia I_x and I_y about the axes OX and OY respectively, as shown in Figure 6.6. Then we choose as our elementary volume, a 'rod' of length $2a$, thickness dy and distance y from the axis OZ, and use the parallel axes theorem.

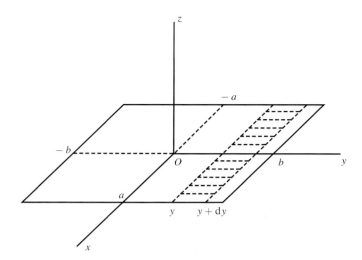

Fig. 6.6 Coordinates for the proof of the perpendicular axes theorem.

We do this in two parts:

(i) First we work out the surface mass density as $\sigma = M/4ab$, and use equation (6.10) to work out the moments of inertia as $I = \sigma \int r^2 \, dS$, for constant σ. Hence:

$$I_x = \sigma \int \int r^2 \, dy \, dx = \sigma \int_{-a}^{a} dx \int_{-b}^{b} y^2 \, dy$$

$$= \sigma \left[x \right]_{-a}^{a} \left[\frac{y^3}{3} \right]_{-b}^{b} = \frac{4\sigma ab^3}{3} = \frac{Mb^2}{3}.$$

Similarly,

$$I_y = \frac{Ma^2}{3}.$$

(ii) Now consider the moment of inertia of an elementary 'rod' of length $2a$ and thickness dy, about a perpendicular axis through its centre. We have, from equation (6.16) with $h = 0$:

$$I_{cm} = \frac{1}{3} M_R a^2,$$

where $M_R = \sigma \cdot 2a \cdot dy$. Now, by the parallel axes theorem, the moment of inertia of the elementary rod about OZ is:

$$I_z = \frac{1}{3} M_R a^2 + M_R y^2 = \frac{1}{3} (\sigma \cdot 2a \cdot dy) a^2 + (\sigma \cdot 2a \cdot dy) y^2.$$

Hence the moment of inertia of the complete lamina about the axis OZ is obtained by integrating with respect to y over the width of the lamina, thus:

$$I_z = \int_{-b}^{b} \frac{2\sigma a^3}{3} \, \mathrm{d}y + \int_{-b}^{b} 2\sigma a y^2 \, \mathrm{d}y = \frac{2\sigma a^3}{3} \cdot 2b + 2\sigma a \cdot \frac{2b^3}{3},$$

$$= \frac{Ma^2}{3} + \frac{Mb^2}{3}$$

$$= I_y + I_x,$$

as required.

6.2.7 Example: *Radius of gyration of a cylinder about its axis and about one of its generators*

A uniform solid cylinder of radius a has mass M. Obtain an expression for the radius of gyration k of the cylinder about its axis. Also find its moment of inertia and the corresponding radius of gyration about one of its generators,[2] as shown in Figure 6.7.

From equation (6.15) and the result (6.17) we have

$$I = Mk^2 = \frac{Ma^2}{2},$$

and so the radius of gyration k is given by

$$k = \frac{a}{\sqrt{2}}.$$

The extension to motion about an axis through one of its generators is obtained by use of the parallel axes theorem in the form of equation (6.18), or:

$$I_p = I_{cm} + Ma^2 = \frac{3Ma^2}{2},$$

and hence in this case $k^2 = 3a^2/2$ and so $k = (\sqrt{3/2})a$.

[2]In the present context, a generator is simply any line parallel to the axis of the cylinder and lying along its surface.

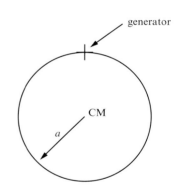

Fig. 6.7 A cylinder suspended from an axis through one of its generators.

6.2.8 *The compound pendulum*

The compound pendulum consists of a solid body, free to swing about a horizontal axis through the point O, as shown in Figure 6.8. If the centre of mass (shown as CM) is displaced away from its equilibrium position immediately below the point O, then there will be a restoring force due to gravity and the body will oscillate around the axis through O. However, we shall use the principle of conservation of energy, as we did for the simple pendulum in Section 2.2.5, to obtain the equation of motion.

Let the moment of inertia about the axis through the point O be denoted by I. Then we may write the kinetic energy of angular motion T as

$$T = \frac{1}{2} I \omega^2 = \frac{1}{2} I \dot{\theta}^2,$$

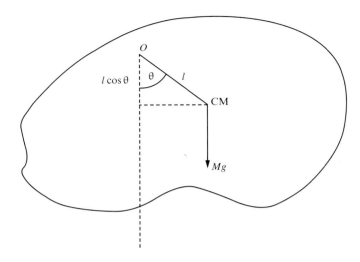

Fig. 6.8 The compound pendulum.

where we refer to the figure for the definition of θ, and as usual the dot stands for differentiation with respect to time. Next, if we take the reference level for the potential energy to be $U = 0$ at the point O, then the potential energy may be written as

$$U = -Mgl\cos\theta.$$

This latter step should be clear after reference to the figure. Now we invoke the principle of conservation of energy in the form:

$$T + U = E,$$

where E is the total energy, and is of course constant. Thus, substituting from above for the kinetic and potential energies, conservation of energy gives us:

$$\frac{1}{2}I\dot{\theta}^2 - Mgl\cos\theta = E. \tag{6.23}$$

We may compare this to the result for the simple pendulum, as derived in Section 2.2.5. There we obtained the equation

$$\frac{1}{2}m(L\dot{\theta})^2 + mgL(1 - \cos\theta) = E,$$

for a simple pendulum of length L and with a bob of mass m. This equation was derived on the assumption that the reference level $U = 0$ was at the lowest point of the swing. In order to get a result comparable with the above equation for the compound pendulum, we need to change the reference point to the point of suspension. Then the equation of motion of the simple pendulum becomes

$$\frac{1}{2}m(L\dot{\theta})^2 - mgL\cos\theta = E. \tag{6.24}$$

In order to make our comparison, we divide both equations (6.23) and (6.24) across by the factor I in the first case and mL^2 in the second. Then equations (6.23) and (6.24) become respectively:

$$\frac{1}{2}\dot{\theta}^2 - \frac{Mgl}{I}\cos\theta = \text{constant}, \qquad (6.25)$$

and

$$\frac{1}{2}\dot{\theta}^2 - \frac{g}{L}\cos\theta = \text{constant}. \qquad (6.26)$$

The two pendulums will oscillate at the same frequency if they each have the same amplitude[3] and, from a comparison of equations (6.25) and (6.26), if

$$\frac{Ml}{I} = \frac{1}{L}.$$

Or, alternatively, the compound pendulum is equivalent to a simple pendulum which has a length given by:

$$L = \frac{I}{Ml}. \qquad (6.27)$$

We can put this in a more helpful form as follows. We express the moment of inertia of the compound pendulum about its centre of mass in terms of its radius of gyration as Mk^2. Then, by the parallel axes theorem, the moment of inertia about the axis through the point O is just:

$$I = Mk^2 + Ml^2.$$

Substitution of this expression into (6.27) yields the equivalent simple-pendulum length as

$$L = \frac{k^2 + l^2}{l}, \qquad (6.28)$$

where k is the radius of gyration of the compound pendulum and l is the distance between the point of suspension and the centre of mass.

6.2.9 Example: *Circular cylinder oscillating about one of its generators*

If a solid cylinder is suspended from, and is free to oscillate about a fixed axis which coincides with one of its generators, it forms a *compound* pendulum. Show that the equivalent *simple* pendulum would have the same frequency of oscillation if its length L were given by

$$L = \frac{3a}{2},$$

where a is the radius of the cylinder.

We follow the treatment of the general compound pendulum, suspended at a point a distance l from the centre of mass, and compare its energy

[3]This restriction would not be necessary if we treated the case of small oscillations, when the frequency would not depend on the amplitude.

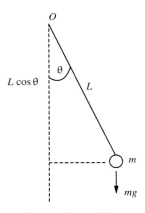

Fig. 6.9 The simple pendulum.

equation to that of a simple pendulum of length L. The actual case we are considering is a right circular cylinder of radius a and mass M. The configuration is illustrated in Figure 6.7, while we remind ourselves of the form of the simple pendulum in Figure 6.9. In both cases, we take O as reference level where the potential energy $U = 0$, and invoke the principle of conservation of energy, viz.,

$$T + U = E.$$

For the simple pendulum we have:

$$\frac{1}{2}ML^2\dot{\theta}^2 - mgL\cos\theta = E$$

$$\implies \frac{1}{2}\dot{\theta}^2 - \frac{g}{L}\cos\theta = \frac{E^2}{mL}.$$

For the compound pendulum we have:

$$\frac{1}{2}I_O\dot{\theta}^2 - Mgl\cos\theta = E,$$

where I_O is the moment of inertia of the cylinder about a horizontal axis through the point O. But, by the parallel axes theorem,

$$I_O = I_{CM} + Ml^2 = Mk^2 + Ml^2,$$

where I_{CM} is the moment of inertia of the cylinder about an axis through its centre of mass (in this case, its axis of symmetry). Hence the equation of motion for the compound pendulum becomes

$$\frac{1}{2}\dot{\theta}^2 - g\left(\frac{l}{k^2 + l^2}\right)\cos\theta = \frac{E}{M(k^2 + l^2)};$$

and by comparison of the left-hand side of this with the left-hand side of the equation for the simple pendulum:

$$L = \frac{k^2 + l^2}{l}.$$

In this case, we have $k = a/\sqrt{2}$ and $l = a$, hence

$$L = \frac{a^2/2 + a^2}{a} = \frac{3a}{2}.$$

6.3 Motion about a fixed point

In this section, we give a more general treatment of rotation and the concept of moment of inertia. Now we consider a 'particle' of the body, with mass m_s, distance $\mathbf{r}_s$ from the fixed point and with linear velocity $\mathbf{v}_s$. If the body rotates with angular velocity ω, then the linear velocity of the particle labelled s is given by

$$\mathbf{v}_s = \boldsymbol{\omega} \times \mathbf{r}_s. \tag{6.29}$$

The total angular momentum L for the system of N particles (i.e. the body) is given by:

$$\mathbf{L} = \sum_{s=1}^{N} \mathbf{r}_s \times \mathbf{p}_s = \sum_{s=1}^{N} m_s(\mathbf{r}_s \times \mathbf{v}_s)$$

$$= \sum_{s=1}^{N} m_s(\mathbf{r}_s \times (\boldsymbol{\omega} \times \mathbf{r}_s)),$$

as $\mathbf{v}_s = \boldsymbol{\omega} \times \mathbf{r}_s$, and so

$$L = \sum_{s=1}^{N} m_s r_s^2 \boldsymbol{\omega} - \sum_{s=1}^{N} m_s(\mathbf{r}_s \cdot \boldsymbol{\omega})\mathbf{r}_s, \tag{6.30}$$

by decomposition of a vector triple product.[4]

Now we write this in component form: we take the x-component,

$$L_x = \sum_{s=1}^{N} m_s r_s^2 \boldsymbol{\omega}_x - \sum_{s=1}^{N} m_s(x_s\omega_x + y_s\omega_y + z_s\omega_z)x_s, \tag{6.31}$$

as an example, while noting that similar expressions can be written for L_y and L_z. First we rearrange the expression for L_x,

$$L_x = \sum_{s=1}^{N} m_s(r_s^2 - x_s^2)\omega_x - \sum_{s=1}^{N} m_s x_s y_s \omega_y - \sum_{s=1}^{N} m_s x_s z_s \omega_z, \tag{6.32}$$

and introduce new symbols for the coefficients of ω_x, ω_y, and ω_z, thus:

$$I_{xx} = \sum_{s=1}^{N} m_s(r_s^2 - x_s^2) = \sum_{s=1}^{N} m_s(y_s^2 + z_s^2) \tag{6.33}$$

$$I_{xy} = -\sum_{s=1}^{N} m_s x_s y_s = I_{yx} \tag{6.34}$$

$$I_{xz} = -\sum_{s=1}^{N} m_s x_s z_s = I_{zx}. \tag{6.35}$$

Hence the equation for L_x can be written as:

$$L_x = I_{xx}\omega_x + I_{xy}\omega_y + I_{xz}\omega_z, \tag{6.36}$$

and similarly,

$$L_y = I_{yx}\omega_x + I_{yy}\omega_y + I_{yz}\omega_z \tag{6.37}$$

$$L_z = I_{zx}\omega_x + I_{zy}\omega_y + I_{zz}\omega_z. \tag{6.38}$$

[4]That is, $\mathbf{A} \times (\mathbf{B} \times \mathbf{C}) = \mathbf{B}(\mathbf{A} \cdot \mathbf{C}) - \mathbf{C}(\mathbf{A} \cdot \mathbf{B})$.

Evidently, by an extension of our previous treatment of rotation about a fixed axis, we can identify the quantities I_{xx}, I_{yy}, I_{zz} as the **moments of inertia** about the x, y and z axes respectively. The new quantities I_{xy}, I_{yz}, I_{zx} are called the **products of inertia**.

6.4 Inertia tensor

The complete array of coefficients appearing in equations (6.36)–(6.38) is called the **inertia tensor**. In general, we relate the angular momentum of a body to its angular velocity by combining all three of these equations into one compact equation, thus:

$$L_i = I_{ij}\omega_j. \tag{6.39}$$

Here the quantity I_{ij} is the inertia tensor and the subscripts $i, j = 1, 2, 3$ or x, y, z. Students who are less familiar with this kind of Cartesian tensor notation should satisfy themselves that equations (6.36)–(6.38) can be recovered from the decomposition of (6.39). In particular, it should be noted that the repeated index (j in this instance) is summed.

The general formula for the inertia tensor is easily deduced from the comparison of equations (6.36)–(6.38) with (6.39). However, before writing it down, we shall make a small change of notation. As we are now using subscripts to denote the coordinate direction, we shall move our label denoting a specific particle up to be a superscript. In order to reinforce this, we shall also change to a Greek letter. That is, the generic particle is taken to be at position $\mathbf{x}^{(\alpha)}$ and with mass $m^{(\alpha)}$. Then the general formula for the components of the inertia tensor may be written as:

$$I_{ij} = \sum_{\alpha=1}^{N} m^{\alpha} \left\{ \left(x_k^{(\alpha)} x_k^{(\alpha)} \right) \delta_{ij} - x_i^{(\alpha)} x_j^{(\alpha)} \right\}, \tag{6.40}$$

where $\mathbf{x}^{(\alpha)} = x_i^{(\alpha)} \mathbf{e}_i$ is the position vector of a mass $m^{(\alpha)}$, in terms of the basis $\mathbf{e}_i$, and the sum over α is over all the elements of the body. As is usual, the symbol δ_{ij} is the Kronecker delta, with the properties $\delta_{ij} = 1$ if $i = j$; but $\delta_{ij} = 0$ if $i \neq j$.

6.4.1 Example: Inertia tensor for four point masses in a plane

A body is constructed from four particles bound together by a light rigid framework, such that two particles of mass μ are situated at points $(a, a, 0)$ and $(-a, -a, 0)$, together with two particles of mass 2μ at points $(a, -a, 0)$ and $(-a, a, 0)$. Show that the non-zero components of the inertia tensor are given by: $I_{11} = 6\mu a^2$, $I_{22} = 6\mu a^2$, $I_{33} = 12\mu a^2$, and $I_{12} = I_{21} = 2\mu a^2$.

From equation (6.40) the components of the inertia tensor are defined by:

$$I_{ij} = \sum_{\alpha=1}^{N} m^{(\alpha)} \left\{ \left(x_k^{(\alpha)} x_k^{(\alpha)} \right) \delta_{ij} - x_i^{(\alpha)} x_j^{(\alpha)} \right\},$$

where $\mathbf{x}^{(\alpha)} = x_i^{(\alpha)}\mathbf{e}_i$ is the position vector of mass $m^{(\alpha)}$ and the sum is over all elements of the body.

In this case $x_3 = 0$ for all particles and the various elements of the inertia tensor can be written down as follows:

$$I_{11} = \sum_\alpha m^\alpha \{(x_1^2 + x_2^2) - x_1^2\} = \sum_\alpha m^\alpha x_2^2$$

$$I_{22} = \sum_\alpha m^\alpha \{(x_1^2 + x_2^2) - x_2^2\} = \sum_\alpha m^\alpha x_1^2$$

$$I_{33} = \sum_\alpha m^\alpha \{(x_1^2 + x_2^2 - 0)\} = \sum_\alpha m^\alpha (x_1^2 + x_2^2)$$

$$I_{12} = \sum_\alpha m^\alpha \{(x_1^2 + x_2^2) \cdot 0 - x_1 x_2\} = \sum_\alpha m^\alpha x_1 x_2 = I_{21}$$

$$I_{13} = I_{31} = -\sum_\alpha m^\alpha x_1 x_3 = 0$$

and similarly

$$I_{23} = I_{32} = 0.$$

Now it is just a matter of adding up the masses of the constituent particles. We shall work out I_{11} as an example:

$$I_{11} = \sum_\alpha m^\alpha x_2^2 = \mu a^2 + \mu(-a)^2 + 2\mu(-a)^2 + 2\mu a^2 = 6\mu a^2.$$

All the other components can be worked out in the same way and we can write the final result as a matrix:

$$I_{ij} = \mu a^2 \begin{pmatrix} 6 & 2 & 0 \\ 2 & 6 & 0 \\ 0 & 0 & 12 \end{pmatrix}.$$

6.4.2 Example: *Verification of the perpendicular axes theorem for a lamina using the inertia tensor*

Write down the definition of the components of the inertia tensor and prove that for a lamina (in the plane $x_3 = 0$), the diagonal components of the inertia tensor are related by $I_{11} + I_{22} = I_{33}$: this is the perpendicular axes theorem for laminae only.

For the given geometry, we have $x_3 = 0$ for all particles (or mass elements) and equation (6.40) reduces to

$$I_{ij} = \sum m\{(x_1^2 + x_2^2)\delta_{ij} - x_i x_j\}.$$

It should be noted here that the summation is a shorthand for an integral over the surface of the lamina. It is easily seen that the moments of inertia about the respective coordinate axes are

$$I_{11} = \sum m x_2^2$$

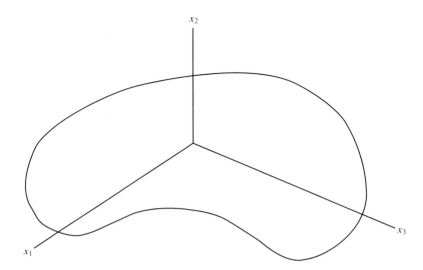

Fig. 6.10 Coordinate system for the perpendicular axes theorem for a lamina.

$$I_{22} = \sum m x_1^2,$$

and hence

$$I_{33} = \sum m\left(x_1^2 + x_2^2\right) = I_{11} + I_{22},$$

which verifies the perpendicular axes theorem as discussed in Section 6.2.6.

It should also be noted that for this geometry $I_{13} = I_{31} = I_{23} = I_{32} = 0$ but $I_{12} = I_{21} \neq 0$ in general.

6.4.3 Example: *Inertia tensor for a thin disc*

Derive the inertia tensor for a thin disc of radius a and mass m with respect to an orthonormal basis at its centre with e_3 normal to plane of the disc.

By symmetry $I_{11} = I_{22}$ and by the perpendicular axes theorem $I_{11} = I_{22} = \frac{1}{2}I_{33}$. Also, mass elements above and below Ox_1 (say) contribute equal and opposite amounts to $\sum m x_1 x_2$ for fixed x_1, so $\sum m x_1 x_2 = 0$.

Consider an elementary hoop between r and $r + \mathrm{d}r$, as illustrated in Figure 6.11.

$$I_{33} = \int_0^a \rho r^2 \, \mathrm{d}V = \int_0^a \left(\frac{m}{\pi a^2}\right)(2\pi r \, \mathrm{d}r) r^2$$

$$= \frac{1}{2} m a^2.$$

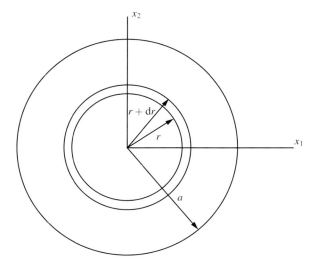

Fig. 6.11 Inertia tensor for a circular disc.

Therefore

$$I_{ij} = ma^2 \begin{pmatrix} \frac{1}{4} & 0 & 0 \\ 0 & \frac{1}{4} & 0 \\ 0 & 0 & \frac{1}{2} \end{pmatrix}.$$

6.5 Exercises

6.1 Find the moment of inertia of a long, slender rod about an axis perpendicular to its length and: (a) at the centre of the rod; and (b) at one end of the rod.

6.2 A thin rectangular plate has sides of length A and B. Obtain its moment of inertia about an axis which is: (a) perpendicular to the plate through its centre; and (b) parallel to the side of length B and through the centre of the plate.

6.3 Show that the moment of inertia of a hollow cylinder, with outer radius A, inner radius B and length l, about the axis of the cylinder is:

$$I = \frac{1}{2} M(A^2 + B^2),$$

where M is the mass of the cylinder.

6.4 Obtain the moment of inertia of a solid sphere about an axis which is: (a) along a diameter; and (b) tangent to the surface.

6.5 A solid sphere rolls without slipping down a slope which makes an angle α with the horizontal. Find the speed of its centre after it has travelled a distance l and compare this value with the result which would be obtained if the sphere had been sliding without rolling over the same distance.

6.6 A sphere of mass M and radius a spins with angular speed ω while sliding over a smooth surface with speed v. What is its kinetic energy?

6.7 A fly-wheel which is free to turn about a horizontal axis supports a mass m which is suspended from a vertical string which is wrapped round a pulley of radius b attached to the fly-wheel. Show that the tension in the string is given by:

$$T = \frac{I}{I + mb^2} mg,$$

where I is the moment of inertia of the fly-wheel.

6.8 A uniform circular disk of radius 0.1 m and mass 1 kg is set rotating about its axis at a rate of 100 revolutions per minute. It is brought to rest in 10 s by a constant frictional force applied tangentially to its rim. What is the magnitude of the force?

6.9 If a rotating planet contracts by cooling, show that when its radius is reduced by a factor λ, the length of day will be reduced by 2λ. The planet may be assumed to be spherical at all times.

6.10 A right circular cylinder of radius a is made of a non-uniform material such that it has its centre of mass at a distance h from its axis. If the cylinder rolls back and forward on a horizontal plane, show that its period of oscillation is given by:

$$T = 2\pi\sqrt{\frac{M(a-h)^2 + I}{Mgh}},$$

where M is the mass of the cylinder and I is its moment of inertia about an axis through its centre of mass and parallel to its generators.

6.11 Repeat the calculation of Example 6.4.1; but now with the particles of mass μ at $(0, 0, 0)$ and $(2a, 2a, 0)$, and the particles of mass 2μ at $(2a, 0, 0)$ and $(0, 2a, 0)$. Show that the non-zero components are of the inertia tensor are:
$I_{11} = 12\mu a^2; I_{22} = 12\mu a^2; I_{33} = 24\mu a^2; I_{12} = I_{21} = -4\mu a^2$.

6.12 A system of four point masses forms the vertices of a regular tetrahedron. Each particle has mass m and they are situated respectively at the points $(a, -a, a)$, $(a, a, -a)$, $(-a, a, a)$ and $(-a, -a, -a)$. Show that the inertia tensor at the origin of coordinates takes the form $8ma^2\delta_{ij}$, where δ_{ij} is the Kronecker delta and each of the subscripts corresponds to the coordinate directions.

6.13 A uniform rectangular block is bounded on three edges by the coordinate axes and by $z \leq 2a$, $y \leq 2b$ and $z \leq 2c$. Find the elements of the inertia tensor at the origin of coordinates.

Non-inertial frames of reference

It is helpful in understanding what constitutes an inertial frame to examine some examples of non-inertial frames. For our purposes here, it will be sufficient to consider two important special cases of non-inertial frames, viz., a coordinate system moving in a straight line with constant acceleration; and a coordinate system rotating with constant angular velocity.

7.1 Coordinate system moving with constant linear acceleration

An interesting example of such a system would be some experimental apparatus in free fall towards the Earth. As shown in Figure 7.1, we take our inertial frame S to be fixed relative to the Earth and let the non-inertial frame S'_a move with constant acceleration a in the direction of the negative z-axis of S. We assume that the origins of S and S'_a coincide at $t = 0$ and that S'_a starts from rest in S. Note that we put the subscript 'a' on S'_a in order emphasize that it is accelerating and hence is non-inertial, and also to distinguish it from our usual second *inertial* frame S'. Ignoring the Earth's rotation (in this section) we may write down transformations of coordinates between the two frames. The two coordinates at right angles to the motion are of course unaffected, thus:

$$x = x' \qquad \text{and} \qquad y = y'.$$

Assuming that $t = t'$ (as in this part of the book we are still dealing with Galilean relativity), the transformations in the z-direction are:

$$z = z' - \frac{at^2}{2} \qquad \dot{z} = \dot{z}' - at \qquad \text{and} \qquad \ddot{z} = \ddot{z}' - a. \qquad (7.1)$$

In S Newton's second law is just

$$m\ddot{z} = F, \qquad (7.2)$$

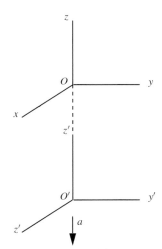

Fig. 7.1 A non-inertial frame accelerating in the negative z-direction.

for a body of mass m, here assumed to be constant. In S'_a, using the acceleration transformation of equation (7.1), we obtain N2 in the form

$$m\ddot{z}' = F + ma. \qquad (7.3)$$

If the experimental apparatus (a box, say) which defines S'_a is in free fall towards the Earth, then the gravitational force inside the box is cancelled by the inertial force ma. That is,

$$m\ddot{z} = -mg + ma, \qquad (7.4)$$

and $a = g$ for S'_a in free fall.[1] Therefore a particle in the box experiences no force (or acceleration) relative to the box. (Although, in S, both box and particle are being accelerated with g towards Earth.)

However, if we move the box to some region of the universe where there is no gravitational force, and accelerate it with $a = g$ in the direction of $\mathbf{k}$, then every particle within the box will experience an apparent force of magnitude mg in the direction of $-\mathbf{k}$. This equivalence of an accelerated frame of reference and a gravitational field is of fundamental importance in general relativity, and we shall discuss this topic further in Chapter 15.

Taking the basis vectors of S to be $(\mathbf{i}, \mathbf{j}, \mathbf{k})$ we may restate all the above results in vector form as

$$\mathbf{x} = \mathbf{x}' - \frac{at^2}{2}\mathbf{k} \qquad (7.5)$$

$$\dot{\mathbf{x}} = \dot{\mathbf{x}}' - at\mathbf{k} \qquad (7.6)$$

$$\ddot{\mathbf{x}} = \ddot{\mathbf{x}}' - a\mathbf{k} \qquad (7.7)$$

$$m\ddot{\mathbf{x}} = \mathbf{F} \qquad (7.8)$$

$$m\ddot{\mathbf{x}}' = \mathbf{F} + ma\mathbf{k} \qquad (7.9)$$

where $\mathbf{x}$ is the general position vector in S of a particle of mass m.

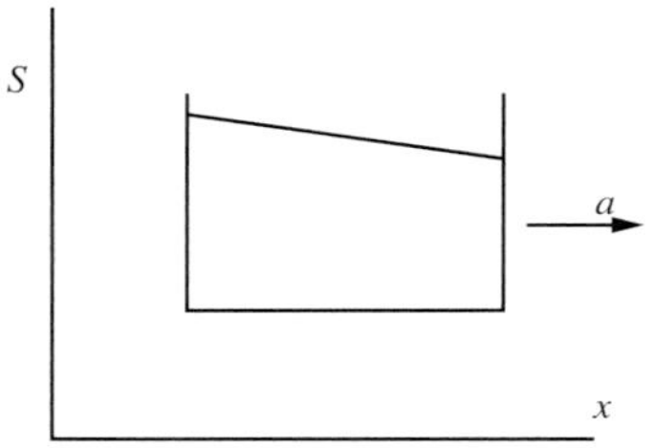

[1] We make the assumption here that the inertial mass and the gravitational mass of a body are the same thing. We have introduced this assumption as the principle of equivalence in Section 1.4.3.

Fig. 7.2 A bucket of water undergoing constant acceleration in the horizontal direction.

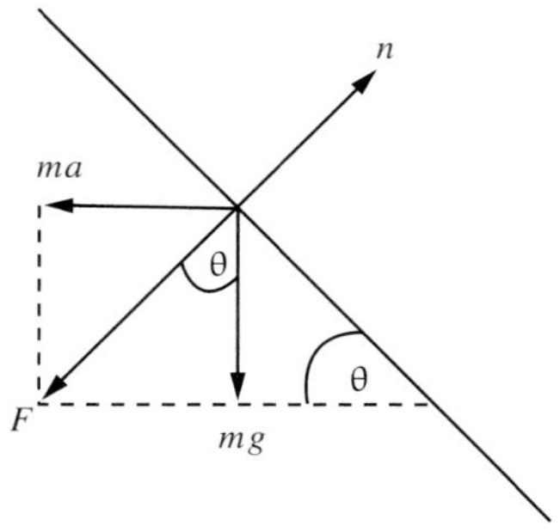

Fig. 7.3 Resolution of forces acting on the free surface of the liquid in the frame S'_a.

7.1.1 Example: *A bucket of water undergoing constant acceleration*

As an example of a coordinate frame undergoing a constant linear acceleration, we may consider a bucket of water being accelerated in a horizontal plane. The situation is as shown in Figure 7.2. We take the direction of acceleration to be the x-direction and the gravitational acceleration is assumed to be in the y-direction. Under these circumstances the free surface of the water in the bucket will no longer be horizontal but instead will make some angle θ with the horizontal. We take the observer's frame to be that of the Earth which is locally inertial and denote this as usual by S. Then we take the frame of reference fixed with respect to the bucket to be S'_a. We would like to calculate the angle θ and to do this we work in the frame S'_a and consider the forces acting on an element of the free surface of mass m. It is clear from Figure 7.3 that the forces acting on the element of mass are mg downwards due to gravity and ma to the left due to the bucket's acceleration to the right. Now the pressure at the free surface is constant (equal to the atmospheric pressure) and so the normal to the surface must lie along the resultant force acting on the surface. It follows from inspection of the diagram that the angle that the surface of the water makes with the horizontal is given by $\theta = \arctan(a/g)$.

7.2 Coordinate system rotating with constant angular velocity

Now let us consider an inertial frame S with basis vectors $(\mathbf{i}, \mathbf{j}, \mathbf{k})$ and a rotating frame S'_{rot} with basis vectors $(\mathbf{i}', \mathbf{j}', \mathbf{k}')$. The origins of the two frames coincide at all times and S'_{rot} rotates with constant angular velocity $\boldsymbol{\omega}$ relative to some axis of rotation which passes through the origin of coordinates.

We begin by reminding ourselves of an important result from vector calculus. If any arbitrary vector $\mathbf{A}$ is rotated about a fixed axis in an inertial frame S, then its rate of change with time in S is:

$$\left[\frac{\mathrm{d}\mathbf{A}}{\mathrm{d}t}\right]_S = \boldsymbol{\omega} \times \mathbf{A}, \tag{7.10}$$

where $\boldsymbol{\omega}$ is the angular velocity about an axis in the direction of the unit vector given by

$$\mathbf{e} = \boldsymbol{\omega}/|\boldsymbol{\omega}|. \tag{7.11}$$

In the next section, we shall make use of this result to transform differential coefficients with respect to time between the two frames of reference.

7.2.1 Rate of change of a vector relative to a rotating coordinate system

First, we shall obtain the required transformation for a position vector $\mathbf{x}$, as the geometric significance in this case is obvious. After that we can generalize to any vector $\mathbf{A}$.

Consider a particle at a point P. Draw a line from the common origin of S and S'_{rot} to P: this line is the position vector $\mathbf{x}$. There is only one position vector $\mathbf{x}$, but its components will be different in different frames. Thus in S and S'_{rot} we have respectively

$$\mathbf{x} = x\mathbf{i} + y\mathbf{j} + z\mathbf{k} = x'\mathbf{i}' + y'\mathbf{j}' + z'\mathbf{k}'. \tag{7.12}$$

Now take the time derivative, as measured in S, of $\mathbf{x}$ in terms of its components in the frame S'_{rot}. Note that $\mathbf{i}'$, $\mathbf{j}'$ and $\mathbf{k}'$ are moving relative to S and are therefore variables. Hence we have,

$$\left[\frac{\mathrm{d}\mathbf{x}}{\mathrm{d}t}\right]_S = \underbrace{\frac{\mathrm{d}x'}{\mathrm{d}t}\mathbf{i}' + \frac{\mathrm{d}y'}{\mathrm{d}t}\mathbf{j}' + \frac{\mathrm{d}z'}{\mathrm{d}t}\mathbf{k}'}_{A} + \underbrace{x'\frac{\mathrm{d}\mathbf{i}'}{\mathrm{d}t} + y'\frac{\mathrm{d}\mathbf{j}'}{\mathrm{d}t} + z'\frac{\mathrm{d}\mathbf{k}'}{\mathrm{d}t}}_{B}. \tag{7.13}$$

In equation (7.13), the terms grouped on the right-hand side by under-braces may be interpreted as follows:

A These terms make up the particle velocity relative to S'_{rot};

B These terms are the components of the velocity of a point fixed in S'_{rot} relative to S.

Since the first group of terms constitutes the time derivative of the position vector with respect to the frame S'_{rot}, accordingly we may write it as

$$\frac{\mathrm{d}x'}{\mathrm{d}t}\mathbf{i}' + \frac{\mathrm{d}y'}{\mathrm{d}t}\mathbf{j}' + \frac{\mathrm{d}z'}{\mathrm{d}t}\mathbf{k}' = \left[\frac{\mathrm{d}\mathbf{x}}{\mathrm{d}t}\right]_{S'}, \tag{7.14}$$

where the subscript S' indicates that the differentiation is carried out in S'_{rot}. Note that, in the interests of clarity, we dispense with the subscript 'rot', when S'_{rot} is itself a subscript.

The second group of terms on the right-hand side of equation (7.13) can be simplified as follows. Recalling equation (7.10), we can write

$$\frac{\mathrm{d}\mathbf{i}'}{\mathrm{d}t} = \boldsymbol{\omega} \times \mathbf{i}', \tag{7.15}$$

and similarly for the other two coordinate directions. Thus we can rewrite equation (7.13) as

$$\left[\frac{\mathrm{d}\mathbf{x}}{\mathrm{d}t}\right]_{S} = \left[\frac{\mathrm{d}\mathbf{x}}{\mathrm{d}t}\right]_{S'} + \boldsymbol{\omega} \times (x'\mathbf{i}' + y'\mathbf{j}' + z'\mathbf{k}'); \tag{7.16}$$

or, more compactly, as

$$\left[\frac{\mathrm{d}\mathbf{x}}{\mathrm{d}t}\right]_{S} = \left[\frac{\mathrm{d}\mathbf{x}}{\mathrm{d}t}\right]_{S'} + \boldsymbol{\omega} \times \mathbf{x}, \tag{7.17}$$

and, in terms of the particle velocities $\mathbf{v}_S$ and $\mathbf{v}_{S'}$, as measured in frames S and S' respectively,

$$\mathbf{v}_S = \mathbf{v}_{S'} + \boldsymbol{\omega} \times \mathbf{x}. \tag{7.18}$$

Obviously this derivation can be repeated for any arbitrary vector $\mathbf{A}$ and so we may write the general result:

$$\left[\frac{\mathrm{d}\mathbf{A}}{\mathrm{d}t}\right]_{S} = \left[\frac{\mathrm{d}\mathbf{A}}{\mathrm{d}t}\right]_{S'} + \boldsymbol{\omega} \times \mathbf{A}. \tag{7.19}$$

7.2.2 Interpretation of acceleration in a rotating frame

Equation (7.18) gives us the relationship between the velocity of a particle in an inertial frame and that in a uniformly rotating frame. In order to obtain the analogous relationship for the acceleration, we merely differentiate (7.18) again with respect to time and apply the rule given in equation (7.19) by setting $\mathbf{A} = \mathbf{v}_S$, thus:

$$
\left[\frac{d\mathbf{v}_S}{dt}\right]_S = \left[\frac{d\mathbf{v}_S}{dt}\right]_{S'} + \boldsymbol{\omega} \times \mathbf{v}_S
$$

$$
= \left[\frac{d}{dt}(\mathbf{v}_{S'} + \boldsymbol{\omega} \times \mathbf{x})\right]_{S'} + \boldsymbol{\omega} \times [\mathbf{v}_{S'} + (\boldsymbol{\omega} \times \mathbf{x})], \qquad (7.20)
$$

where we have used equation (7.18) to substitute for $\mathbf{v}_S$. Then, remembering that $\boldsymbol{\omega}$ is constant, we may further write this as

$$
\left[\frac{d\mathbf{v}_S}{dt}\right]_S = \left[\frac{d\mathbf{v}_{S'}}{dt}\right]_{S'} + \boldsymbol{\omega} \times \left[\frac{d\mathbf{x}}{dt}\right]_{S'} + \boldsymbol{\omega} \times \mathbf{v}_{S'} + \boldsymbol{\omega} \times (\boldsymbol{\omega} \times \mathbf{x}); \qquad (7.21)
$$

or, grouping terms, and substituting back for the velocities $\mathbf{v}_S$ and $\mathbf{v}_{S'}$, we have

$$
\left[\frac{d^2\mathbf{x}}{dt^2}\right]_S = \left[\frac{d^2\mathbf{x}}{dt^2}\right]_{S'} + 2\boldsymbol{\omega} \times \left[\frac{d\mathbf{x}}{dt}\right]_{S'} + \boldsymbol{\omega} \times (\boldsymbol{\omega} \times \mathbf{x}). \qquad (7.22)
$$

Alternatively, we may give this equation a rather simpler appearance by writing it in terms of the particle accelerations and the particle velocity in the moving frame, thus:

$$
\mathbf{a}_S = \mathbf{a}_{S'} + 2\boldsymbol{\omega} \times \mathbf{v}_{S'} + \boldsymbol{\omega} \times (\boldsymbol{\omega} \times \mathbf{x}),
$$

where the terms in this version of the equation may be interpreted as follows:

$\mathbf{a}_S \equiv$ acceleration of particle relative to S;

$\mathbf{a}_{S'} \equiv$ acceleration of particle relative to S'_{rot};

$2\boldsymbol{\omega} \times \mathbf{v}_{S'} \equiv$ Coriolis acceleration;

$\boldsymbol{\omega} \times (\boldsymbol{\omega} \times \mathbf{x}) \equiv$ acceleration in S of a point fixed to S'_{rot}.

We shall discuss these interpretations further at a later stage, but at this point, it may be useful to remind ourselves that, if we were dealing with the more usual case in this book, where S and S' are *both* inertial, then there would be no difference between $\left[d^2\mathbf{x}/dt^2\right]_S$ and $\left[d^2\mathbf{x}/dt^2\right]_{S'}$ according to Galilean relativity. As we shall see in the second part of this book, this would not be the case in special relativity.

7.2.3 Newton's second law in a rotating frame: the centrifugal and Coriolis forces

Note that the last term on the right-hand side of equation (7.21) is more familiar as the *centripetal* acceleration; or as the acceleration associated with the *centrifugal* force. For instance, in the simple case of circular motion, we may put $\mathbf{x} \equiv r\hat{\mathbf{r}}$, where r is the radius and $\boldsymbol{\omega}$ is just the angular speed ω, to recover the familiar elementary form $\omega^2 r$ of the centripetal acceleration.

Next we turn to a consideration of N2 in a rotating frame of reference. We already have this in the inertial frame as

$$\mathbf{F} = m\frac{\mathrm{d}^2\mathbf{x}}{\mathrm{d}t^2},\qquad(7.23)$$

where we are treating the mass of the body as constant. Now from (7.22) we may write N2 as

$$F = m\left[\frac{\mathrm{d}^2\mathbf{x}}{\mathrm{d}t^2}\right]_{S'} + 2m\boldsymbol{\omega}\times\left[\frac{\mathrm{d}\mathbf{x}}{\mathrm{d}t}\right]_{S'} + m\boldsymbol{\omega}\times(\boldsymbol{\omega}\times\mathbf{x}).\qquad(7.24)$$

Or, rearranging as an equation of motion:

$$m\left[\frac{\mathrm{d}^2\mathbf{x}}{\mathrm{d}t^2}\right]_{S'} = \mathbf{F} - 2m\boldsymbol{\omega}\times\left[\frac{\mathrm{d}\mathbf{x}}{\mathrm{d}t}\right]_{S'} - m\boldsymbol{\omega}\times(\boldsymbol{\omega}\times\mathbf{x}).\qquad(7.25)$$

Thus the acceleration in S'_{rot} is determined not only by the externally applied force $\mathbf{F}$, but also by, respectively, the Coriolis force and the centrifugal force. We shall discuss these forces further when we have considered some applications of the theory.

7.2.4 Example: *A bucket of water in a rotating frame*

Consider a bucket of water rotating about its vertical axis. As is well known — perhaps, in more everyday terms, from stirring a cup of tea or coffee — the free surface is not flat, but curved; with a depression in the middle. We can work out the profile of a cross-section through the free surface of the liquid in the rotating bucket as follows.

An observer at rest in the rotating frame of the bucket observes an element of mass as being accelerated radially outwards under the action of centrifugal force. At the same time, the mass is subject to the acceleration due to gravity in (say) the z-direction.

Let $\mathrm{d}l$ be a line element of the free surface in a plane of symmetry through the bucket. Let m be the mass of the line element. This mass is subject to both centrifugal and gravitational forces, as shown in Figure 7.4. Clearly from the geometry of the figure we may write

$$\tan\theta = \frac{m\omega^2 r}{mg} = \frac{\mathrm{d}z}{\mathrm{d}r},\qquad(7.26)$$

and so the slope of the free surface is given by

$$\frac{\mathrm{d}z}{\mathrm{d}r} = \frac{\omega^2 r}{g}.\qquad(7.27)$$

The equation of the free surface may be obtained by integrating both sides of this equation with respect to r, with the result

$$z = \frac{\omega^2 r^2}{2g} + C,\qquad(7.28)$$

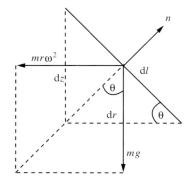

Fig. 7.4 Forces acting on an element of the free surface of the liquid in a rotating frame.

where C is the constant of integration. Clearly the value of C depends on where we choose to measure z from.

7.2.5 Example: *A particle falling freely to earth*

We treat the Earth as a perfect sphere (i.e. we neglect polar flattening and any other irregularities). We take our inertial frame S to have its origin at the centre of the Earth and its z-axis along the Earth's axis of rotation. The z'-axis of S'_{rot} is taken to be along a radius of the idealized Earth at latitude ϕ (measured from the equatorial plane) and the positive sense corresponding to that of a normal pointing outwards from the Earth. The x'-axis is taken to point south and the y'-axis to point east.

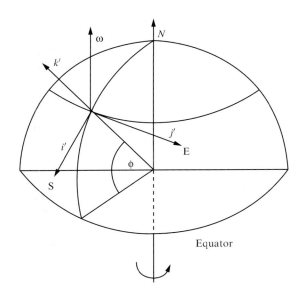

Fig. 7.5 Coordinate system based on the northern hemisphere of a perfectly spherical Earth.

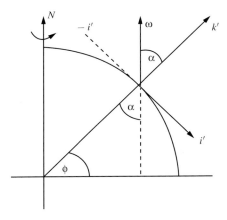

Fig. 7.6 Cartesian coordinates for a particle at latitude ϕ. The j'-axis is normal to the page. Note that the angle α is known as the colatitude.

From (7.25), the equation of motion can be written as

$$m\left[\frac{d^2\mathbf{x}}{dt^2}\right]_{S'} = -mg\mathbf{k}' - 2m\boldsymbol{\omega} \times \left[\frac{d\mathbf{x}}{dt}\right]_{S'}. \qquad (7.29)$$

Note that when we set

$$\mathbf{F} = -mg\mathbf{k}',$$

on the right-hand side of equation (7.29) we are also taking the centrifugal force $\boldsymbol{\omega} \times (\boldsymbol{\omega} \times \mathbf{x}) \equiv$ into account, as g is the *resultant* acceleration towards Earth at the given latitude. It is easily shown that the angular velocity $\boldsymbol{\omega}$ has components in S'_{rot} as follows:

$$\omega_1 = -\omega \sin\alpha = -\omega \sin\left(\frac{\pi}{2} - \phi\right) = -\omega \cos\phi \qquad (7.30)$$

$$\omega_2 = 0 \qquad (7.31)$$

$$\omega_3 = \omega \cos\alpha = \omega \cos\left(\frac{\pi}{2} - \phi\right) = \omega \sin\phi. \qquad (7.32)$$

Thus if we write (7.29) out in component form, we obtain the three equations

$$m\frac{d^2x'}{dt^2} = 2m\omega \sin\phi \frac{dy'}{dt} \qquad (7.33)$$

$$m\frac{d^2y'}{dt^2} = -2m\omega \sin\phi \frac{dx'}{dt} - 2m\omega \cos\phi \frac{dz'}{dt} \qquad (7.34)$$

$$m\frac{d^2z'}{dt^2} = -mg + 2m\omega \cos\phi \frac{dy'}{dt}. \qquad (7.35)$$

Now, due to the rotation of the Earth, there will be some variation of x', y' during the free fall. However, we shall neglect dx'/dt, dy'/dt compared with the 'vertical velocity' dz'/dt. Hence the above equations take the simplified forms:

$$\frac{d^2x'}{dt^2} = 0 \qquad (7.36)$$

$$\frac{d^2y'}{dt^2} = -2m\omega \cos\phi \frac{dz'}{dt} \qquad (7.37)$$

$$\frac{d^2z'}{dt^2} = -g. \qquad (7.38)$$

Equation (7.36) shows that there is no deviation in the north–south direction, while (7.38) is the usual equation which we consider in elementary treatments where the rotation of the Earth is neglected. Integrating this twice, and taking initial conditions to be $z' = dz'/dt = 0$ at $t = 0$, we find

$$\frac{dz'}{dt} = -gt, \qquad z' = -\frac{1}{2}gt^2. \qquad (7.39)$$

If we substitute the derivative into equation (7.37), we obtain for the y'-deviation

$$\frac{d^2 y'}{dt^2} = 2\omega g t \cos \phi, \tag{7.40}$$

and, integrating twice with initial conditions $y' = dy'/dt = 0$ at $t = 0$, we further obtain

$$y' = \frac{1}{3} \omega g t^3 \cos \phi. \tag{7.41}$$

Thus the eastward deviation of falling bodies is proportional to the cube of the time of fall, which has been confirmed by experiment.

7.2.6 Example: *Foucault's pendulum*

We have previously considered the simple pendulum in Section 4.2.2. Now we consider such a pendulum where the bob, of mass m, is free to move in any of the three coordinate directions. We again invoke N2, as given by equation (7.25), and make the same reduction of the full equations of motion as we did in the previous section. However, this time the force **F** is, in addition to mg in the vertical direction, determined by the tension in the string T. For a string of length l, we have the appropriate three-dimensional generalization of the results of Section 4.2.2 as

$$\mathbf{F} = \left(\frac{Tx'}{l}\right)\mathbf{i}' + \left(\frac{Ty'}{l}\right)\mathbf{j}' + \left(\frac{Tz'}{l} - mg\right)\mathbf{k}'. \tag{7.42}$$

From (7.25), the equations of motion can then be written out in component form, thus:

$$\frac{d^2 x'}{dt^2} - 2\omega \sin \phi \frac{dy'}{dt} - \frac{Tx'}{ml} = 0 \tag{7.43}$$

$$\frac{d^2 y'}{dt^2} + 2\omega \sin \phi \frac{dx'}{dt} + 2\omega \cos \phi \frac{dz'}{dt} - \frac{Ty'}{ml} = 0 \tag{7.44}$$

$$\frac{d^2 z'}{dt^2} - 2\omega \cos \phi \frac{dz'}{dt} - \frac{Tz'}{ml} + mg = 0, \tag{7.45}$$

where in each case we have divided across by m.

We now make the usual approximations for a simple pendulum undergoing motions of small amplitude. First we neglect motion in the z'-direction and set $z' = l$. Thus $d^2 z'/dt = 0$, $dz'/dt = 0$ and the equations of motion reduce to the form:

$$\frac{d^2 x'}{dt^2} - 2\omega \sin \phi \frac{dy'}{dt} - \frac{Tx'}{ml} = 0 \tag{7.46}$$

$$\frac{d^2 y'}{dt^2} + 2\omega \sin \phi \frac{dx'}{dt} - \frac{Ty'}{ml} = 0 \tag{7.47}$$

$$-\frac{Tz'}{ml} + mg = 0. \tag{7.48}$$

The last of these immediately gives the tension as

$$T = mg, \tag{7.49}$$

thus allowing the further simplification:

$$\frac{d^2 x'}{dt^2} - 2b \frac{dy'}{dt} + n^2 x' = 0 \tag{7.50}$$

$$\frac{d^2 y'}{dt^2} + 2b \frac{dx'}{dt} + n^2 y' = 0, \tag{7.51}$$

where:

$$n^2 = g/l, \tag{7.52}$$

and

$$b = \omega \sin \phi. \tag{7.53}$$

While this coupled set of equations may be solved in general, by the usual methods for simultaneous ordinary differential equations, it is sufficient for our purposes here to note that, if we choose suitable initial conditions, a particularly simple solution takes the form:

$$x' = A \cos bt \cos \Omega t \tag{7.54}$$
$$y' = -A \sin bt \cos \Omega t, \tag{7.55}$$

where $\Omega^2 = n^2 + b^2$.

This result can be more easily interpreted when expressed in plane polar coordinates (r, θ), where:

$$r = A \cos \Omega t \tag{7.56}$$
$$\theta = \arctan(y'/x') = \arctan(-\tan bt) = -bt. \tag{7.57}$$

That is, the bob of the pendulum executes simple harmonic motion with angular frequency $\Omega = (n^2 + b^2)^{1/2}$ along a line which is itself rotating with angular velocity $b = \omega \sin \phi$ in the horizontal plane.

In this case, the period of oscillation of the pendulum differs from that for the simple pendulum because of the effect of the rotation of the Earth. If we denote the period of Foucault‚s pendulum by τ_F, then it is easily shown that

$$\tau_F = 2\pi \left[\frac{l}{g + l\omega^2 \sin^2 \phi} \right]^{1/2}. \tag{7.58}$$

Thus a measurement of the period of oscillation of Foucault's pendulum yields the rotational velocity of the Earth as measured at latitude ϕ.

7.3 Inertial and fictitious forces

The term *inertial force* is often used to refer to the force experienced by a particle within an accelerating frame. This terminology is usually applied to all such forces, which arise purely as a consequence of the non-inertial nature of the frame of reference in which the force is experienced. Naturally, it can be applied to the Coriolis and centrifugal forces too.

Moreover, it also quite usual to refer to such forces as fictitious forces. This usage is obviously well justified when one considers the Coriolis force. For instance, in the example of a particle falling freely to Earth, as discussed in Section 7.2.5, the effect of the Coriolis force is to displace a falling body in a westward direction, according to an observer on Earth. However, according to an observer in a nearby inertial frame, the body falls without deviation and instead the surface of the Earth rotates eastward beneath it. So there is no actual force acting on the body in the westward direction.

However, in contrast, centrifugal force is arguably a real physical effect. For instance, if we continue with the example of a body falling to Earth, the gravitational acceleration which the body experiences towards the Earth is reduced by the centrifugal force due to the Earth's rotation. Similarly, one can think of many practical consequences of centrifugal force ranging from the action of a centrifuge to the disintegration of an unbalanced flywheel. Certainly, anyone who has driven a motor car which needs to have its wheels balanced will be familiar with both the reality of centrifugal force and the phenomenon of resonance!

Perhaps it is purely a matter of taste whether or not one includes centrifugal force in with the other inertial forces, in order to describe all such forces as fictitious. Nevertheless, some physicists prefer to make the distinction, and the present writer is among them.

7.4 Exercises

7.1 A simple pendulum hangs vertically from the roof of a car which is moving in a straight line at constant speed. What angle does the pendulum make with the vertical, if the car accelerates at a uniform rate a? Discuss what is happening from the point of view of an inertial observer outside the car and a non-inertial observer inside the car.

7.2 A simple pendulum hangs vertically from the roof of a railway carriage which is moving with a constant speed of $30\,\mathrm{m\,s^{-1}}$. When the brakes are applied, the pendulum swings through an angle of 3 degrees. How long will it take for the train to come to rest, if the braking force is taken to be constant?

7.3 A simple pendulum hangs from the ceiling of a laboratory in a spaceship. If the spaceship takes off vertically from Earth with an acceleration equivalent to $8g$, what will be the effect on the period of the pendulum at that time?

7.4 A particle rotates with angular velocity $\boldsymbol{\omega}$ about an axis fixed in an inertial frame S and passing through the origin. Verify that its instantaneous linear velocity is given by $\mathbf{v} = \boldsymbol{\omega} \times \mathbf{x}$, where $\mathbf{x}$ is the position vector of the particle.

7.5 Consider a rotating frame of reference S'_{rot}, with the same origin as S, and which rotates with angular velocity $\boldsymbol{\omega}$ relative to an axis which is fixed in S and passes through the origin. If a particle is moving with respect to both sets of axes, use an instantaneous Galilean transformation to deduce the general result:

$$\left[\frac{d\mathbf{x}}{dt}\right]_S = \left[\frac{d\mathbf{x}}{dt}\right]_{S'} + \boldsymbol{\omega} \times \mathbf{x},$$

where $\mathbf{x}$ is the position vector of the particle, and the subscripts on the square brackets have the same significance as in Section 7.2.1.

7.6 A particle is released from a point which is at a height h above the surface of the Earth and at a latitude of ϕ. Show that the easterly deviation of the point of impact is given by

$$y' = \frac{1}{3}\omega\left(\frac{8h^3}{g}\right)^{1/2}.$$

7.7 Describe the events of the previous exercise, as viewed by an observer on Earth and by an observer in a nearby inertial frame.

7.8 A projectile is launched from the Earth's surface at an angle of elevation β and heading due south. If the angular speed of the Earth is ω, show that the range and time of flight are unaffected by the Earth's rotation, to order ω; but that the point of impact is displaced to the west by a distance

$$y' = \frac{4\omega V^3}{3g^2}\sin^2\beta(\cos\phi\sin\beta + 3\sin\phi\cos\beta),$$

where ϕ is the latitude of the launch site and V is the speed of the projectile.

Variable-mass problems

8

In this chapter we consider some situations where the mass of a moving body may change with time because it actually loses or gains matter.[1] Apart from reminding us of the more general form of Newton's second law, this topic also has the salutary effect of making us think about the reference frame we are working in, and in the case of rocket motion, brings in the use of the instantaneous comoving frame.

We shall find it helpful to distinguish between two general cases.

In the first case, the process by which the particle gains or loses mass does not directly affect its motion. That is, the effect on the motion is an indirect one, through the change of the mass with time. In this category we place processes involving, for example, condensation, evaporation and sublimation. For instance, one could consider the fall of a raindrop under gravity, in which the raindrop gains mass by condensation or loses mass by evaporation, as it falls, depending on its immediate environment. Such situations will be described as **accretion-of-mass** problems.

In the second case, the process by which a particle gains or loses mass exerts a direct force on the particle. In this category are physical systems involving propulsion, such as rockets, jet engines and even the recoil of guns. We shall take the rocket as an example and discuss this case under the heading of the **rocket-propulsion equation**.

We begin by deriving a general equation of motion which will be appropriate for both these cases.

8.1 Derivation of the general equation of motion from $d\mathbf{p}/dt = \mathbf{F}$

For situations in which the mass of a body varies with time, we must consider N2 in the form:

$$\mathbf{F} = \frac{d\mathbf{p}}{dt} \equiv \frac{d}{dt}(m\mathbf{v}). \tag{8.1}$$

This is of course N2 for *one* body. Variable mass problems are, in effect, many-body problems, and at the very least we have to consider a system consisting of two bodies which may be moving with different velocities. This point will become more obvious as we consider some specific

[1] We shall see later that one interpretation of Einstein's theory of special relativity is that the mass of a particle depends on its speed, with appreciable changes occurring for speeds approaching the speed of light. We shall not be concerned with such effects in this chapter.

examples. For the moment, we formulate the general problem, in which we assume that a particle of mass m, moving with velocity $\mathbf{v}$ at time t, gains (or loses) mass δm, over an interval δt.

We obtain the *equation of motion* most conveniently by integrating both sides of N2, as given by (8.1) over the time interval δt, thus:

$$\int_t^{t+\delta t} \mathbf{F}\,dt = \mathbf{p}(t + \delta t) - \mathbf{p}(t),$$

where the quantity on the left-hand side is known as the **impulse** of the force $\mathbf{F}$ and is sometimes written as:

$$I = \int_t^{t+\delta t} \mathbf{F}\,dt.$$

We shall restrict our attention to cases where either the force $\mathbf{F}$ is constant, or at least so slowly varying with time that it is effectively constant over the interval δt. Then the above expression for the change in momentum of the system may be written as:

$$\mathbf{F}\delta t = \mathbf{p}(t + \delta t) - \mathbf{p}(t). \tag{8.2}$$

In the following sections we shall apply this result to various physical systems. However, at this stage it should be noted that in the present chapter $\mathbf{F}$ is always the externally applied force which acts on the system, such as the force due to gravity or to air resistance.

8.2 Accretion-of-mass problems

This is the first of our two categories of problem, in which a particle, initially of mass m, gains mass δm, becoming a new particle with mass $m + \delta m$. Note at this stage the two-body nature of the problem. There are, in effect, two masses which coalesce. The situation is illustrated in Figure 8.1. We consider the total momentum of the system at successive instants t and $t + \delta t$.

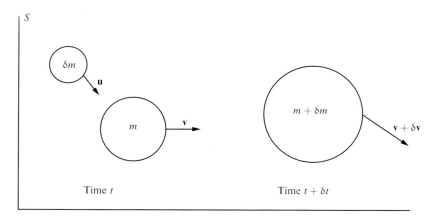

Fig. 8.1 A moving particle gaining mass from its surroundings.

At time t, the total momentum of the two-body system is:

$$\mathbf{p}(t) = m\mathbf{v} + \delta m\mathbf{u}. \tag{8.3}$$

At time $(t + \delta t)$, the total momentum has changed to

$$\mathbf{p}(t + \delta t) = (m + \delta m)(\mathbf{v} + \delta\mathbf{v}). \tag{8.4}$$

Therefore, from (8.2), the equation of motion takes the form:

$$\mathbf{F}\delta t = \mathbf{p}(t + \delta t) - \mathbf{p}(t) \tag{8.5}$$
$$= (m + \delta m)(\mathbf{v} + \delta\mathbf{v}) - m\mathbf{v} - \delta m\mathbf{u} \tag{8.6}$$
$$= m\mathbf{v} + m\delta\mathbf{v} + \delta m\mathbf{v} + O(\delta m\delta v) - m\mathbf{v} - \delta m\mathbf{u}. \tag{8.7}$$

Note that we neglect the product of the differentials as being of the second order of small quantities. Next we cancel the $m\mathbf{v}$ terms, divide across by δt and take the limit as $\delta t \to 0$, thus:

$$\mathbf{F} = m\frac{d\mathbf{v}}{dt} + (\mathbf{v} - \mathbf{u})\frac{dm}{dt}, \tag{8.8}$$

where $\mathbf{F}$ is the total force acting on the system. In the following section, we shall consider a simple one-dimensional application of this equation.

8.2.1 A raindrop falling through a cloud

We assume for the sake of simplicity that the water in the cloud is stationary in the *inertial frame* in which the raindrop velocity is $\mathbf{v}$, hence we set $\mathbf{u} = 0$ in equation (8.8). Note that this is a one-dimensional problem as $\mathbf{v}$ and $\mathbf{F} = \mathbf{g}$ are in the same direction, as shown in Figure 8.2. Take this direction—the vertical direction—to be the z-direction, and also take the downwards direction as positive, so that we may put $\mathbf{F} = (0, 0, mg)$ and $\mathbf{v} = (0, 0, v)$ in equation (8.8). Then, working only with the z-components, we have the scalar equation

$$mg = m\frac{dv}{dt} + v\frac{dm}{dt}. \tag{8.9}$$

Then, rearranging and dividing across by m, we obtain

$$\frac{dv}{dt} = g - \frac{v}{m}\frac{dm}{dt}. \tag{8.10}$$

Therefore, given $m(t)$ or dm/dt, and the initial conditions, this expression can be integrated to yield $v(t)$ and hence, with a further integration, $x(t)$.

It is worth noting that, if $dm/dt = 0$, then the acceleration of the drop is just equal to g. This is, of course, the expected result for a falling body which neither loses nor gains mass.

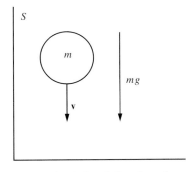

Fig. 8.2 A raindrop falling through a cloud.

8.2.2 Example: *Falling raindrop with given rate of mass increase*

A raindrop falls from rest through a stationary cloud, its mass m increasing by absorption of water vapour from the cloud. Find the velocity after a time t for the two cases:

1. the rate of increase of mass is proportional to the mass (i.e. $\dot{m} = km$);

2. the mass increases at a constant rate (i.e. $m = m_0 + \alpha t$, where m_0 and α are constants).

The general equation of motion for a falling raindrop is, as given by equation (8.9):

$$mg = m\frac{\mathrm{d}v}{\mathrm{d}t} + v\frac{\mathrm{d}m}{\mathrm{d}t}.$$

We consider the two cases in turn.

Case 1: $\mathrm{d}m/\mathrm{d}t = km$ The equation of motion becomes:

$$mg = m\frac{\mathrm{d}v}{\mathrm{d}t} + vkm.$$

Cancelling m across, and rearranging:

$$\frac{\mathrm{d}v}{\mathrm{d}t} = g - kv.$$

The solution may be found by separation of variables and, as $v(0) = 0$, takes the form:

$$v = \frac{g}{k}(1 - \mathrm{e}^{-kt}),$$

as may be verified by direct substitution.

Case 2: $m = m_0 + \alpha t$ From the given relationship, we have the rate of change of mass as

$$\frac{\mathrm{d}m}{\mathrm{d}t} = \alpha.$$

The equation of motion (8.9) then becomes

$$(m_0 + \alpha t)\frac{\mathrm{d}v}{\mathrm{d}t} + \alpha v = (m_0 + \alpha t)g$$

$$\implies \frac{\mathrm{d}}{\mathrm{d}t}[(m_0 + \alpha t)v] = (m_0 + \alpha t)g$$

with $v(0) = 0$. The solution of this equation is readily obtained by integrating both sides, with the result

$$[(m_0 + \alpha t)v]_0^t = \left(m_0 t + \frac{1}{2}\alpha t^2\right)g.$$

8.3.4 Example: *Final speed of a two-stage rocket*

A two-stage rocket of total mass M is used to accelerate from rest a payload of mass m, in circumstances where the effects of gravity may be neglected. The rocket motor of each stage ejects gases at a constant speed V (relative to the rocket) and the fuel mass of the stage is ε times the total mass of the stage ($0 < \varepsilon < 1$). If M_2 is the mass of the second stage, show that the final speed of the rocket is

$$v_f = V \ln \left[\frac{(M + m)(M_2 + m)}{(m + M - \varepsilon M + \varepsilon M_2)(m + M_2 - \varepsilon M_2)} \right].$$

Hence calculate v_f for $M = 99m$, $\varepsilon = \frac{5}{6}$ and $M_2 = M/2$.

This question may seem rather complicated, but reference to Figure 8.6 should make matters clear. Let us now consider the data for the two stages in turn. We are given the following information:

Stage 1: For the first stage we have:
- mass of fuel $= \varepsilon(M - M_2)$;
- mass of case $= (1 - \varepsilon)(M - M_2)$.

Stage 2: For the second stage we have:
- mass of fuel $= \varepsilon M_2$;
- mass of case $= (1 - \varepsilon)M_2$.

From (8.14) the rocket equation is

$$\mathbf{F} = m\frac{d\mathbf{v}}{dt} - \mathbf{V}\frac{dm}{dt},$$

and for this particular case we may put

$$\mathbf{F} = 0, \qquad \mathbf{v} = v\mathbf{i}, \qquad \mathbf{V} = -V\mathbf{i},$$

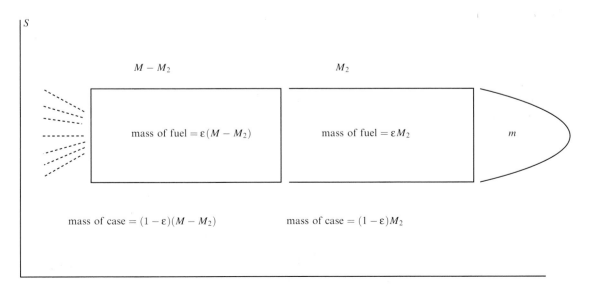

Fig. 8.6 Two-stage rocket with a payload m.

where **i** is the unit vector in the direction of flight. With these forms substituted, the rocket equation becomes

$$\frac{dv}{dt} = -V\frac{\dot{m}}{m} \Rightarrow v_2 - v_1 = V\ln\left(\frac{m_1}{m_2}\right).$$

For the first stage we have $v_1 = 0$, as the rocket takes off from rest. So the final speed of this stage is given by

$$v_2 = V\ln\{(M+m)/[M+m-\varepsilon(M-M_2)]\}.$$

At the end of the first burn, the first stage rocket is detached, so the second burn gives

$$v_f = v_2 + V\ln\{(M_2+m)/(M_2+m-\varepsilon M_2)\}.$$

Then, substituting from the preceding equation for v_2, the desired result follows. Direct numerical substitution then yields:

$$v_f = V\ln 9.3 \simeq 2.23V.$$

In fact it can be shown that an optimum choice of M_2 gives $v_f = 2.77V$, and this is the subject of Exercise 8.5, at the end of this chapter.

8.4 Exercises

8.1 An electrostatically charged sphere is launched vertically from the surface of the Earth through a cloud of dust. As it climbs, it attracts dust particles, so that its mass increases at a rate equal to km, where m is the mass of the sphere and k is a constant. Show that, if the constant k may be taken to be small, the time τ taken for it to reach its maximum height is given by

$$\tau = \frac{V_0}{k+g} + O(k^2),$$

where g is the gravitational acceleration and V_0 is the initial velocity of the sphere.

8.2 A gun of mass M fires a shell of mass m. Show that the less the gun recoils, the greater the speed of the shell. In particular, show that the speed v of the shell is given by:

$$v = \sqrt{\frac{2ME}{m(m+M)}},$$

where E is the energy liberated by the explosion of the propellant.

8.3 A rocket is launched from the surface of the Earth at an angle θ to the horizontal. Show that at any later time,

the angle α which its trajectory makes with the horizontal is related to its initial angle by:

$$\tan\alpha = \tan\theta\left[1 - \frac{gt}{V\cos\theta\ln(m_0/m)}\right],$$

where V is the speed of the exhaust gas relative to the rocket, m_0 is the initial mass of the rocket, and m is its mass at any time.

8.4 For test purposes, a rocket is mounted on horizontal rails, and fired from rest. If it experiences a resistive force $F = kv$, show that its speed v at any time is given by:

$$\frac{RV - kv}{RV} = \left[\frac{m_0}{m}\right]^k,$$

where $R = -dm/dt$ is the reaction rate of the rocket motor, m_0 is the initial mass and m is the mass at any time.

8.5 In the problem discussed in Example 8.3.4, assume that both M and m are fixed, but that M_2 may be varied. Show that v_f is a maximum if M_2 is a solution of $M_2^2 + 2mM_2 - mM = 0$, independent of ε, and write down the solution. Calculate M_2 and the final velocity v_f for $M = 99m$ and $\varepsilon = 5/6$.

Collisions

In Section 2.4, we discussed the two-body problem for the special case of one-dimensional motion and introduced concepts like the centre-of-mass (CM) frame of reference. Now we shall take a more general look at two-body collisions. However, first we make a few remarks about what we mean by the terms 'collisions' and 'particles'; and then we shall extend the work of Section 2.4 by considering collisions which are two-dimensional, before going on to discuss the general case.

9.1 Elastic collisions

Let us start by considering what we mean by a body or a particle. To begin with, we can fall back on our usual expedient of a point mass. This will be adequate for many purposes. However, in some cases, it will turn out that the finite extent of the body or particle is important. In these circumstances we shall take the body under consideration to be a sphere. Also, as we do not wish to consider frictional effects, we shall take the sphere to be smooth.

Now we come to the single most important concept in two-body collisions: elastic and inelastic collisions. This is so important that we shall highlight the definition of an elastic collision, thus:

Definition: A collision is said to be elastic if the total kinetic energy of all the particles before the collision is the same as that after.

In general, collisions between macroscopic bodies are *inelastic*, as in practice energy is lost in the form of heat or sound. In this book we shall restrict our attention to **elastic** collisions involving **two bodies**.

9.1.1 Example: *Elastic collision of identical particles: particles move off at right angles to each other*

A particle moving with constant speed v in the x-direction makes an elastic collision with an identical particle which is initially at rest. The incident particle is scattered at an angle θ_1 to the x-axis, with speed v_1, while the target particle recoils with speed v_2 at an angle θ_2 to the x-axis, as shown in Figure 9.1. By considering conservation of energy and momentum, show that the two particles move off at right angles to each other.

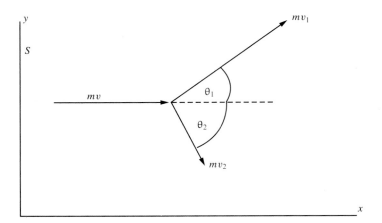

Fig. 9.1 Elastic collision of two identical particles.

First we apply the principle of conservation of energy, which implies:

$$v^2 = v_1^2 + v_2^2.$$

Next we invoke the principle of conservation of momentum, but we resolve into components in the x and y directions, and consider each separately.

Conservation of x-momentum implies:

$$v = v_1 \cos \theta_1 + v_2 \cos \theta_2,$$

and conservation of y-momentum implies:

$$0 = v_1 \sin \theta_1 + v_2 \sin \theta_2.$$

Now square both sides of each of these equations, to obtain:

$$v^2 = v_1^2 \cos^2 \theta_1 + 2 v_1 v_2 \cos \theta_1 \cos \theta_2 + v_2^2 \cos^2 \theta_2,$$

and

$$0 = v_1^2 \sin^2 \theta_1 + 2 v_1 v_2 \sin \theta_1 \sin \theta_2 + v_2^2 \sin^2 \theta_2,$$

respectively. Then we add the two squared momentum equations, thus:

$$v^2 + 0 = v_1^2(\cos^2 \theta_1 + \sin^2 \theta_1) + v_2^2(\cos^2 \theta_2 + \sin^2 \theta_2) \\ + 2 v_1 v_2 (\cos \theta_1 \cos \theta_2 + \sin \theta_1 \sin \theta_2),$$

and, using the trigonometric formula for the compound angle $\theta_1 + \theta_2$, which is

$$v^2 = v_1^2 + v_2^2 + 2 v_1 v_2 \cos(\theta_1 + \theta_2).$$

we find that comparison of this result with that obtained by conservation of energy yields the requirement for consistency:

$$2 v_1 v_2 \cos(\theta_1 + \theta_2) = 0.$$

Hence, for $v_1, v_2 \neq 0, \theta_1 + \theta_2 = \pi/2$, which is the required result.

9.1.2 Example: *Relationship between initial and final speeds of an elastically scattered particle*

A particle of mass m_1 and initial velocity v_{1i} collides with a stationary particle of mass m_2, and is scattered through an angle θ_1. If the collision is elastic, show that the initial and final speeds of the scattered particle are related by:

$$\left(\frac{v_{1f}}{v_{1i}}\right)^2 - \frac{2m_1}{m_1 + m_2}\left(\frac{v_{1f}}{v_{1i}}\right)\cos\theta_1 + \frac{m_1 - m_2}{m_1 + m_2} = 0,$$

where v_{1f} is the final velocity of the scattered particle.

As in the previous problem, we invoke conservation of energy and momentum, but this time we use vector notation in the momentum equation.

Conservation of energy:

$$\frac{1}{2}m_1 v_{1i}^2 = \frac{1}{2}m_1 v_{1f}^2 + \frac{1}{2}m_2 v_{2f}^2,$$

and rearranging to give an expression for $m_2^2 v_{2f}^2$:

$$m_2^2 v_{2f}^2 = m_1 m_2 v_{1i}^2 - m_1 v_{1f}^2.$$

Now, in vector form, we have conservation of momentum as:

$$m_1 \mathbf{v}_{1i} = m_1 \mathbf{v}_{1f} + m_2 \mathbf{v}_{2f},$$

and rearranging and squaring both sides we get:

$$m_2^2 v_{2f}^2 = m_1^2 v_{1i}^2 + m_1^2 v_{1f}^2 - 2m_1^2 v_{1i} v_{1f}\cos\theta_1.$$

Now, equating the two expressions for $m_2^2 v_{2f}^2$:

$$m_1 m_2 v_{1i}^2 - m_1 m_2 v_{1f}^2 = m_1^2 v_{1i}^2 + m_1^2 v_{1f}^2 - 2m_1^2 v_{1i} v_{1f}\cos\theta_1,$$

and rearranging:

$$(m_1^2 + m_1 m_2)v_{1f}^2 - 2m_1^2 v_{1i} v_{1f}\cos\theta_1 + (m_1^2 - m_1 m_2)v_{1i}^2 = 0.$$

Lastly, we divide across by $v_{1i}^2(m_1^2 + m_1 m_2)$ for the general result:

$$\left(\frac{v_{1f}}{v_{1i}}\right)^2 - \frac{2m_1}{m_1 + m_2}\left(\frac{v_{1f}}{v_{1i}}\right)\cos\theta_1 + \frac{m_1 - m_2}{m_1 + m_2} = 0.$$

Then, for the particular case $m_1 = m_2$, we have

$$\left(\frac{v_{1f}}{v_{1i}}\right)^2 - (v_{1f}/v_{1i})\cos\theta_1 = 0,$$

with solutions

$$\frac{v_{1f}}{v_{1i}} = \cos\theta_1 \quad \text{or} \quad 0.$$

For the case where the scattering angle is $\theta_1 = \pi/2$, it follows that $v_{1f} = 0$ for finite $v_{1i} \neq 0$, and all the energy is given to the recoiling particle.

9.2 Conservation laws for a general two-body collision

In the examples just considered, we did not pay any particular attention to the frame of reference and this habit of taking it for granted is quite usual in dynamics. However, from now on, we shall be more particular. We consider a general collision between particle 1 (say) and particle 2, as depicted in the left-hand side of Figure 9.2. For a general *elastic* collision, momentum and kinetic energy are conserved in the sense:

$$\text{initial state} = \text{final state.}$$

For conservation of energy, this means that we add up all the kinetic energy in the initial state (that is, before the collision) and all the kinetic energy in the final state (that is, after the collision) and equate them, thus:

$$\frac{p_{1i}^2}{2m_1} + \frac{p_{2i}^2}{2m_2} = \frac{p_{1f}^2}{2m_1} + \frac{p_{2f}^2}{2m_2}. \tag{9.1}$$

Similarly for linear momentum, we add up the momenta of the initial state and equate the result to the sum of the individual momenta of the final state, thus:

$$\mathbf{p}_{1i} + \mathbf{p}_{2i} = \mathbf{p}_{1f} + \mathbf{p}_{2f}. \tag{9.2}$$

Now, at this stage we note an important fact. If we consider the most general form of collision, with both particles moving relative to some

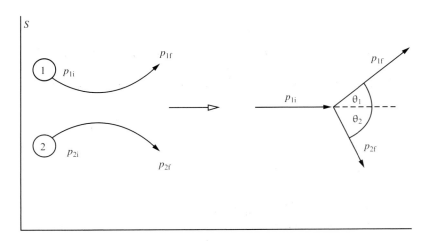

Fig. 9.2 Sketch defining the laboratory (LAB) frame of reference S.

inertial frame, then even when we are given *all* the initial conditions, we are unable to predict the final state. It is easily seen that the above equations only predict four scalar variables of the final state and hence only four components of final momentum: not all six.

In order to cope with this, we introduce the Laboratory or LAB frame (also known as the target frame), which is the frame in which particle 2 is initially at rest. Then we may further simplify the theoretical problems by working in the centre-of-mass frame, a concept which we have already met briefly in Section 2.4. We now formulate conservation laws in both these frames and then consider the transformations between the frames.

9.3 LAB and centre-of-mass (CM) frames

The general two-body collision, as observed in the LAB frame, is shown on the right-hand side of Figure 9.2, and we describe it as follows:

- Particle 1 is **incident** then **scattered**.
- Particle 2 is the stationary **target**, then it **recoils**.

We have in fact already used this terminology in the preceding examples and from now on we shall use it consistently. For instance, referring again to Figure 9.2, the angle θ_1 is called the scattering angle, whereas θ_2 is the angle of recoil.

In practice there is no hardship in working in the LAB frame, as this is usually where measurements are made. However, as we shall see, it can be easier to solve problems in the centre-of-mass frame. Accordingly we shall show how to set up conservation laws in both frames. Then some variant of the Galilean transformation—albeit a little complicated by the trigonometry—will allow us to transform results obtained in the CM frame to the LAB frame for comparison with experiment.

Formally we define the CM frame for two-body collisions as follows:

The CM frame of reference is a coordinate system in which the centre of mass of the two bodies is at rest.

The concept is illustrated in Figure 9.3, where we denote the CM frame by S', to remind us that an observer at rest in the CM frame is moving (with the velocity of the centre of mass) in S. Note that we denote variables in the CM frame by primes, so that the scattering angle becomes θ', and from the figure it may be seen that, in this frame of reference, the scattering angle equals the angle of recoil.

It may not be so intuitively obvious, but to an observer at rest in the CM frame, the initial momenta of the two particles are equal and opposite, as indeed are the final momenta. We shall show this presently, but for the moment it is worth noting that the centre of momentum of the two particles is also fixed in this frame, giving us an alternative meaning for the term 'CM frame'. We shall find this interpretation of some interest when we go on to special relativity in the second part of the book.

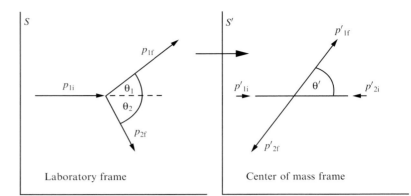

Fig. 9.3 Sketch defining the centre-of-mass frame.

9.3.1 Conservation of energy and momentum in the LAB frame

Setting up the conservation laws in the LAB frame is quite straight-forward. In this frame the initial velocity of particle 2 (the target particle) is zero, and so is its momentum, hence we simply put $\mathbf{p}_{2i} = 0$ in equation (9.1) for conservation of energy, to obtain

$$\frac{p_{1i}^2}{2m_1} = \frac{p_{1f}^2}{2m_1} + \frac{p_{2f}^2}{2m_2}.$$
(9.3)

Similarly, from equation (9.2) for conservation of momentum, we have

$$\mathbf{p}_{1i} = \mathbf{p}_{1f} + \mathbf{p}_{2f}.$$
(9.4)

However, in practice these relationships are simply constraints that have to be satisfied by the theoretical calculations which we shall carry out in the CM frame.

9.3.2 Conservation of energy and momentum in the CM frame

We begin this section by examining some of the relationships between position vectors and velocities in the CM frame, and end it by stating the appropriate formulations of the conservation laws.

As we have already seen in Section 2.4, in the CM system the positions of the particles are given by the *relative* coordinates $\mathbf{r}_1$ and $\mathbf{r}_2$ such that

$$m_1\mathbf{r}_1 + m_2\mathbf{r}_2 = 0.$$
(9.5)

Also, the separation $\mathbf{r}$ of the particles is given by

$$\mathbf{r} = \mathbf{r}_1 - \mathbf{r}_2 = \mathbf{x}_1 - \mathbf{x}_2,$$
(9.6)

which is, of course, the same in both frames. Or, in other words, the particle separation is an invariant under Galilean transformation.

We may use these equations to express either of the relative position vectors $\mathbf{r}_1$ or $\mathbf{r}_2$ in terms of the separation vector $\mathbf{r}$. For instance, if we substitute $\mathbf{r}_2 = \mathbf{r}_1 - \mathbf{r}$ in equation (9.5) we obtain

$$m_1\mathbf{r}_1 + m_2\mathbf{r}_1 - m_2\mathbf{r} = 0,$$

and so

$$\mathbf{r}_1 = \frac{m_2}{m_1 + m_2}\mathbf{r}, \tag{9.7}$$

or

$$\mathbf{r}_1 = \frac{\mu}{m_1}\mathbf{r}, \tag{9.8}$$

where the reduced mass μ is given by equation (2.44), which we reproduce here, thus:

$$\mu = \frac{m_1 m_2}{m_1 + m_2}.$$

Similarly, for the second relative position coordinate we have,

$$\mathbf{r}_2 = -\frac{\mu}{m_2}\mathbf{r}. \tag{9.9}$$

We may obtain useful relationships between the particle velocities (and hence their momenta) if we differentiate both these relationships for $\mathbf{r}_1$ and $\mathbf{r}_2$, with respect to time:

$$\dot{\mathbf{r}}_1 = \frac{\mu}{m_1}\dot{\mathbf{r}} \quad \text{and} \quad \dot{\mathbf{r}}_2 = -\frac{\mu}{m_2}\dot{\mathbf{r}}. \tag{9.10}$$

But, from the differentiation of all terms in (9.6) with respect to time, we obtain

$$\dot{\mathbf{r}} = \dot{\mathbf{r}}_1 - \dot{\mathbf{r}}_2 = \dot{\mathbf{x}}_1 - \dot{\mathbf{x}}_2 = \mathbf{u}, \tag{9.11}$$

where $\mathbf{u}$ is the relative velocity of the two particles. (Or, just like the relative position in equation (9.6), relative velocity is also a Galilean invariant.) Therefore the equations in (9.10) become

$$m_1\dot{\mathbf{r}}_1 = \mu\mathbf{u} = -m_2\dot{\mathbf{r}}_2. \tag{9.12}$$

This is characteristic of the CM frame: linear momenta are equal and opposite, in both initial and final states. It follows that in the CM frame, conservation laws take the form:

Energy:

$$\frac{\mathbf{p}_{1i}'^2}{2\mu} = \frac{\mathbf{p}_{1f}'^2}{2\mu} = \frac{1}{2}\mu u^2. \tag{9.13}$$

Momentum:

$$\mathbf{p}_{1i}' + \mathbf{p}_{2i}' = 0 = \mathbf{p}_{1f}' + \mathbf{p}_{2f}'. \tag{9.14}$$

9.4 Galilean transformation between the LAB and CM frames

We start by examining the general relationships between LAB and CM frames for the final state after collision, as shown in Figure 9.4. First, we have the straightforward vector results for the LAB to CM translation of particle positions in the two coordinate systems:

$$\mathbf{x}_1 = \mathbf{R} + \mathbf{r}_1 \qquad \text{and} \qquad \mathbf{x}_2 = \mathbf{R} + \mathbf{r}_2. \qquad (9.15)$$

Now differentiate both of these equations with respect to time to get velocities:

$$\mathbf{v}_{1f} = \dot{\mathbf{x}}_1 = \dot{\mathbf{R}} + \dot{\mathbf{r}}_1 = \dot{\mathbf{R}} + \mathbf{v}'_{1f} = \dot{\mathbf{R}} + \frac{\mu}{m_1}\mathbf{u}, \qquad (9.16)$$

and

$$\mathbf{v}_{2f} = \dot{\mathbf{x}}_2 = \dot{\mathbf{R}} + \dot{\mathbf{r}}_2 = \dot{\mathbf{R}} + \mathbf{v}'_{2f} = \dot{\mathbf{R}} - \frac{\mu}{m_2}\mathbf{u}, \qquad (9.17)$$

(remembering that

$$\dot{\mathbf{r}} = \dot{\mathbf{r}}_1 - \dot{\mathbf{r}}_2 = \dot{\mathbf{x}}_1 - \dot{\mathbf{x}}_2 = \mathbf{u}).$$

Lastly, the velocity of the centre of mass $\dot{\mathbf{R}}$ can be obtained from the conservation of momentum, thus:

$$m_1\mathbf{v}_{1i} = (m_1 + m_2)\dot{\mathbf{R}}, \qquad (9.18)$$

and so

$$\dot{\mathbf{R}} = \frac{\mu}{m_2}\mathbf{v}_{1i}. \qquad (9.19)$$

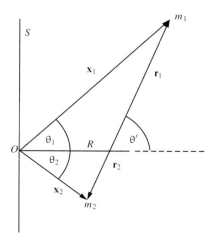

Fig. 9.4 Final state after the collision in the LAB and CM frames.

Next, we shall consider the question of how to relate the scattering angle in the CM frame to the scattering angle and the angle of recoil in the LAB frame. It turns out that it is easier, if less natural, to deal with the recoil angle first.

9.4.1 Relationship between θ_2 and θ'

The situation after the collision is shown in Figure 9.5, in terms of the various momenta. The individual triangles represent the possible statements of conservation of momentum and we explain and justify our construction as follows.

1. Triangle ABC represents

$$(m_1 + m_2)\dot{\mathbf{R}} = m_1\dot{\mathbf{x}}_1 + m_2\dot{\mathbf{x}}_2. \tag{9.20}$$

We can verify this by taking the defining relationship for the centroid coordinate, viz.,

$$\mathbf{R} = \frac{m_1\mathbf{x}_1 + m_2\mathbf{x}_2}{m_1 + m_2},$$

and differentiating each term with respect to time.

2. Triangle ABP represents

$$m_1\dot{\mathbf{R}} + m_1\dot{\mathbf{r}}_1 = m_1\dot{\mathbf{x}}_1. \tag{9.21}$$

We may verify this by taking the defining equation for the relative coordinate $\mathbf{R} + \mathbf{r}_1 = \mathbf{x}_1$, multiplying through by m_1 and differentiating each term with respect to time.

3. Triangle PBC represents

$$m_2\dot{\mathbf{R}} - m_1\dot{\mathbf{r}}_1 = m_2\dot{\mathbf{x}}_2. \tag{9.22}$$

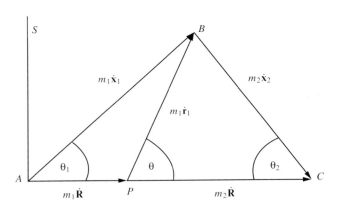

Fig. 9.5 Relationship between the recoil angle θ_2 in the LAB frame and the scattering angle θ' in the CM frame.

We verify this at first in the same way as for the preceding case, although with coordinates for particle 2, thus $m_2(\mathbf{R} + \mathbf{r}_2 = \mathbf{x}_2)$, leading, upon differentiation of each term with respect to time, to

$$m_2\dot{\mathbf{R}} + m_2\dot{\mathbf{r}}_2 = m_2\dot{\mathbf{x}}_2.$$

Then the usual condition on the relative coordinates, when differentiated, gives

$$m_1\mathbf{r}_1 + m_2\mathbf{r}_2 = 0 \implies m_1\dot{\mathbf{r}}_1 + m_2\dot{\mathbf{r}}_2 = 0,$$

therefore

$$m_2\dot{\mathbf{r}}_2 = -m_1\dot{\mathbf{r}}_1,$$

and so

$$m_2\dot{\mathbf{R}} - m_1\dot{\mathbf{r}}_1 = m_2\dot{\mathbf{x}}_2,$$

as required.

Now we use our construction as follows:

$$\text{the length } PC = |m_2\dot{\mathbf{R}}| = \mu|\mathbf{v}_{1i}| \tag{9.23}$$
$$\text{the length } PB = |m_1\dot{\mathbf{r}}_1| = \mu|\mathbf{u}| = \mu|\mathbf{v}_{1i} - \mathbf{v}_{2i}| = \mu|\mathbf{v}_{1i}|, \tag{9.24}$$

as $|\mathbf{v}_{2i}| = 0$. Therefore we have that the length of PB is equal to the length of PC, hence PBC is an isosceles triangle, and so

$$\angle PBC = \angle BCP = \theta_2.$$

As the angles of triangle PBC must add up to π, it follows from the diagram that:

$$\theta' + \theta_2 + \theta_2 = \pi,$$

or

$$\theta_2 = \frac{\pi - \theta'}{2}, \tag{9.25}$$

which is the required relationship.

9.4.2 *Relationship between* θ_1 *and* θ'

We now consider how to relate the scattering angles in the LAB and CM frames. We have already done the basic groundwork in the previous section and now all we have to do is consider just triangle ABP of the previous figure, which we present here as Figure 9.6. From this figure, it is clear that

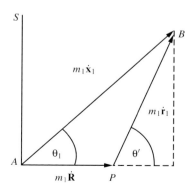

Fig. 9.6 Relationship between the scattering angles in the LAB and CM frames.

$$\tan\theta_1 = \frac{m_1\dot{\mathbf{r}}_1 \sin\theta'}{m_1\dot{\mathbf{r}}_1 \cos\theta' + m_1\dot{\mathbf{R}}} = \frac{\dot{\mathbf{r}}_1 \sin\theta'}{\dot{\mathbf{r}}_1 \cos\theta' + \dot{\mathbf{R}}}, \tag{9.26}$$

where we have cancelled m_1. Now use the results:

$$\dot{\mathbf{r}}_1 = \frac{\mu}{m_1}\dot{\mathbf{r}} = \frac{\mu}{m_1}\mathbf{v}_{1i},$$

from (9.24), and:

$$\dot{\mathbf{R}} = \frac{\mu}{m_2}\mathbf{v}_{1i},$$

from (9.19), to get:

$$\frac{\dot{\mathbf{r}}_1}{\dot{\mathbf{R}}} = \frac{m_2}{m_1}.$$

Therefore

$$\tan\theta_1 = \frac{\sin\theta'}{\cos\theta' + \dot{\mathbf{R}}/\dot{\mathbf{r}}} \qquad (9.27)$$

or

$$\tan\theta_1 = \frac{\sin\theta'}{\cos\theta' + m_1/m_2}, \qquad (9.28)$$

which, along with (9.25), completes the process of relating the angles of scattering and recoil in the LAB and CM frames.

9.4.3 Relation between scattering angles in elastic collision of equal-mass particles as a special case

Taking the special case $m_1 = m_2$ makes this the same problem as in Section 9.1.1 and this is a special case of some importance. From equation (9.28), the relationship between the scattering angles in the two coordinate systems becomes:

$$\tan\theta_1 = \frac{\sin\theta'}{\cos\theta' + 1} = \tan\frac{\theta'}{2} \qquad (9.29)$$

$$\theta_1 = \frac{\theta'}{2}. \qquad (9.30)$$

When taken in conjunction with the relationship between θ_2 and θ', as given by equation (9.25), this allows us to verify the result of Section 9.1.1, that two particles of equal mass move off at right angles to each other following an elastic collision. A demonstration of this point is the subject of Exercise 9.2.

9.5 Exercises

9.1 In an elastic collision between two particles of mass m_1 and m_2, the total kinetic energy is conserved. However, during the collision, the incident particle loses energy to the target particle. Show that the initial and final velocities of the incident particle satisfy the relationship:

$$\left(\frac{v_{1f}}{v_{1i}}\right)^2 - 2\left(\frac{v_{1f}}{v_{1i}}\right)\frac{\mu}{m_2}\cos\theta_1 - \left(\frac{m_2 - m_1}{m_2 + m_1}\right) = 0,$$

where μ is the reduced mass. If the two particles have the same mass, under what circumstances will the scattered particle transfer all its energy to the recoiling particle?

9.2 Using the results of Section 9.4, show that in an elastic collision of two particles of equal mass, the scattering angle and the angle of recoil add up to a right angle in the LAB frame.

9.3 If a particle of mass m_1 undergoes an elastic collision with a particle of mass m_2, show that the scattering angles in the LAB and CM frames become identical as the ratio m_1/m_2 tends to zero. Give a physical interpretation of this result.

9.4 If a particle undergoes an elastic collision with another particle of the same mass then, as we have shown in Exercise 9.2, the angles of scattering and recoil must add up to a right angle in the LAB frame. What determines the relative magnitude of the two angles?

9.5 Two smooth spherical particles collide elastically. By taking the y-axis along the line joining the centres of the particles, and considering conservation of momentum and energy, obtain the following relationship:

$$\dot{y}_1 - \dot{y}_2 = -(\dot{y}_1' - \dot{y}_2').$$

Here the primes refer to after the collision and the subscript to particle 1 and particle 2.

9.6 In this book, we only consider elastic collisions, but in many practical situations collisions are inelastic, and the total kinetic energy of the two-particle system is less after the impact than before it. For the special case of spherical particles, we may take this loss of mechanical energy into account by introducing the coefficient of restitution e such that the relationship obtained in the previous exercise becomes modified to the form:

$$e(\dot{y}_1 - \dot{y}_2) = -(\dot{y}_1' - \dot{y}_2'),$$

where $0 \leq e \leq 1$ and the y and y' axes are as defined in the preceding exercise. If a ball is projected horizontally with speed U at a height h above a smooth horizontal table, show that the horizontal distance x_m before the ball stops bouncing is given by

$$x_m = \left(\frac{2h}{g}\right)^{1/2}\left(\frac{1+e}{1-e}\right)U,$$

where e is the coefficient of restitution between the ball and the table.

Scattering of particles

10

The subject of scattering is essentially statistical in character, in that we consider a lot of particles which are incident on some target. Typically, we talk of a **beam of particles**. We then consider the effect of the target on the incident particles in terms of the probability of particles being scattered through some given angle. In experiments, this probability is determined by counting the number of particles scattered in the given direction per unit time. This measure of probability is known as a **scattering cross-section** and we shall give it a formal definition shortly. However, for the moment, we should note that the scattering cross-section will depend on both the nature of the incident particles and also on the nature of the target. In this chapter we shall use the theory of collisions developed in the previous chapter in order to work out some cross-sections from first principles.

In order to do this, we shall make the same restrictions as in the previous chapter. Particles will be either point masses or smooth spheres and all collisions will be assumed to be elastic. In addition, an incident beam will be assumed to be made up of one type of particle, of the one size, all having the same energy and being uniformly distributed over the beam. In the order given, such a beam may be referred to as *homogeneous*, *monodisperse*, *monoenergetic* and *uniform*.

Incident particles may be scattered by other particles or by a centre of force (e.g. gravity, or electrostatic attraction or repulsion). We shall assume that any force falls off to zero at large distances from the centre. This allows us to specify the nature of the incident and scattered beams asymptotically: that is, at large distances from the target. In this way, the target can be regarded as the traditional black box, with input and output beams being connected by the scattering cross-section.

We should note that particle scattering is of immense importance in physics, being the basic tool for investigating microscopic structure. Nowadays the emphasis is on probing the deepest subnuclear structure of matter using high-energy particle beams. In such cases we have to consider a theoretical treatment which is fully compatible with special relativity and we shall return to this topic later on in this book. However, once we have dealt with the Galilean or Newtonian theory, we shall round off our consideration of this topic with a brief discussion of the classic experiment usually described as 'Rutherford scattering', by which the nuclear structure of matter was discovered.

Lastly, we shall also give some attention to the situation where the forces between the incident and target particles are attractive, and the question then is whether or not the incident particle is captured by the target. The probability of this happening is known as the **capture cross-section**. We shall close this chapter with a brief discussion of capture processes and the calculation of a capture cross-section.

10.1 Scattering from a fixed centre of force

For convenience, in a discussion of the general picture of particle scattering, we shall consider the case where the scattering centre does not recoil. In practice this is often the case where there is a fixed centre of force, or a target particle is a fixed (in a lattice, say) or is so massive compared to incident particles that its recoil will be small enough to neglect. In addition, as we saw in Section 2.4.4, the two-body problem[1] can be reduced to an equivalent one-body problem, in which a fictitious particle of mass equal to the reduced mass is a distance **r** from a fixed centre of force, where **r** is the separation vector of the two particles.

Also, as we discussed the problem of relating the LAB and CM frames in the previous chapter, we should note here an important fact: if the target particle does not recoil, then there is no difference between the LAB and CM frames. It is left as an exercise for the reader to verify this.

Now we may summarize the general aspects of particle scattering as follows:

1. A uniform beam of monoenergetic identical particles coming from infinity is incident upon a centre of force.
2. The incident beam is characterized by its intensity I, where: $I =$ The number of particles crossing unit area (normal to the beam) in unit time.
3. As each particle in the beam approaches the centre of force, it will be either attracted or repelled, and its orbit will deviate from the initial straight line.
4. As each particle passes the centre of force and moves off to infinity, the force acting on it will diminish, and its orbit will revert to a straight line.
5. The kinetic energy and angular momentum of each particle is conserved during the scattering process.

10.2 Impact parameter and scattering angle

We have previously touched on the use of an impact parameter in Section 3.5.2, where we treated some aspects of α-particle scattering; and of course we have discussed the concept of scattering angle in the previous chapter. Referring to Figure 10.1, which is the definition sketch for scattering from a fixed centre, we may define the impact parameter b formally as follows:

[1] Note the requirement that the two bodies must interact along a line joining their centres, so that this is effectively a central force problem.

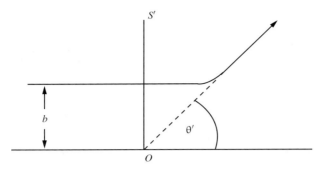

Fig. 10.1 Definition sketch for scattering from a fixed centre of force.

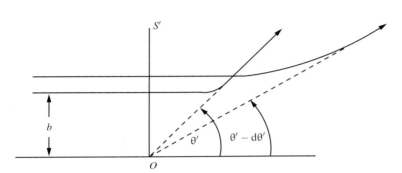

Fig. 10.2 Relationship between scattering angle and impact parameter.

The impact parameter is the distance between the line of approach of an incident particle and a parallel line through the scattering centre.

The scattering angle is the angle by which the scattered line of flight deviates from the incident line of flight.

However, in the latter case, note that we are working with θ', which was defined in Chapter 9 as the scattering angle in the CM frame. We do this to remind ourselves of the need to transform our results to the LAB frame in those cases where the target particle is not fixed.

In Figure 10.2, we illustrate the effect on the scattering angle of varying the impact parameter. Clearly if we *increase* the impact parameter, the incident particle does not pass the scattering centre so closely and is therefore less affected. Accordingly, the greater the impact parameter, the smaller the scattering angle, and this inverse relationship should be borne in mind when we work out the scattering cross-section in the next section.[2]

Lastly, we note that a conceptual problem sometimes arises, when we treat scattering as we do here, due to the fact that there are two symmetries involved. That is:

1. The incident beam has cylindrical symmetry and is invariant under rotations round an axis through the scattering centre.

2. The scattering centre (with the restrictions we impose here) has spherical symmetry and the scattering force is invariant under all rotations.

[2]In this discussion we are implicitly assuming that the strength of the force falls off with increasing distance from its centre.

The conceptual problem which can arise is that the elementary solid angle which we have to consider in order to calculate the scattering cross-section is defined by the intersection of these two symmetries. As this can be a cause of difficulty, we shall devote some attention to this particular point in the next section, once we have given the formal definition of cross-section.

10.3 Differential and total cross-sections for scattering

The cross-section for scattering is denoted by σ and is defined as follows:

$\sigma(\Omega)d\Omega$

$$= \frac{\text{Number of particles scattered into solid angle } d\Omega \text{ per unit time}}{\text{Incident intensity}}$$

(10.1)

where $d\Omega$ is an element of solid angles. The quantity $\sigma(\Omega)$ is known as the **differential cross-section**, while its integral over all solid angles is known as the **total cross-section** σ_T, or:

$$\sigma_T = \int \sigma(\Omega)\, d\Omega.$$

Where there is complete symmetry around the axis of the incident beam (as assumed here), the element of solid angle $d\Omega$ depends only on the scattering angle. As before, we take the scattering angle to be θ_1 in the LAB frame and θ' in the CM frame. The *total* cross section σ_T is just obtained by integration, thus:

$$\sigma_T = \int \sigma(\Omega)\, d\Omega = 2\pi \int \sigma(\theta') \sin \theta'\, d\theta', \qquad (10.2)$$

in the CM frame.

10.3.1 Calculation of the differential cross-section in the CM frame

We will evaluate σ in the CM frame, and then show how to transform the result to the LAB frame.

Referring to Figure 10.3, we see that this is a two-dimensional slice through the real picture. The full three-dimensional version is obtained by rotating the diagram about the axis of symmetry AB through an angle of 2π. This is made clear in Figure 10.4, where we have a view of the scattering experiment looking from point B in the direction of A.

In order to calculate the cross-section from its definition, as given by equation (10.1), we need to calculate that fraction of the incident intensity I of particles which is scattered into an elementary solid angle $d\Omega$. In order to define the elementary solid angle, we consider those particles passing through an annulus with impact parameters between b

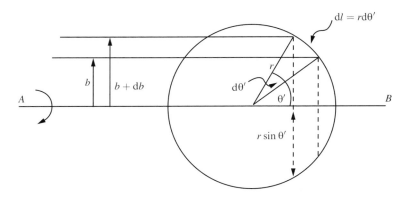

$\mathrm{d}l = r\mathrm{d}\theta'$

Fig. 10.3 Elementary solid angle for scattering from a fixed centre: side view.

and $b + \mathrm{d}b$. As shown in Figure 10.3, this variation in impact parameter will produce a variation in scattering angle, such that (in two dimensions) the corresponding incident particles will all pass through an arc of the circle of length $\mathrm{d}l = r\,\mathrm{d}\theta'$. The full element of area in three dimensions is then obtained by rotating the elementary arc $\mathrm{d}l$ through an angle 2π about the axis AB. From Figure 10.3, it is clear that the element $\mathrm{d}l$ is a distance $r\sin\theta'$ from the axis of symmetry and from Figure 10.4 it may be seen that the rotation generates an elementary area given by:

$$\mathrm{d}A = 2\pi r\sin\theta' \times r\,\mathrm{d}\theta' = 2\pi r^2\sin\theta'\,\mathrm{d}\theta'.$$

Then, from Appendix B, where we discuss the subject of solid angles, we have the corresponding solid angle as

$$\mathrm{d}\Omega = \frac{\mathrm{d}A}{r^2} = 2\pi\sin\theta'\,\mathrm{d}\theta', \tag{10.3}$$

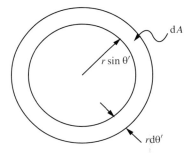

Fig. 10.4 Elementary solid angle for scattering from a fixed centre: front view.

and, as expected from the symmetry, the cross-section depends only on the scattering angle, viz.,

$$\sigma(\Omega) \equiv \sigma(\theta'). \tag{10.4}$$

Now let $\mathrm{d}N$ be the number of particles per unit time passing through the annulus defined by b and $b + \mathrm{d}b$; thus:

$$\mathrm{d}N = 2\pi b\,\mathrm{d}b \times I, \tag{10.5}$$

which is the cross-sectional area of the annulus multiplied by the intensity I of the beam. Then $\mathrm{d}N$ particles per unit time must be scattered into the solid angle $\mathrm{d}\Omega$, as defined by the scattering angles θ' and $\theta' + \mathrm{d}\theta'$, so we also have

$$\mathrm{d}N = I\sigma\,\mathrm{d}\Omega, \tag{10.6}$$

by the definition of cross section given in equation (10.1).

Hence, equating these two expressions for dN, and substituting for $d\Omega$ from (10.3), we obtain

$$2\pi b\, db \times I = -I\sigma 2\pi \sin\theta'\, d\theta' \qquad (10.7)$$

or, cancelling the common factors, and rearranging,

$$\sigma = -\frac{b}{\sin\theta'}\frac{db}{d\theta'}, \qquad (10.8)$$

where we note that the minus sign is because, as b increases θ' decreases.

But $\sigma(\theta')$ is the *probability* of a particle being scattered through θ', and therefore must be positive. So we tidy this up by taking the modulus:

$$\sigma = \frac{b}{\sin\theta'}\left|\frac{db}{d\theta'}\right|. \qquad (10.9)$$

10.3.2 *Transformation to the LAB frame*

Now we consider the transformation of this result to give us $\sigma_L(\theta_1)$ in the LAB frame. In order to do this, we note the basic conservation requirement:

The number of particles scattered in unit time into solid angle $2\pi \sin\theta'\, d\theta'$ in the CM frame

must equal

The number of particles scattered in unit time into the corresponding solid angle $2\pi \sin\theta_1\, d\theta_1$ in the LAB frame.

In other words, it follows that we have to have:

$$I\sigma_L(\theta_1)2\pi \sin\theta_1\, d\theta_1 = I\sigma(\theta')2\pi \sin\theta'\, d\theta'; \qquad (10.10)$$

and hence, with some rearrangement, and cancellation,

$$\sigma_L(\theta_1) = \sigma(\theta')\frac{\sin\theta'd\theta'}{\sin\theta_1 d\theta_1}. \qquad (10.11)$$

Alternatively, using the result that follows from differentiation of $\cos\theta$, we may write the expression for the differential cross-section in the compact form:

$$\sigma_L(\theta_1) = \sigma(\theta')\frac{d(\cos\theta')}{d(\cos\theta_1)}. \qquad (10.12)$$

(This may look a little unusual, but if we renamed $\cos\theta'$ as y and $\cos\theta_1$ as x, then it would be no more than a perfectly familiar differential coefficient dy/dx!)

In order to make use of this result, we simply invoke our earlier formula for the relationship between θ' and θ_1: see equation (9.28).

10.4 Hard-sphere scattering as a special case

This is a very important special case and we begin by considering very small spheres of mass m being scattered elastically by a fixed smooth sphere of mass M and radius R. Then we move on to consider the scattering of one sphere by an identical one, while drawing a distinction between the two extreme situations, where the target particle is fixed in space or is free to recoil.

However, before going on to do this, it is of interest to note that hard-sphere scattering corresponds to a potential:

$$V(r) = 0 \quad \text{for} \quad r \geq R$$
$$V(r) = \infty \quad \text{for} \quad r \leq R.$$

This form is of considerable importance in statistical and many-body physics.

10.4.1 Example: *Scattering of a small sphere by a large, fixed sphere*

Rigid spheres, of mass m and negligible radius, are projected with speed V and impact parameter b towards a target made up of large spheres of mass M and radius R, which are fixed in position. Find the scattering angle and the differential and total scattering cross-sections in the CM frame, on the assumption that the collisions are elastic.

Referring to Figure 10.5, we take the speed of the incident particle to be U after the collision, θ' is the scattering angle, α, β are the angles of incidence and reflection, the dotted line is a radius produced to define the normal to the surface, O is the centre of the large sphere, and P is the point of contact. As the target particles do not recoil, the CM and LAB frames are identical. Hence, in Figure 10.5 we emphasize this by labelling the frame of reference with both S and S'.

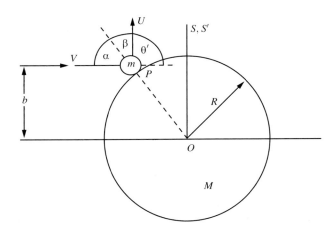

Fig. 10.5 Small sphere being scattered by a large sphere.

We begin by making use of our two given conditions:

- Elastic collision: the component of incident velocity normal to the surface of the target sphere is conserved, thus:

$$V \cos \alpha = U \cos \beta;$$

- Smooth collision: the component of incident velocity tangential to the surface of the target sphere is conserved, thus:

$$V \sin \alpha = U \sin \beta;$$

and solving these two equations simultaneously leads to:

$$U = V \quad \text{and} \quad \alpha = \beta.$$

In other words, the speed of the incident particle is unaffected by the collision and its reflection is specular. That is, the angles of incidence and reflection are equal.

Our immediate objective is to obtain a relationship between the impact parameter and the scattering angle, such that we may use equation (10.9) to calculate the scattering cross-section. When we consider the situation, clearly there are two possibilities. If $b > R$ then there is no collision and hence $\theta' = 0$. On the other hand, if $b < R$, then there is a collision and by elementary trigonometry we can show that the impact parameter and the angle of incidence are related by

$$b = R \sin \alpha.$$

The latter instance is the non-trivial case, and is therefore the one which we shall analyse. Our next step is to re-express this result in terms of the scattering angle. From the figure, $\theta' = \pi - (\alpha + \beta)$, and hence,

$$\theta' = \pi - 2\alpha,$$

therefore

$$\alpha = \frac{\pi}{2} - \frac{\theta'}{2},$$

and, using the trigonometrical identity for $\sin(A - B)$,

$$b = R \cos \frac{\theta'}{2}, \tag{10.13}$$

is the desired relationship between the impact parameter and the scattering angle. We have previously shown—see equation (10.9)—that:

$$\sigma = \frac{b}{\sin \theta'} \left| \frac{\mathrm{d}b}{\mathrm{d}\theta'} \right|.$$

So, substituting into this from (10.13) for b, doing the differentiation, and making use of the identity $2 \sin A \cos A = \sin 2A$, gives:

$$\sigma(\theta') = \frac{R \cos(\theta'/2)}{\sin \theta'} \frac{R}{2} \sin \frac{\theta'}{2} = \frac{R^2}{2} \frac{1}{\sin \theta'} \frac{1}{2} \left(2 \sin \frac{\theta'}{2} \cos \frac{\theta'}{2} \right) = \frac{1}{4} R^2.$$

$$(10.14)$$

The total scattering cross-section is found by integration over the solid angle thus:

$$\sigma_{\mathrm{T}} = \int \sigma(\Omega) \, d\Omega = 2\pi \int_0^\pi \frac{1}{4} R^2 \sin \theta' \, d\theta'$$

$$= \frac{\pi R^2}{2} \int_{-1}^1 d(\cos \theta') = \pi R^2. \qquad (10.15)$$

Note that the total cross-section is just the geometrical cross-section (that is, the projected area) of the target particle in this case.

10.4.2 Example: *Scattering by identical spherical particles which are fixed in space*

Smooth, rigid spheres, of radius R are projected with velocity V and impact parameter b at a target made up of identical spherical particles. Find the scattering angle θ_1 in the **LAB** frame, as a function of R and b, when the target sphere is held fixed and it is assumed that the collisions are elastic. Also, obtain the differential cross-section σ_{L} and the total cross-section σ_{T}.

The situation is illustrated in Figure 10.6, where we again assume that the speed of the incident particles after the collision is U. The target

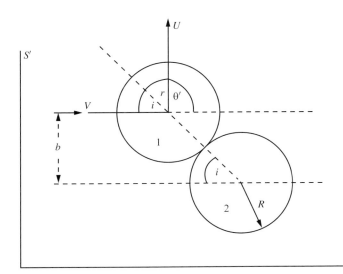

Fig. 10.6 Scattering of identical spheres.

sphere is held fixed, while the incoming sphere is incident at angle i, and is reflected at angle r (measured with respect to the line joining their centres). We take the frame of reference to be the CM frame, although in this case, as the target is fixed, this is identical to the LAB frame. However, as the same figure will serve for the example in the following section, where the target particle is free to recoil, we have labelled the frame of reference as S' in order to denote the moving CM frame. We begin by analysing the collision in the same way as in the previous example.

Since the spheres are smooth, we have

$$V \sin i = U \sin r.$$

Since the collision is elastic, we also have

$$V \cos i = U \cos r.$$

Then, solving these two equations simultaneously, we again conclude that:

$$U = V \quad \text{and} \quad i = r,$$

just as in the previous example. Now, from inspection of Figure 10.6, we note that:

$$\sin i = b/2R \quad \text{and} \quad \theta' = \pi - 2i.$$

Hence, the impact parameter may be expressed in term of the scattering angle, thus:

$$b = 2R \sin\left(\frac{\pi}{2} - \frac{\theta'}{2}\right) = 2R \cos\frac{\theta'}{2},$$

and so, from equation (10.9), the differential cross-section in the LAB frame is:

$$\sigma_L(\theta') = \frac{b}{\sin\theta'}\left|\frac{db}{d\theta'}\right| = \frac{2R\cos\theta'/2}{\sin\theta'} \cdot R\sin\frac{\theta'}{2} = R^2.$$

Then the total cross-section is simply obtained by integrating over the total solid angle to give:

$$\sigma_T = 2\pi \int_0^\pi \sigma(\theta') \sin\theta' \, d\theta' = 4\pi R^2,$$

which is, of course, the geometrical cross-section for one particle of radius R to collide with another particle of the same radius.

For this case, where the target particle is fixed, we reiterate that there is no difference between the CM and LAB frames; and hence we have

$$\theta_1 = \theta' = 2\cos^{-1}(b/2R),$$

as required.

10.4.3 Example: *Scattering by identical spherical particles which are free to recoil*

Re-calculate the previous problem with the change that the target sphere is free to recoil.

The first part of the calculation is the same for both cases, and we still have $\sigma(\theta') = R^2$, in the CM frame.

However, for this case, where the target particles are free to move, we have a less trivial transformation between the two coordinate systems. From equation (9.28), θ_1 and θ' are related by

$$\tan \theta_1 = \frac{\sin \theta'}{\cos \theta' + m_1/m_2} = \frac{\sin \theta'}{\cos \theta' + 1} = \tan \frac{\theta'}{2},$$

for $m_1 = m_2$, where the last step is just one of the standard trigonometric half-angle formulae.

From this result, it follows immediately that $\theta_1 = \theta'/2$, and hence the angle of scattering in the LAB frame is given by:

$$\theta_1 = \theta'/2 = \cos^{-1}(b/2R).$$

Note also that the maximum angle of scattering in the CM frame is π and that the above relationship indicates that, in this case where the target particle is free to recoil, the maximum angle of scattering in the LAB frame is $\pi/2$. We shall need to bear this in mind when we come to work out the total cross-section.

Now, from equation (10.12), we transform differential cross-sections from the CM frame to the LAB frame by

$$\sigma_L(\theta_1) = \sigma(\theta') \frac{d(\cos 2\theta_1)}{d(\cos \theta_1)},$$

where we have substituted $\theta' = 2\theta_1$. In order to do the differentiation, we employ the double-angle formula $\cos 2A = 2\cos^2 A - 1$, and hence

$$\sigma_L(\theta_1) = 4\cos \theta_1 \sigma(\theta') = 4R^2 \cos \theta_1.$$

Then the total scattering cross-section is given by the integral

$$\sigma_T = 2\pi \int_0^{\pi/2} 4R^2 \cos \theta_1 \sin \theta_1 \, d\theta_1 = 4\pi R^2,$$

where we note that the maximum angle of scattering is half of that in the CM frame, as we saw above. It is left as an exercise—see Exercise 10.2

at the end of this chapter—for the reader to explain on physical grounds why this result for the total cross-section is exactly the same as in the previous case where the target particle was fixed.

10.5 Scattering in a central-force field

In Chapter 3, we have given a general treatment of the motion of bodies under the influence of a central-force field, and this material can be adapted to allow us to work out the scattering cross-section, where required, for a given force. The case of greatest importance, whether one is considering astronomical scales where gravity dominates, or atomic and molecular scales where electrostatic forces dominate, is the inverse-square law. Accordingly we conclude this chapter on scattering by working out the scattering cross-section for the case of the inverse-square law. We should note that we have already studied various aspects of the motion of a particle under a repulsive inverse-square law in Section 3.5.2, and we now make use of these results.

10.5.1 Repulsive inverse-square law of force

Consider an inverse-square law force, with centre at the point O. A particle approaches from $r = \infty$, with speed V. We assume that its initial parameters have been chosen to be such that its orbit is a hyperbola. In the notation of Chapter 3, we write the force per unit mass as

$$f(r) = -\frac{k}{r^2}.$$

From the results of Section 3.5.2, we have a relationship between the scattering angle and the given data for the problem, thus:

$$\tan \frac{\theta'}{2} = \frac{k}{mbV^2}. \tag{10.16}$$

In order to work out the cross-section, we need to differentiate the impact parameter with respect to the scattering angle, and accordingly we rearrange this equation to get:

$$b = \frac{k}{mV^2} \cot \frac{\theta'}{2},$$

and hence, since $d\cot A / dA = -\csc^2 A$,

$$\frac{db}{d\theta'} = -\frac{k}{2mV^2} \csc^2 \frac{\theta'}{2}.$$

Then, substituting into equation (10.9) for the cross-section, we have:

$$\sigma(\theta') = \frac{b}{\sin \theta'} \left| \frac{db}{d\theta'} \right| = \frac{1}{2} \left(\frac{k}{mV^2} \right)^2 \frac{\cot \theta'/2 \csc^2 \theta'/2}{2 \sin \theta'/2 \cos \theta'/2}$$

$$= \left(\frac{k}{2mV} \right)^2 \frac{1}{\sin^4 \theta'/2}. \tag{10.17}$$

It is worth noting that if we go on and try to calculate the corresponding total scattering cross-section, in this case we should get an infinite result. Physically, this is because the Coulomb force is of infinitely long range and so all particles (in principle, at least) are scattered to some extent. This type of divergence is often encountered in physics and arises because we are employing an idealized analysis in an attempt to describe a complicated physical situation. In practice, we have to take into account factors like the presence of a screening cloud of electrons around the nucleus or the many-body effect of the presence of other scattering centres in the material specimen being studied. Thus, in the actual case, the Coulomb potential is correct at short distance; but is screened or cut-off at long distances, thus eliminating the divergence.

10.5.2 Rutherford's model of the atom

Rutherford studied atoms by scattering α-particles from thin metal foils. For the Coulomb interaction, we may identify the constant in the law of force

$$f = \frac{-k}{r^2},$$

as being

$$k = z_1 z_2 e^2,$$

where e is the electronic charge, z_1 is the atomic number for α-particles ($= 2$), and z_2 is the atomic number of the metal in the foil being studied. He found that the distance of closest approach was bounded by 10^{-14} m and concluded that the charge was concentrated within this region.

This observation led to the concept of the **nucleus**. It is difficult for us—having grown up in an age where the existence of the nucleus is an everyday fact—to imagine the surprise felt by Rutherford when his α-particles were scattered through large angles. Picturesquely he described it to be as if a sixteen inch shell from a naval gun had been fired at a sheet of tissue paper and had rebounded!

10.6 Attractive scattering forces and capture processes

There are many situations where the interaction between the incoming particle and the target involves attractive forces. One may think of examples ranging from the microscopic, such as neutrons being absorbed by nuclei, up to the macroscopic (and, indeed, industrial) such as chemical processes, powder technology and electrostatic paint-spraying. As we have pointed out at the beginning of this chapter, it is conventional to express the probability of the incident particles being captured in terms of a **capture cross-section**. In practice this can be obtained experimentally by measuring the attenuation of a particle beam as it passes through the scattering medium. In order to have a simple example of a calculation of a capture cross-section, we shall turn to an astronomical problem, and this is the subject of the next section.

10.6.1 Example: *A planet passing through a cloud of meteors*

An airless planet passes through a cloud of meteors with relative speed V. Show that the planet captures those meteors which lie within a cylinder, with its axis along the path of the planet and its radius given by $R\sqrt{1 + 2gR/V^2}$, where R is the planet's radius and g is the acceleration due to gravity at its surface.

To an observer on the planet, the meteors seem to approach with velocity V, and those meteors with impact parameter b, such that $b < b_0$ (say), will be captured by the planet. Now, referring to Figure 10.7, we fix the value of the impact parameter which determines whether or not meteors are captured as follows. Consider a meteor with impact parameter b_0 which has velocity V_0 as it grazes the planet's surface at $r = R$.

We invoke the principle of conservation of energy at that point and this gives:

$$\frac{1}{2}mV^2 = \frac{1}{2}mV_0^2 - \frac{GmM}{R},$$

where m is the mass of the meteor and M is the mass of the planet. Then, with some rearrangement, we obtain an expression for the grazing velocity in the form:

$$V_0 = V\left(1 + \frac{2GM}{RV^2}\right)^{1/2},$$

and as the acceleration due to gravity at the surface of the planet must satisfy

$$mg = \frac{mMG}{R^2} \implies \frac{GM}{R} = gR$$

we may substitute as appropriate to obtain:

$$V_0 = V\left(\frac{1 + 2gR}{V^2}\right)^{1/2}.$$

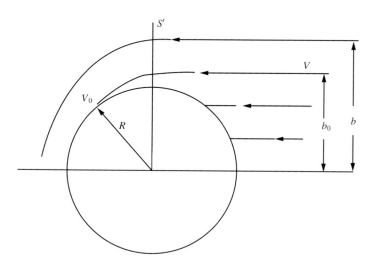

Fig. 10.7 Meteorites captured by a planet.

Conservation of angular momentum then gives us:

$$mb_0 V = mRV_0,$$

and hence the limiting impact parameter for capture takes the value:

$$b_0 = \frac{RV_0}{V} = R\left(1 + \frac{2gR}{V^2}\right)^{1/2},$$

as required.

If only short-range forces were involved, then from purely geometric considerations the capture cross-section would be:

$$\sigma = \pi R^2.$$

Instead, due to the long-range nature of the gravitational attraction, we have

$$\sigma = \pi b_0^2 = \pi R^2\left(1 + \frac{2gR}{V^2}\right),$$

and clearly the effect of the gravitational attraction is to increase the probability of capture.

10.7 Exercises

10.1 In Section 10.4.1, we showed that the total cross-section for scattering of particles with negligible radius by particles of radius R was $\sigma_T = \pi R^2$. In Section 10.4.2, we found that the corresponding result when the incident particles also were of radius R, was $4\pi R^2$. What is the physical explanation for the difference between these two results?

10.2 In Sections 10.4.2 and 10.4.3, we established that the total cross-section for the mutual scattering of identical particles with radius R was equal to $4\pi R^2$, irrespective of whether or not the target particle was fixed or free to recoil. Why, on physical grounds, should these two results be the same?

10.3 Smooth, rigid spheres of radius r are projected with speed V and impact parameter b at a target made up from similar particles, but with radius R, where $R > r$. Obtain the total scattering cross-section and discuss its form in the limiting cases $r \to 0$ and $r \to R$.

10.4 A particle of mass m is repelled from a fixed centre by a central force of magnitude κ/r^3. It can be shown that if such a particle is projected from a large distance, with velocity V and impact parameter b, then it will be scattered through an angle given by

$$\theta_1 = \pi\left[1 - \left(1 + \frac{\kappa}{mb^2 V^2}\right)^{-1/2}\right],$$

as measured in the LAB frame. Show that the differential cross-section is:

$$\sigma(\theta_1) = \frac{\pi^2 \kappa(\pi - \theta_1)}{mV^2(2\pi - \theta_1)^2\theta_1^2 \sin\theta_1}.$$

10.5 A sphere of radius R falls with a constant speed V through a cloud of dust. The dust consists of a uniform distribution of small particles, each of mass m. Obtain an expression for the capture of dust particles by the sphere, under the following circumstances:

(a) the sphere is covered in a sticky substance which will cause any particle which touches its surface to adhere; or:

(b) the sphere is electrostatically charged and attracts any dust particle with a force given by

$$f = -\frac{\lambda}{r^2},$$

where r is the distance of the particle from the centre of the sphere.

11 Special relativity

In this chapter, we begin our 'modern physics' approach to the subject of dynamics. We show that in order to satisfy Einstein's second axiom—that the speed of light *in vacuo* is the same in all inertial frames—we have to modify the Galilean transformation to take the form of the Lorentz transformation. In the process, we are forced to give up the Newtonian postulate of the universality of time. Instead, we now must treat time as a variable which transforms between different inertial frames. It is from this one aspect that all the bizarre and fascinating consequences of special relativity flow.

11.1 The need for a new principle of relativity

We shall begin this chapter by considering how the necessity for a new principle of relativity became recognized. We start by reviewing *Galilean* relativity. As before, we take two inertial frames S and S' in standard configuration. The Galilean transformations between the coordinates of the same event, as measured in the two different frames, take the form discussed in Chapter 1, thus:

$$x' = x - Vt \qquad y' = y \qquad z' = z \qquad t' = t;$$

where the last transformation embodies the Newtonian postulate of a universal time. Also, differentiating with respect to time, we have the additional relations of Galilean relativity, viz.,

- addition of velocities:

$$\dot{x}' = \dot{x} - V;$$

- invariance of accelerations:

$$\ddot{x}' = \ddot{x},$$

where the dots denote differentiation with respect to time.
Now we remind ourselves of the basic principle of Galilean relativity:

The laws of mechanics are the same in all inertial frames.

This property is known as Galilean invariance, and we speak of Newton's laws as being 'Galilean invariant'.

By the end of the nineteenth century, there was a complacent view that physics as a subject was complete. The natural philosophy of the universe was thought to be well understood. However, there were some cracks in this façade, which many people at the time thought could be patched up, but which were ultimately to turn into large fissures which would destroy the whole structure. The result was the intellectual revolution that gave us both quantum theory and special relativity. It is the latter topic which concerns us here, and so we shall only discuss those particular problems which led to special relativity. There were two of these.

Problem 1: The laws of propagation of electromagnetic waves (that is, Maxwell's equations) are not Galilean invariant.

Problem 2: The Michelson–Morley experiment implied that the velocity of light, c, (*in vacuo*) is independent of the frame of reference. This result violates the Galilean law of addition of velocities.

These difficulties with classical physics prompted a number of *ad hoc* explanations and attempts to resolve the situation. Two such theories are worthy of note here.

Firstly, the FitzGerald contraction was postulated in an attempt to explain the 'null' result of the Michelson–Morley experiment. The postulate was simply that moving bodies contract in their direction of motion. Thus the arm of the Michelson interferometer which pointed in the direction of the Earth's motion, was supposed to be shorter than the arm at right angles.

Secondly, the Lorentz transformations were introduced as *modified* Galilean transformations to make Maxwell's equations the same in S and S'. As we shall see, Einstein was to unite both these ideas in the theory of special relativity and to do this, he needed to postulate only two axioms.

11.2 Einstein's axioms

Faced, like other physicists of the time, with the need to choose between two mutually exclusive choices, viz.,

- the correctness of Newton's laws and Galilean invariance on the one hand; and
- the correctness of Maxwell's equations and Lorentz invariance on the other;

Einstein made his choice and enunciated it boldly in the form of two now famous axioms:

Axiom 1: The laws of **physics** are the same in all inertial reference frames.

Axiom 2: The velocity of light *in vacuo* (c) is the same in all inertial frames.

A corollary can be appended to these axioms, as follows:

Corollary: No **physical** experiment can be used to tell whether an inertial frame is moving or at rest (with respect to any other frame).

These axioms are the basis of the theory of special[1] relativity. If we accept them, then, as we shall see, we must equally accept the following implications:

1. Time is not universal.
2. Simultaneity is relative.
3. Time is dilated: 'moving clocks run slow'.
4. Length is contracted: a moving body contracts in its direction of motion.
5. Mass and energy are equivalent.[2]

These consequences can all be demonstrated to be the results of the change from the Galilean transformations between inertial frames, which we have been using up until now, to the Lorentz transformations. These are derived in the next section.

[1] The use of the word 'special' means that gravitational effects are excluded. This imposes a restriction that the inertial frames S, S', S'' etc. must all be at the same gravitational potential. In Chapter 15 we shall give a very brief introduction to general relativity which includes a discussion of the limitations of the concept of a *locally* inertial frame.

[2] A conventional but perhaps now somewhat old-fashioned view is that the mass of a moving particle is increased in value over its rest mass. In this book we shall relegate this concept to an appendix.

11.3 Derivation of the Lorentz transformations

Let us consider two frames S and S' in standard configuration, as illustrated in Figure 11.1. We begin by noting that the transformations between any two frames must be linear, in order to ensure that Newton's first law holds in all inertial frames. An obvious starting point is some generalization of the Galilean transformations. However, it is better just to say that in each case we write the transformations such that the coordinates of an event in one frame are linear combinations of the coordinates in the other frame.

Thus we assume that the following relationships hold

$$x' = ax + bt \tag{11.1}$$

and

$$x = a'x' + b't', \tag{11.2}$$

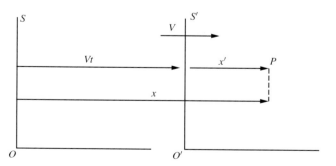

Fig. 11.1 Inertial frames in standard configuration for derivation of the Lorentz transformations.

where a, b, a' and b' are coefficients to be determined, and both sets of coordinates refer to the same event.

Now the origin of the frame S' is at $x' = 0$, thus equation (11.1) reduces to

$$0 = ax + bt,$$

for this particular case and hence we must have the relationship

$$-ax = bt,$$

and so

$$-\frac{b}{a} = \frac{x}{t} = V, \tag{11.3}$$

for constant velocity V.

Similarly, set $x = 0$ for the position of O in equation (11.2) and we obtain the corresponding result

$$0 = a'x' + b't',$$

and hence the primed coefficients are related by

$$-a'x' = b't',$$

and so

$$-\frac{b'}{a'} = \frac{x'}{t'} = -V. \tag{11.4}$$

Then, substituting from (11.3) and (11.4) for b and b' in equations (11.1) and (11.2), we obtain

$$x' = a(x - Vt), \tag{11.5}$$

and

$$x = a'(x' + Vt'), \tag{11.6}$$

as our embryonic transformations. Note that the requirement of symmetry under the interchange of S and S' in turn requires $a = a'$. In other words, going from S to S' should be similar to going from S' to S.

So far we have still to use either of Einstein's two basic axioms of special relativity. In order to determine the coefficient a (and hence a'), we now make use of the second axiom. Let us suppose that we send out a light signal at $t = t' = 0$ when $x = x' = 0$. The signal propagates in both S and S' satisfying

$$x' = ct', \tag{11.7}$$

and

$$x = ct. \tag{11.8}$$

That is, the light signal propagates at speed c in both frames, which is just Einstein's second axiom.

Next we substitute from (11.7) and (11.8) for x and x' in equations (11.5) and (11.6) to obtain

$$ct' = a(ct - Vt);$$

$$ct = a(ct' + Vt'),$$

after which we multiply both left-hand sides and both right-hand sides together:

$$c^2tt' = a^2tt'(c^2 - V^2),$$

and cancel the common factor tt' across, to obtain

$$a^2(c^2 - V^2) = c^2.$$

Then, rearranging,

$$a^2 = \frac{c^2}{c^2 - V^2} = \frac{1}{1 - V^2/c^2},$$

and taking the square root of both sides, we find for the coefficient a:

$$a = \frac{1}{\sqrt{(1 - V^2/c^2)}} \equiv \gamma(V). \tag{11.9}$$

Having obtained the form of the coefficient a which is compatible with Einstein's second axiom, we note that a is usually written as γ or $\gamma(V)$, and this notation will be used throughout the rest of this book.

Now we obtain the non-trivial Lorentz transformation for the space coordinates by substituting (11.9) for $a = a'$ into equations (11.5) and (11.6), thus:

$$\underbrace{x' = \frac{x - Vt}{\sqrt{(1 - V^2/c^2)}}}_{\text{direct}} \qquad \underbrace{x = \frac{x' + Vt'}{\sqrt{(1 - V^2/c^2)}}}_{\text{reverse}}.$$

These are the Lorentz transformations for x. We also have the trivial transformations

$$y' = y \qquad y = y'$$

$$z' = z \qquad z = z'$$

as before, in the Galilean case.

In order to obtain the time transformations, we rewrite (11.6) as an equation for t':

$$aVt' = x - ax',$$

and so

$$t' = \frac{x}{aV} - \frac{x'}{V}. \tag{11.10}$$

Next, we substitute from (11.5) for x':

$$t' = \frac{x}{aV} - \frac{a}{V}(x - Vt),$$

and rearrange:

$$t' = at + \frac{x(1 - a^2)}{aV}. \tag{11.11}$$

Lastly, we substitute from (11.9) for the coefficient a to obtain

$$\underbrace{t' = \frac{t - Vx/c^2}{\sqrt{(1 - V^2/c^2)}}}_{\text{direct}} \qquad \underbrace{t = \frac{t' + Vx/c^2}{\sqrt{(1 - V^2/c^2)}}}_{\text{reverse}}.$$

That is, the Lorentz transformations for the time coordinate.

Now let us summarize the standard[3] Lorentz transformations, as follows:

[3]So-called because they are for inertial frames in *standard* configuration.

Direct	**Reverse**	
$x' = \dfrac{x - Vt}{\sqrt{(1 - V^2/c^2)}}$	$x = \dfrac{x' + Vt'}{\sqrt{(1 - V^2/c^2)}}.$	(11.12)
$y' = y$	$y = y'.$	
$z' = z$	$z = z'.$	
$t' = \dfrac{t - Vx/c^2}{\sqrt{(1 - V^2/c^2)}}$	$t = \dfrac{t' + Vx'/c^2}{\sqrt{(1 - V^2/c^2)}}.$	(11.13)

We can also introduce the γ factor, as defined in equation (11.9), thus:

$$\gamma \equiv \frac{1}{\sqrt{(1 - V^2/c^2)}}.$$

Therefore an alternative form for the standard Lorentz transformations is:

$$x' = \gamma(x - Vt) \qquad x = \gamma(x' + Vt'). \tag{11.14}$$

$$t' = \gamma(t - Vx/c^2) \qquad t = \gamma(t' + Vx'/c^2). \tag{11.15}$$

The reader should note the resemblance to the Galilean transform and should verify that as $V/c \to 0$, then $\gamma \to 1$, and the Lorentz transformation reduces to the Galilean transformation.

11.4 Implications of replacing Galilean with Lorentz invariance

We have already seen one immediate implication of replacing the requirement that the laws of mechanics be Galilean invariant with the new requirement that they should be Lorentz invariant:

We no longer have a universal time.

Time, like all the other coordinates of the space-time point which is an event, must undergo transformation from one inertial frame to another. In fact, as we spelled out briefly at the beginning of this chapter, there are several counter-intuitive consequences of the change to Lorentz invariance and in this chapter we shall look at some of them.

11.4.1 Relativity of simultaneity

The first (and somewhat surprising) casualty is the concept of simultaneity. One might think that this is a qualitative concept and that it either holds in all frames or it does not. In fact, as we shall now see, the concept of simultaneity depends very much on the frame in which the observations are carried out.

Let us consider events at the space-time points (x_1, t_1) and (x_2, t_2) in S. Suppose they are simultaneous, so that $t_1 = t_2$, in S. Will they be observed to be simultaneous in S'? In S', the Lorentz transformation gives us

$$t_1' = \frac{t_1 - Vx_1/c^2}{\sqrt{1 - V^2/c^2}} \tag{11.16}$$

$$t_2' = \frac{t_2 - Vx_2/c^2}{\sqrt{1 - V^2/c^2}}. \tag{11.17}$$

Hence, in S', the time interval between the two events is given by:

$$t_1' - t_2' = \frac{V(x_2 - x_1)/c^2}{\sqrt{1 - V^2/c^2}}, \tag{11.18}$$

as $t_1 = t_2$ (that is, they are simultaneous in S).

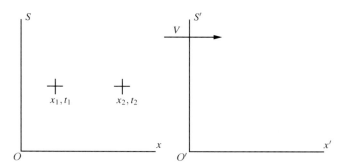

Fig. 11.2 Events in S transformed to S'.

Therefore, simultaneity in S does **NOT** imply simultaneity in S', unless both events are at the same place, in which case $x_1 = x_2$, and equation (11.18) indicates that the time interval between the two events is zero in S' as well. However, note what happens if $V/c \to 0$! For a velocity V small compared to the speed of light, the two events are simultaneous in both frames.

11.4.2 Lorentz–Fitzgerald contraction

Now let us consider a measuring rod at rest in S', lying parallel to the x-axis with one end at x_1', and the other at x_2'. Then we have

$$L_0 = x_2' - x_1', \qquad (11.19)$$

where L_0 is the proper length of the rod in its own rest frame.

In S, the rod moves with speed V. We use the Lorentz transformation to transform the end coordinates of the rod into S, thus:

$$L_0 = x_2' - x_1' = \frac{(x_2 - x_1) - V(t_2 - t_1)}{\sqrt{1 - V^2/c^2}}. \qquad (11.20)$$

Now we must measure x_2 and x_1 *simultaneously* in S. This precaution is not needed in S', where the rod is at rest.[4] So we take $t_1 = t_2$ and hence

[4]Refer back to Exercise 1.6!

$$L_0 = \frac{(x_2 - x_1)}{\sqrt{1 - V^2/c^2}} = \frac{L}{\sqrt{1 - V^2/c^2}}. \qquad (11.21)$$

Then, rearranging, we find

$$L = L_0 \sqrt{1 - V^2/c^2}, \qquad (11.22)$$

where L is the length of the rod in S. Thus, for $V > 0$, this expression indicates that $L < L_0$ and the moving rod is reduced in length in the frame of a stationary observer.

Lastly, we should note the reciprocal effect arising from the fact that V^2 is positive. Thus, if we positioned the rod at rest in S and transformed its end coordinates to S', then we would come to exactly the same conclusion, despite the change from V to $-V$.

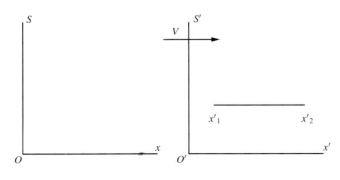

Fig. 11.3 Contraction of length in direction of motion.

11.4.3 Example: *Rod at an angle to the x-axis moving in the x-direction*

Use the Lorentz transformations to show that a straight rod of proper length L_0, appears to have a length

$$L = L_0\sqrt{1 - V^2/c^2}$$

to an observer moving with speed V parallel to the axis of the rod.

A straight rod makes an angle θ with the x-axis. What angle does the rod make with this axis, according to an observer moving with speed V in the x-direction?

The first part of this example is just the reverse of the case considered in the previous section and hence makes the point about reciprocity. Evidently we have for the length of the rod as measured in S:

$$L_0 = x_2 - x_1 = \frac{x_2' + Vt_2'}{\sqrt{1 - V^2/c^2}} - \frac{x_1' + Vt_1'}{\sqrt{1 - V^2/c^2}},$$

by Lorentz transformation. Now we must measure x_1' and x_2' simultaneously in S', therefore we set

$$t_1' = t_2',$$

thus:

$$L_0 = \frac{x_2' - x_1'}{\sqrt{1 - V^2/c^2}} = \frac{L}{\sqrt{1 - V^2/c^2}},$$

and rearranging

$$L = L_0\sqrt{1 - V^2/c^2},$$

as expected.

Now we turn our attention to the case of the rod moving with its axis at an angle to the x-axis. According to an observer in S', there is a Fitzgerald contraction of the rod in the x-direction but not the y-direction. We resolve the length of the rod in the x and y directions.

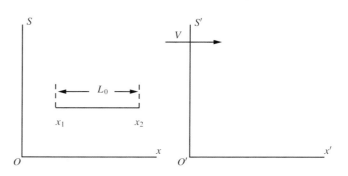

Fig. 11.4 A rod moving parallel to its own axis.

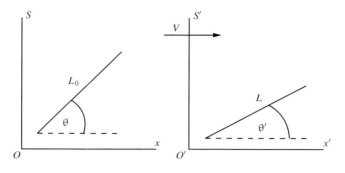

Fig. 11.5 Rod at an angle to the x-axis.

In S,

x-direction: $L \cos \theta' = (L_0 \cos \theta)\sqrt{1 - V^2/c^2}$

y-direction: $L \sin \theta' = (L_0 \sin \theta)$.

Thus, taking the ratio of the y and x projections in the two frames to define the tangents of the angles, we have for the angle in S'

$$\tan \theta' = \tan \theta / \sqrt{1 - V^2/c^2}.$$

11.4.4 Example: *Moving particle track at an angle to the x-axis in S relative to S'*

A particle is emitted at a point P in S. It travels along a straight line in the (x, y) plane, reaching the origin of coordinates at time $t = 0$. Its path is of length r, its velocity is u, making an angle α with the x-axis. Write down the coordinates of the event P in four-dimensional space-time.

Hence show that if an observer in S' measures the angle of incidence of the particle track to be α', then this can be related to the value in S by

$$\tan \alpha' = \frac{\tan \alpha}{\gamma(1 + (V/u) \sec \alpha)}.$$

By considering a star situated at P, obtain a formula which will allow an astronomer to compensate for the stellar aberration which results when one observes a fixed star from a moving Earth.

A particle is emitted at $t = -r/u$ and reaches $r = 0$ at $t = 0$.
   In S, the coordinates of the event P are $(r \cos \alpha, r \sin \alpha, 0, -r/u)$.
   In S', Lorentz transformation gives:

$$x' = (x - Vt)\gamma \qquad \text{and} \qquad y' = y.$$

Therefore

$$x' = \gamma(r \cos \alpha + Vr/u) \qquad \text{and} \qquad y' = r \sin \alpha.$$

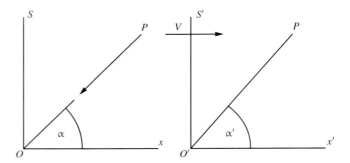

Fig. 11.6 Aberration of a moving particle track due to the motion of the observer.

Hence

$$\tan \alpha' = \frac{y'}{x'} = \frac{r \sin \alpha}{\gamma(r \cos \alpha + Vr/u)} = \frac{\tan \alpha}{\gamma(1 + (V/u) \sec \alpha)}.$$

For stellar aberration, just put $u = c$ and the equation becomes

$$\tan \alpha' = \frac{\tan \alpha}{\gamma[1 + (V/c) \sec \alpha]}.$$

This is indeed the result for stellar aberration, as we shall see in the next chapter.

11.5 Time dilation

Let us consider a clock situated at x' in S' and suppose that t'_1 and t'_2 are two successive instants (i.e. $t'_1 < t'_2$) recorded by an observer in S'. Therefore the corresponding time interval in S' is $\Delta t' = t'_2 - t'_1$.

Now, an observer in S measures these instants as t_1 and t_2, and by Lorentz transformation relates them to the measurements in S' as

$$t_1 = \frac{t'_1 + Vx'/c^2}{\sqrt{1 - V^2/c^2}} \tag{11.23}$$

and

$$t_2 = \frac{t'_2 + Vx'/c^2}{\sqrt{1 - V^2/c^2}}. \tag{11.24}$$

Therefore the time interval according to the observer in S is just the difference between the two times as given by equations (11.23) and (11.24), thus:

$$\Delta t = t_2 - t_1 = \frac{t'_2 - t'_1}{\sqrt{1 - V^2/c^2}},$$

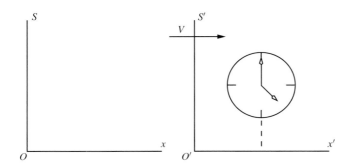

Fig. 11.7 Clock moving relative to observer.

or, alternatively,

$$\Delta t = \frac{\Delta t'}{\sqrt{1 - V^2/c^2}}, \qquad (11.25)$$

is the relationship between the measured time intervals in S and S'. Thus we may conclude that a clock moving at speed V in S runs slow to an observer at rest in S. We may also express this in a slightly more general way by rewriting equation (11.25) as

$$\Delta t = \gamma \Delta t_0, \qquad (11.26)$$

where t_0 is the proper time interval measured in the rest frame of the clock. In this form, we have the more general statement: moving clocks run slow.

11.5.1 Experimental evidence

Unlike the effect of length contraction, the phenomenon of time dilation has received direct experimental verification. For many years the classic evidence for the effect was that muons created high in the Earth's atmosphere by incoming cosmic rays could only reach the surface if one postulated that their decay times were slowed down by their speed, and measurements were made which verified this hypothesis.[5] More recently, experiments in accelerators at CERN have confirmed this phenomenon under laboratory conditions.[6] Also, intriguingly, it is now possible to measure the effect by taking atomic clocks on a commercial air flight![7] Other direct and indirect evidence is available, especially in the case of the transverse Doppler effect, and we shall mention some of this where appropriate.

[5]B. Rossi and D.B. Hall *Phys. Rev.* **59**, 223 (1941).

[6]J. Bailey *et al. Nature* **268**, 301 (1977).

[7]J.C. Hafele and R. Keating *Science* **177**, 166 (1972).

11.5.2 Example: *Decay of high-energy pions*

A pion decays into a muon and a neutrino. Its lifetime is obtained from the radioactive decay law for pions at rest, as

$$\tau_0 = 2.6 \times 10^{-8}\,\text{s}.$$

In a high-energy accelerator, a pion beam is produced with a velocity V so close to c that $\gamma(V) = 1400$. What percentage of the beam has been lost by pion decay by the time it travels 300 m from the point of production? What would be the corresponding Galilean prediction?

The decay law is:

$$N(t) = N(0)e^{-t/\tau}.$$

In this case we have

$$t = \frac{300\,\text{m}}{V} \simeq \frac{300\,\text{m}}{c} = 10^{-6}\,\text{s},$$

where

$$\tau = \frac{\tau_0}{\sqrt{1 - V^2/c^2}} = \gamma\tau_0 = 3.64 \times 10^{-5}\,\text{s}.$$

The fraction lost is then:

$$\frac{N(0) - N(t)}{N(0)} = 1 - e^{-1/36.40} \simeq 1 - e^{-0.03} \simeq 0.03 \sim 3\%.$$

The Galilean prediction (no time dilation) is:

$$N(t) = N(0)e^{-t/\tau_0} \simeq N_0 e^{-38},$$

and hence the fraction remaining is:

$$\frac{N(t)}{N(0)} = e^{-38} \simeq 10^{-17}.$$

Evidently the effect of time dilation is of considerable relevance to experiments in high-energy accelerators.

11.6 The twins paradox

The so-called twins paradox is a well-known bone of contention in relativity and seems to crop up every few years in the correspondence columns of some journal or other. Even now, it can be relied upon to inject some real enthusiasm into an undergraduate class, and debate can become quite heated. It is worth quoting W. Rindler[8] on the subject. He writes:

'It is quite easily resolved, but seems to possess some hidden emotional content that makes it the subject of interminable debate among dilettantes of relativity.'

Perhaps it is the idea of twins no longer being the same age that arouses unease or perhaps it is just the very subtlety inherent in choices between only two alternatives. Whatever the reason may be, we shall drop the unfortunate twins and resort to an emotionally neutral scientific experiment which is carried out with clocks!

[8]W. Rindler, *Introduction to Special Relativity*, Oxford University Press (1991).

We shall discuss this topic under three headings in turn. These are:

1. The experiment.
2. The paradox or apparent contradiction.[9]
3. The resolution of the paradox.

11.6.1 *The experiment*

- Two physicists, A and B, agree to carry out an experiment to test the idea of time dilation.

- They synchronize their identical clocks. This means that they carry out a comparison of their clocks in the same frame of reference. This is of the essence in this type of experiment.

- A takes off in a rocket and is accelerated up to a constant speed V. This speed is maintained for a time $T_A/2$, then A turns round and returns home. On A's clock the whole trip takes time T_A and we assume that all periods of acceleration are brief compared to the periods of free flight at constant speed V.

- On return to Earth, A compares clocks with B. Can B predict the result? Note that because it is the square of V which occurs in the γ factor, from the point of view of time dilation, there is no difference between the trip out and the trip back.

It is always vital in relativity to be specific about frames. Here B says that Earth is the usual frame S, while the usual moving frame S' is the rest frame of A's rocket. Once B has made this identification it is just a matter of applying our existing theory and in particular equation (11.26) for time dilation.

From equation (11.26), B predicts the result of the experiment to be:

$$T_B = \frac{T_A}{\sqrt{1 - V^2/c^2}} = \gamma T_A.$$

That is, B has to multiply A's time for the trip T_A by a factor $\gamma > 1$ to make it equal to B's own time which is T_B. According to special relativity, B measures the time interval between the clock comparisons as being longer than the time interval measured by A. (If A and B were twins, at this stage A is now younger than B!)

11.6.2 *The paradox or apparent contradiction*

According to the analysis just given, B asserts that A's clock ran slow. The apparent contradiction or paradox arises when A says that it is the other way round, and that it was B who went away with velocity $-V$ and came back with velocity $+V$. Everyone knows, says A, that moving clocks run slow. B's clock was moving relative to A's and so it ran slow. Accordingly, A predicts the result of the same experiment to be: $T_A = \gamma T_B$. This is the apparent contradiction. So which of them is correct in predicting the result of the experiment?

[9]There is a widespread and longstanding confusion over the meaning of the word paradox in physics. A paradox is defined to be an apparent contradiction. There is no such thing as an 'apparent paradox'; or, if there were, its consideration would be a waste of time. Once a paradox has been resolved, it does not go away. It is still a paradox!

11.6.3 The resolution of the paradox

A clue to the fact that this is only an apparent contradiction lies in the fact that the experiment will have a definite outcome which must inevitably decide between the opposing contentions of A and B. The plausibility of the paradox lies in the apparent symmetry of the arguments put forward by A and B. However, just as the experimental result must inevitably destroy that symmetry, so also do theoretical considerations. At the fundamental level, the symmetry between A and B is broken by the fact that A accelerated and this is not purely relative. The observer A physically experienced forces that B did not. Hence it was A who moved and A's clock which ran slow. It is worth noting that once the symmetry is broken, the consequences can be as small or as large as one pleases (or is physically able to make them).

11.7 Example: Analogy of Lorentz transformation with rotations

S and S' are two systems of rectangular Cartesian coordinates in a plane with common origin O. If a point P has coordinates (w, x) in S and (w', x') in S', show that

$$w' = w \cos \phi + x \sin \phi$$

$$x' = -w \sin \phi + x \cos \phi,$$

where ϕ is the angle between Ow and Ow'. Also, verify that

$$x^2 + w^2 = x'^2 + w'^2,$$

and give a simple geometrical interpretation of this result. Further, show that if we put $w = ict$, where $i = \sqrt{-1}$, then the standard Lorentz transformations can be interpreted as a rotation of axes through an imaginary angle ϕ which satisfies

$$\cos \phi = \gamma \qquad \sin \phi = -\frac{iV\gamma}{c}.$$

We begin by resolving the vector $\mathbf{r}$ into its components in both frames of reference. Thus:

In S the coordinates of P are:

$$w = r \cos \theta \qquad x = r \sin \theta.$$

In S' the coordinates of P are:

$$w' = r \cos(\theta - \phi) \qquad x' = r \sin(\theta - \phi).$$

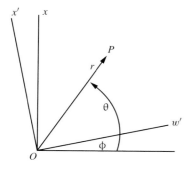

Fig. 11.8 Analogy between Lorentz transformation and rotation of axes.

Using the standard trigonometric identities for the compound angle, we may expand out the last two relationships as:

$$w' = r\cos\theta\cos\phi + r\sin\theta\sin\phi = w\cos\phi + x\sin\phi \qquad (11.27)$$

$$x' = r\sin\theta\cos\phi + r\cos\theta\sin\phi = -w\sin\phi + x\cos\phi. \qquad (11.28)$$

Squaring both sides of each of these equations and adding them together yields:

$$r^2 = x'^2 + w'^2 = (x^2 + w^2)(\sin^2\phi + \cos^2\phi) = x^2 + w^2.$$

This quantity is invariant under the transformation from S to S' because it is simply the square of the distance of the point P from the origin, and this length is unaffected by rotation of axes.

The idea of invariance under rotation of axes will be very useful to us later on, but for the moment we note that we can turn equations (11.27) and (11.28) into the Lorentz transformations, in the form of equation (11.14), by making the substitutions $w = ict$ and $w' = ict'$. Starting with (11.27), substitution for w gives us:

$$x' = x\cos\phi - ict\sin\phi. \qquad (11.29)$$

In the case of (11.28), we substitute for both w and w', and divide across by ic to get:

$$t' = t\cos\phi + \frac{x}{ic}\sin\phi. \qquad (11.30)$$

Comparison of these results with the Lorentz transformations as given by (11.14) and (11.15), yields equivalence if:

$$\gamma = \cos\phi \qquad \text{and} \qquad ic\sin\phi = \gamma V.$$

From the latter condition it follows that:

$$\sin\phi = \frac{\gamma V}{ic} = \frac{-i\gamma V}{c}.$$

We should note that as $\cos\phi = \gamma > 1$, then the angle ϕ must be imaginary.

11.8 General Lorentz transformation

Our results in this book will all be for Lorentz transformations between frames in standard configuration: that is, **standard** Lorentz transformations. However, a generalization is possible. Any law of physics, which is invariant under:

1. standard Lorentz transformation,
2. spatial rotations and translations,
3. time translations

is also invariant between *any* two inertial frames.

A *general* transformation between any two inertial frames whose co-ordinates are standard, but whose configuration is not, consists of the following:

1. A spatial rotation and translation to make the x-axis of S coincide with the line of motion of S'.
2. A time translation to make their origins coincide at $t = 0$.
3. A standard Lorentz transformation.
4. Another spatial rotation and time translation to arrive at the co-ordinates of S'.

The last four steps make up a general Lorentz transformation. Evidently any physical law which is invariant under the preceding steps, including the standard Lorentz transformation, will also be invariant under a general Lorentz transformation. This justifies our use of inertial frames in standard configuration throughout as being without loss of generality. Also note that the steps just given provide an answer to Exercise 1.7!

11.9 Exercises

11.1 Plot a graph of γ against V/c in steps of $V/c = 0.1$.

11.2 If the non-trivial coordinates of an event in S, that is x and ct, are taken as the components of a two-dimensional column vector, show that the corresponding coordinates in S' are given by a matrix equation, where the transformation matrix takes the form:

$$\gamma \begin{pmatrix} 1 & -\beta \\ -\beta & 1 \end{pmatrix},$$

where $\beta = V/c$.

11.3 An astronaut leaves Earth and travels outward at a speed V for a time $T_A/2$, as measured by a clock on the spaceship, turns round and returns to base at the same speed. Show that the duration of the round trip as measured on a clock kept at the base is given by

$$T_B = \frac{T_A}{\sqrt{1 - V^2/c^2}}.$$

Periods of acceleration may be neglected.

11.4 The distance from Earth to a certain star is about 10^5 light years. Explain how it would be possible in principle to go there, and return, within a human lifetime. What speed would be required?

11.5 Show that two clocks which are fixed in S, with separation L, and synchronized in that frame, appear out of synchronization to an observer in S' by a time

$$\delta \simeq -\frac{LV}{2c^2},$$

where V is, as usual, the speed of S' relative to S.

11.6 A spaceship passes the Earth in the direction of the Moon, which is a distance 3.84×10^8 m away, at a speed $V = 3c/5$ relative to the Earth. As it passes, the astronauts synchronize their clock with one at their base on Earth, both clocks being set to zero. An event then occurs on the Moon, which is recorded at Earth base as taking place 0.5 seconds after the spaceship had passed. At what time do the astronauts record the event as having happened? At what time is the event **seen** on (a) Earth? (b) the spaceship?

11.7 A farmer has twin daughters who aspire to represent their country at archery in the Olympic Games. In the interests of safety, he insists that they practise by shooting only in the direction of a large barn; and, to make doubly sure that no flying fragment can possibly injure a bystander, he places the target one arrow length inside the door of the barn. However, the twins (who are reading physics at university) begin to argue, as is their wont. 'Father', says the first. 'Don't you know that Professor Einstein has shown that moving objects contract in their direction of motion? It simply isn't necessary to put the target so far into the barn!' Hastily, the farmer moves the target, so that it is only one half of an arrow length inside the barn. But alas, the second sister chimes in. 'Oh Father, my sister has the right theory, but has applied it wrongly. To the arrow it will appear that the barn is moving, and hence the distance between the door and the target will be contracted and so the arrow will not fit in. Better to have left it as it was!' At this stage the farmer goes away muttering under his breath, and the resourceful twins construct an apparatus which will fire an arrow such that it has a gamma factor of $\gamma = 2$, and proceed to put the matter to the test. The question now is: what happens at the moment of impact? Does the arrow just fit in? Or, does three-quarters of it protrude?

12 Relativistic kinematics

In accepting Einstein's second axiom, namely that the velocity of light *in vacuo* is the same in all inertial frames, we have had to give up the Galilean law for the addition of velocities which states that the velocity of anything (including light) must transform between our usual two inertial frames, S and S', according to

$$\dot{x} = \dot{x}' - V,$$

(where as usual the dots denote differentiation with respect to time) and therefore velocities must have different values in the different frames. In this chapter we show that a new velocity transformation, based on the Lorentz transformations, is compatible with Einstein's second axiom. We then explore some of the consequences of using this method of adding velocities, including the transverse Doppler effect. This latter phenomenon is of particular importance, since it has no classical analogue and hence its existence provides even more experimental evidence for the validity of special relativity.

12.1 Lorentz transformation of intervals between events

As a preliminary step, we consider the Lorentz transformation of the space-time interval between events. In particular, if two events have space-time coordinates (x_1, y_1, z_1, t_1) and (x_2, y_2, z_2, t_2) respectively, in frame S, we wish to obtain the corresponding space and time intervals in frame S', where the two frames are in standard configuration, as illustrated in Figure 11.1. In S, we have the spatial intervals in the different component directions, along with the time interval, as

$$dx = x_2 - x_1, \qquad dy = y_2 - y_1, \qquad dz = z_2 - z_1, \qquad dt = t_2 - t_1.$$

We begin by transforming the interval in the x-direction, and to do this we make use of the Lorentz transformation, as given by equation (11.14), thus:

$$dx' = x'_2 - x'_1 = \gamma(x_2 - Vt_2) - \gamma(x_1 - Vt_1),$$

and, with some rearrangement,

$$dx' = \gamma[(x_2 - x_1) - V(t_2 - t_1)] = \gamma[dx - Vdt]. \qquad (12.1)$$

The transformations of the other two space intervals, are just $dy' = dy$ and $dz' = dz$, and we are left with the time interval:

$$dt' = t'_2 - t'_1 = \gamma\left(t_2 - \frac{Vx_2}{c^2}\right) - \gamma\left(t_1 - \frac{Vx_1}{c^2}\right),$$

where we have again used the Lorentz transformation as given by (11.15). Then, with some rearrangement, we find

$$dt' = \gamma\left[dt - \frac{Vdx}{c^2}\right]. \qquad (12.2)$$

In the following section we shall use equations (12.1) and (12.2) to formulate a transformation law for velocities.

12.2 Velocity transformations

Next we consider a body moving in *both* S and S'. In this section we shall change our notation and use a formal Cartesian tensor notation for the velocity vector, such that we denote components by $\mathbf{u} \equiv (u_1, u_2, u_3)$, rather than (u, v, w). This temporary change allows us to avoid a proliferation of primes!

In S:

$$\mathbf{u} \equiv (u_1, u_2, u_3) = \left(\frac{dx}{dt}, \frac{dy}{dt}, \frac{dz}{dt}\right). \qquad (12.3)$$

In S':

$$\mathbf{u}' \equiv (u'_1, u'_2, u'_3) = \left(\frac{dx'}{dt'}, \frac{dy'}{dt'}, \frac{dz'}{dt'}\right). \qquad (12.4)$$

All we need to do, in order to transform velocities, is to transform their constituent space and time intervals. From the Lorentz transformations, or from equations (12.1) and (12.2), we can write:

$$dx' = \frac{dx - Vdt}{\sqrt{1 - V^2/c^2}} \equiv \gamma(dx - Vdt) \qquad (12.5)$$

$$dy' = dy \qquad (12.6)$$

$$dz' = dz \qquad (12.7)$$

$$dt' = \frac{dt - Vdx/c^2}{\sqrt{1 - V^2/c^2}} \equiv \gamma(dt - Vdx/c^2). \qquad (12.8)$$

Now, from the definition of the velocity in S', we have for the component in the x_1-direction:

$$u'_1 = \frac{dx'}{dt'} = \frac{dx - Vdt}{dt - Vdx/c^2} = \frac{dx/dt - V}{1 - (V/c^2)dx/dt}, \quad (12.9)$$

where we have used (12.5) and (12.8) for the Lorentz transformation of the intervals, cancelled common factors of γ, and divided above and below by dt. Hence, from the definition of the velocity in S we have

$$u'_1 = \frac{u_1 - V}{1 - Vu_1/c^2}. \quad (12.10)$$

It is worth emphasizing that, due to the cancellation, the transformation of u_1 to u'_1 does not involve γ. This is not the case for the other two components of the velocity. For the component in the x_2-direction, we have from (12.6) and (12.8),

$$u'_2 = \frac{dy'}{dt'} = \frac{dy}{\gamma(dt - Vdx/c^2)} = \frac{u_2}{\gamma(1 - Vu_1/c^2)}, \quad (12.11)$$

with a similar result for u'_3.

We may summarize these results as follows:

Velocity transformations

$$u'_1 = \frac{u_1 - V}{1 - Vu_1/c^2} \quad (12.12)$$

$$u'_2 = \frac{u_2}{\gamma(1 - Vu_1/c^2)} \quad (12.13)$$

$$u'_3 = \frac{u_3}{\gamma(1 - Vu_1/c^2)}. \quad (12.14)$$

Reverse velocity transformations

$$u_1 = \frac{u'_1 + V}{1 + Vu'_1/c^2} \quad (12.15)$$

$$u_2 = \frac{u'_2}{\gamma(1 + Vu'_1/c^2)} \quad (12.16)$$

$$u_3 = \frac{u'_3}{\gamma(1 + Vu'_1/c^2)}. \quad (12.17)$$

The following points are worth highlighting:

1. The velocity transformation in the x_1-direction is modified for two reasons. The space interval undergoes Fitzgerald contraction in the direction of its motion and the time interval suffers time dilation. The two associated γ factors cancel.

2. The transformations of the two components of velocity at right angles to the relative motion of the reference frames are modified only

by the time dilation and so the γ factor appears in these transformations.

3. As all three components of velocity are affected by time dilation, the component u_1 appears in all three transformation laws.

12.2.1 Composition of velocities and verification of the Lorentz invariance of the speed of light

The above formulae also give the relativistic addition of velocities $(V, 0, 0)$ and (u_1, u_2, u_3), and for this reason we sometimes refer to them as 'addition of velocities' or 'composition of velocities' laws. If the velocities are collinear, e.g. $(V, 0, 0)$ and $(U, 0, 0))$, then equation (12.12) reduces to:

$$U' = \frac{U - V}{1 - VU/c^2},$$
(12.18)

with a corresponding result for equation (12.15):

$$U = \frac{U' + V}{1 + VU'/c^2}.$$
(12.19)

These are the relationships which are mainly used in problems. However, before tackling some problems, let us consider whether or not we have succeeded in finding a velocity transformation which satisfies Einstein's second axiom.

If we take the velocity of light in S to be $U = c$, what value will it have in S'? We make use of equation (12.18) for the composition of velocities, thus:

$$U' = \frac{c - V}{1 - Vc/c^2} = \frac{c - V}{1 - V/c} = c\frac{c - V}{c - V} = c,$$
(12.20)

as required.

12.2.2 Example: *A problem involving two spaceships and the Earth*

A rocket is moving with speed $0.9c$, relative to the Earth. A second rocket overtakes with a speed of $0.4c$ relative to the first rocket. What is the

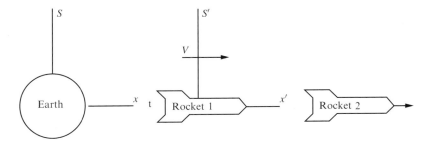

Fig. 12.1 Rocket 2 moves relative to rocket 1 which moves relative to the Earth.

speed of the second rocket, relative to the Earth, assuming use of (a) Galilean transformations and (b) Lorentz transformations?

We take the frames of reference to be as follows:

- $S \equiv$ Earth;
- $S' \equiv$ Rocket 1, $V = 0.9c$.

Rocket 2 has a speed of $0.4c$ relative to Rocket 1, therefore it has a speed $U' = 0.4c$ in S'.

Galilean transformation   Speed of Rocket 2 in S:

$$U_G = U' + V = 1.3c.$$

Lorentz transformation   Speed of Rocket 2 in S:

$$U_L = \frac{U' + V}{1 + U'V/c^2} = \frac{1.3c}{1 + 0.4 \times 0.9c^2/c^2} = \frac{1.3c}{1.36} = 0.9559c.$$

Obviously the Galilean result is unphysical.

12.2.3 Example: *Another problem involving two spaceships and the Earth*

An observer on the Earth seeing two spaceships approaching from opposite directions, measures their speeds of approach and erroneously employs Galilean velocity addition to give them a 'relative speed' of $7c/5$. However, an observer on one of the spaceships measures the relative speed of the other as $35c/37$. Find the speeds of the two spaceships relative to the Earth.

As always, we identify our frames of reference:

- $S \equiv$ Earth;
- $S' \equiv$ Rocket 1, with speed V relative to S.

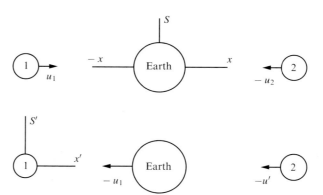

Fig. 12.2 Two spaceships approaching the Earth from opposite directions.

In S, Galilean velocity addition yields:

$$U_1 + U_2 = 7c/5. \tag{12.21}$$

In S', noting that $V =$ (velocity of S' relative to S) $= U_1$, Lorentzian velocity composition gives the velocity of rocket 2 in S' as:

$$U' = \frac{(-U_2) - V}{1 - V(-U_2)/c^2} = \frac{-(U_1 + U_2)}{1 + U_1 U_2/c^2} = \frac{35c}{37}.$$

From this result and equation (12.21) we obtain:

$$\frac{U_1 U_2}{c^2} = \frac{12}{25}. \tag{12.22}$$

Then, multiplying equation (12.21) by either U_1 or U_2 and substituting for $U_1 U_2$ from (12.22) leads to a quadratic equation for U_1 or U_2. For sake of simplicity, we let U stand for either U_1 or U_2, so that we have:

$$U^2 - \frac{7c}{5} U + \frac{12}{25} c^2 = 0 \Rightarrow \left(U - \frac{3c}{5}\right)\left(U - \frac{4c}{5}\right) = 0.$$

The two roots of this equation can then be identified as

$$U_1 = \frac{4c}{5} \quad \text{and} \quad U_2 = \frac{3c}{5},$$

or the other way round!

12.2.4 Fresnel drag coefficient

This is an old problem in optics which predated special relativity. The idea was that a moving fluid could to some extent drag light along with it. A phenomenological theory due to Fresnel can be recovered from the Einstein velocity composition formula. Referring to Figure 12.3, the problem can be stated in our usual terminology as follows.

Consider a transparent liquid flowing down a tube with speed V relative to the tube. Given that light moves at speed U' relative to the liquid, what is the speed of light in the flowing liquid, relative to the tube? We just apply relativistic addition of velocities in the form of equation (12.19):

$$U = \frac{U' + V}{1 + VU'/c^2}. \tag{12.23}$$

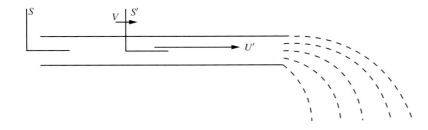

Fig. 12.3 Light moving relative to the water which moves relative to the tube.

Now, Fresnel (*c.* 1818) postulated that light was dragged along by the moving medium such that

$$U = U' + kV, \qquad (12.24)$$

where k is called the Fresnel drag coefficient.

We may compare the two formulae using the fact that $V \ll U' \approx c$. Expanding out in powers of V/c, we can show that equation (12.23) becomes

$$U = \frac{U' + V}{1 + VU'/c^2} \cong U' + V\left(1 - \frac{U'^2}{c^2}\right) + O\left(\frac{V^2}{c^2}\right), \qquad (12.25)$$

and comparison with equation (12.24), due to Fresnel, gives

$$k = 1 - \frac{U'^2}{c^2} = 1 - \frac{1}{n^2}, \qquad (12.26)$$

where $n = c/U' \equiv$ the refractive index of the liquid. The derivation of this result, which was confirmed by experiment, was seen as a direct verification of Einstein's law of addition of velocities.

One point which the reader should consider it this: does this treatment of light in a flowing liquid imply that Einstein's second axiom has been violated? Here we seem to be saying that the speed of light is different in different frames. The answer should be quite obvious, but we shall return to this question at the end of the next section.

12.2.5 *Stellar aberration*

We have already met stellar aberration as a special case in Section 11.4.4. However, formally, we should note that this is a more important topic than such a brief mention might indicate. Stellar aberration is a normal consideration in astronomy, where some allowance has to be made for it. We may define it as follows:

Stellar aberration is the change in the incident angle of light from a star due to the Earth's motion.

We may pose the problem thus: if the angle of incidence is θ in S, what is the angle of incidence θ' in S'?

We begin by resolving the velocity of the incoming light ray in the x and y directions:

$$U_x = -c\cos\theta \qquad \text{and} \qquad U_y = -c\sin\theta,$$

where the negative sign indicates that the beam is coming in towards the origin. Corresponding equations can be written down for the ray of light in S', and reference should be made to Figure 12.4. Now we use addition of velocities for the x and y directions. From (12.12) we have

$$U'_x = \frac{U_x - V}{1 - VU_x/c^2},$$

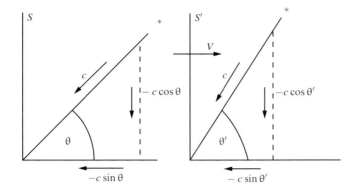

Fig. 12.4 The angle at which light is received from a star is affected by the motion of the Earth.

and substituting for U_x and cancelling factors of $-c$ across, we obtain

$$\cos \theta' = \frac{\cos \theta + V/c}{1 + V \cos \theta/c}. \tag{12.27}$$

Similarly, equation (12.13) yields

$$U'_y = \frac{U_y}{\gamma(1 - VU_x/c^2)},$$

and hence

$$\sin \theta' = \frac{\sin \theta}{\gamma(1 + V \cos \theta/c)}. \tag{12.28}$$

Lastly we divide (12.27) into (12.28) to get:

$$\tan \theta' = \frac{\sin \theta'}{\cos \theta'} = \frac{\sin \theta}{\gamma(\cos \theta + V/c)}, \tag{12.29}$$

and hence

$$\tan \theta' = \frac{\tan \theta}{\gamma[1 + (V/c) \sec \theta]}. \tag{12.30}$$

At this stage, we return to the question posed at the end of the last section; namely, are we infringing Einstein's second axiom by these transformations? The answer to this is that Einstein's axiom refers to the ultimate speed: the speed of light in a vacuum. Here, as in the last section, we are dealing with smaller velocities which do have a non-trivial transformation. In the case of drag in a flowing liquid, the speed of light in the fluid is taken as $U' < c$ and in the present case we are working with U_x and U_y, both of which are by definition less than c.

As we pointed out at the beginning of this section, stellar aberration is a classical effect, and special relativity only provides a correction to the classical result. Arising out of this observation, there are two further points which are of some interest.

Firstly, long before the advent of special relativity, this phenomenon provided a basis for an early measurement of the speed of light. By measuring the angle of aberration, and taking V to be the velocity of the Earth in its orbit, Bradley used stellar aberration to measure the speed of light.

Secondly, when considering such aberration, we are dealing with the apparent change in angular position of a point source due to the relative motion of the observer. However, if we consider an extended source, say a moving body, then each point on the body will act as a point source, but the amount of aberration will not necessarily be equal for all points. Accordingly, this raises the possibility that a fast moving body (in the observer's frame of reference) may change its appearance. A study of such effects leads, among other things, to the surprising conclusion that it is not possible to *see* a Lorentz–Fitzgerald contraction (although it is possible, in principle, to measure it). A full analysis of this effect would take us beyond the scope of the present work, but in Appendix D we provide a brief introduction to this fascinating topic, under the heading of the 'Penrose–Terrell rotation'.

12.2.6 Doppler effect

In Section 4.4.6 we considered the motion of sound waves in moving reference frames and discussed the classical Doppler effect due to relative motion of source and receiver. Now we discuss the effect of taking special relativity into account when we consider electromagnetic (and, in particular, light) waves.

In Figure 12.5, we show a source of radiation moving with velocity **u** relative to an observer. We take S to be the frame of observer and S' to be the frame comoving with the source. We resolve the velocity of the source into two components at right angles, viz.,

1. $u_r \equiv$ radial component of velocity;
2. $u_T \equiv$ transverse component of velocity,

where

$$u^2 = u_r^2 + u_T^2.$$

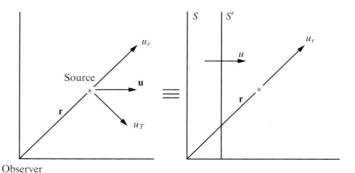

Fig. 12.5 The frequency shift in radiation due to relative motion of source and observer.

In S': the source emits pulses with frequency $\nu_0 = 1/T_0$ ($T_0 =$ time between pulses in S').

In S: the pulses are received with frequency $\nu = 1/T$ ($T =$ time between pulses in S).

Now after emitting a pulse, in time T_0, the source then moves a distance $u_r T_0$ further away from the observer, and so the second pulse starts further away and as a result the time between the arrival of the first and second pulses in S is:

$$T = T_0 + u_r T_0/c = T_0(1 + u_r/c). \qquad (12.31)$$

This is sufficient to account for the classical Doppler effect.

The Lorentz transformation tells us that a moving source suffers time dilation. If an observer in S' measures T_0, then an observer in S measures

$$\frac{T_0}{\sqrt{1 - u^2/c^2}}. \qquad (12.32)$$

Therefore in S we have:

$$T = \frac{T_0}{\sqrt{1 - u^2/c^2}} \left(1 + \frac{u_r}{c}\right). \qquad (12.33)$$

However, the frequency depends inversely on the time, so we have:

$$\frac{T}{T_0} = \frac{\nu_0}{\nu},$$

and hence

$$\frac{\nu_0}{\nu} = \frac{(1 + u_r/c)}{\sqrt{1 - u^2/c^2}}. \qquad (12.34)$$

It is important to note that even if $u_r = 0$ (when classically there is *no* Doppler effect) then we still have

$$u^2 = u_r^2 + u_T^2 = u_T^2, \qquad (12.35)$$

and so

$$\frac{\nu_0}{\nu} = \frac{1}{\sqrt{1 - u_T^2/c^2}}. \qquad (12.36)$$

This is known as the **transverse Doppler effect**.

12.2.7 Thermal Doppler effect

The transverse Doppler effect has been experimentally verified in various ways and of course this constitutes further experimental evidence for the occurrence of time dilation. One particular manifestation of this effect is sufficiently interesting to be worth a brief mention. The situation

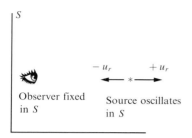

Fig. 12.6 Thermal agitation of ions in a hot crystal causes a Doppler shift in the emitted radiation.

arises when radiating ions located in a hot crystal oscillate back and forward due to the thermal lattice movement. The experimental set-up is shown schematically in Figure 12.6.

From equation (12.34), the instantaneous frequency measurement will give:

$$\frac{\nu_0}{\nu} = \frac{(1 \pm u_r/c)}{\sqrt{1 - u^2/c^2}}. \tag{12.37}$$

The numerator represents the classical effect while the denominator is due to the effect of time dilation. Now, if we average over one complete cycle, we obtain

$$\left\langle \frac{\nu_0}{\nu} \right\rangle = \frac{1}{\sqrt{1 - u^2/c^2}}, \tag{12.38}$$

where $\langle \ldots \rangle$ represents the average value. The correction in the numerator vanishes on the average, because the term u_r/c is as often positive as negative and averages to zero. However, the term u^2/c^2 in the denominator is always positive and hence does not vanish when averaged. Thus the average thermal Doppler effect is due entirely to time dilation.

12.3 Rapidity

In this section we introduce a new variable called the rapidity. Formally the rapidity α associated with velocity V is defined by:

$$\alpha = \tanh^{-1} V/c. \tag{12.39}$$

An important reason for introducing this variable is that it gives us a neat method of doing velocity compositions. However, before using it in this way, we shall consider a couple of examples based on the Lorentz transformations which will show how the concept of the rapidity arises.

12.3.1 Example: *Lorentz transformations as hyperbolic functions*

Prove that the two non-trivial equations of the standard Lorentz transformation may be written as

$$x' = x \cosh \alpha - ct \sinh \alpha$$
$$ct' = -x \sinh \alpha + ct \cosh \alpha,$$

where $\tanh \alpha = V/c$.

In Section 11.7, we showed that the standard Lorentz transformations for x and t could be written as

$$x' = x \cos \phi - ict \sin \phi$$
$$ict' = ict \cos \phi + x \sin \phi,$$

where $\tan \phi = -iV/c$. Using the well-known relationship between trigonometric and hyperbolic functions, we put $\phi = i\alpha$ and invoke the identities

$$\cosh \alpha = \cos i\alpha$$
$$-i \sinh \alpha = \sin i\alpha.$$

Then substitution of $\phi = i\alpha$ in the equations for x' and t' gives

$$x' = x \cosh \alpha - ict(-i \sinh \alpha) = x \cosh \alpha - ct \sinh \alpha$$
$$ict' = ict \cosh \alpha - ix \sinh \alpha$$
$$\Longrightarrow ct' = ct \cosh \alpha - x \sinh \alpha.$$

Hence

$$\tan \phi = -i \tanh \alpha = -iV/c,$$

and so

$$\tanh \alpha = V/c. \qquad (12.40)$$

Evidently the inverse of this result is just (12.39).

12.3.2 Example: *Transitive property of the Lorentz transformations*

Derive the Lorentz transformations in the form:

$$ct' + x' = e^{-\alpha}(ct + x)$$
$$ct' - x' = e^{\alpha}(ct - x).$$

Hence, show that the result of two successive Lorentz transformations, from S to S' and then from S' to S'' is the same as going directly from S to S''.

From the previous section, we have:

$$x' = x \cosh \alpha - ct \sinh \alpha$$
$$ct' = -x \sinh \alpha + ct \cosh \alpha$$

First we add these two equations:

$$ct' + x' = (\cosh \alpha - \sinh \alpha)(ct + x) = e^{-\alpha}(ct + x).$$

Next we subtract the two equations:

$$ct' - x' = (ct - x)(\cosh \alpha + \sinh \alpha) = e^{\alpha}(ct - x),$$

giving the required results.

Two successive transformations:

$$ct' - x' = e^{\alpha_1}(ct - x),$$

then

$$ct'' - x'' = e^{\alpha_2}(ct' - x') = e^{\alpha_2}e^{\alpha_1}(ct - x) = e^{\alpha_1 + \alpha_2}(ct - x),$$

is equivalent to the Lorentz transformation from S to S''.

12.3.3 *Velocity composition in terms of rapidities*

We can obtain the relativistic addition of velocities V and U by simply adding their rapidities: that is, if we rewrite equation (12.9) as

$$W = \frac{U + V}{1 + UV/c^2}, \tag{12.41}$$

where we now denote the resultant velocity by W, then this is equivalent to

$$\alpha_W = \alpha_U + \alpha_V, \tag{12.42}$$

where $\alpha_W = \tanh^{-1} W/c$ and similarly for α_U and α_V. The proof of this result is the subject of Exercise 12.6.

12.3.4 Example: *Speed of multi-stage rocket relative to Earth*

A spaceship is launched from Earth and when it reaches velocity V, with respect to the Earth, it launches a second spaceship, which accelerates until it reaches velocity V with respect to the first spaceship, and so on, all spaceships moving in the same direction. By using the rapidity variable, show that the velocity of the nth spaceship relative to the Earth is given by

$$V_n = c \tanh(n \tanh^{-1} \beta)$$

where $\beta = V/c$.

Rapidities are additive in collinear motion, so

$$\alpha_2 = \alpha_0 + \alpha_1,$$

and as

$$\alpha_0 = \alpha_1 = \alpha \quad \text{(say)},$$

we may put

$$\alpha_2 = 2\alpha,$$

where

$$\alpha = \tanh^{-1}(V/c) \equiv \tanh^{-1} \beta.$$

By definition:

$$\frac{V_2}{c} = \tanh \alpha_2 = \tanh(2\alpha) = \tanh(2\tanh^{-1} \beta).$$

Hence

$$V_2 = c \tanh(2\tanh^{-1}\beta),$$

and thus by induction

$$V_n = c \tanh(n\tanh^{-1}\beta),$$

as required.

12.3.5 Example: *Relationship between gamma factors for a rocket moving relative to S', where S' is moving relative to S*

(a) A moving body has velocity u relative to one inertial frame and velocity u' relative to a second, which has velocity V relative to the first, in the same direction. Prove that

$$\gamma(u') = \gamma(u)\gamma(V)\left(1 - \frac{uV}{c^2}\right),$$

where

$$\gamma(V) = (1 - V^2/c^2)^{-1/2}.$$

(b) Two inertial frames of reference have relative velocity $4c/5$. A body moving in their direction of relative motion is seen, by observers at rest in the two frames, to have velocities of equal magnitudes but opposite directions. What is the velocity of the body, relative to either frame?

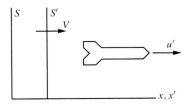

Fig. 12.7 A rocket moves relative to a frame S' which in turn moves relative to S.

(a) From equation (12.18) for the addition of velocities we have:

$$u' = \frac{u - V}{1 - uV/c^2}.$$

Next we define rapidities as follows:

$$u/c = \tanh\alpha, \quad u'/c = \tanh\alpha', \quad V/c = \tanh\beta,$$

and hence the law of addition of velocities may be written as:

$$\alpha' = \alpha - \beta.$$

Now

$$\gamma(u) = \left(1 - \frac{u^2}{c^2}\right)^{-1/2} = (1 - \tanh^2\alpha)^{-1/2} = \cosh\alpha,$$

using a standard identity for hyperbolic functions. So

$$\gamma(u') = \cosh\alpha' = \cosh(\alpha - \beta) = \cosh\alpha\cosh\beta - \sinh\alpha\sinh\beta$$
$$= \cosh\alpha\cosh\beta(1 - \tanh\alpha\tanh\beta)$$
$$= \gamma(u)\gamma(V)(1 - uV/c^2),$$

as required.

(b) Since we have

$$(u')^2 = u^2,$$

it follows that

$$\gamma(u') = \gamma(u),$$

and from the result of part (a),

$$\gamma(u') = \gamma(u)\gamma(V)\left(1 - \frac{uV}{c^2}\right),$$

leading to

$$1 = \gamma(V)\left(1 - \frac{uV}{c^2}\right);$$

therefore, rearranging,

$$\frac{u}{c} = \frac{c}{V}\left(1 - \sqrt{1 - \frac{V^2}{c^2}}\right) = \frac{5}{4}\left(1 - \frac{3}{5}\right) = \frac{1}{2},$$

as we are given $V = 4c/5$.

12.4 Acceleration transformation and proper acceleration

If a particle moves non-uniformly, that is its velocity $u \equiv u(t)$, then we need to know how to transform its acceleration. Suppose a particle moves along x in S with velocity $u(t)$. In S' it moves along x' with velocity $u'(t')$. How do we relate:

$$du'/dt' \equiv \text{acceleration in } S'$$

to

$$du/dt \equiv \text{acceleration in } S?$$

Let us define the instantaneous rest frame of a particle as S', for some instant $t = t_1$ when $u(t_1) = V$. Therefore at $t = t_1$ we have $u'(t_1) = 0$. In other words, at this instant the particle is moving with speed V and therefore has zero speed relative to S'.

We carry out the transformation in two steps. First we transform d/dt', and then we use the velocity composition to transform u'.

1. By time dilation, for events at $t = t_1$:

$$dt = \left[\frac{dt'}{\sqrt{1 - V^2/c^2}}\right]_{V=u(t_1)} = \frac{dt'}{\sqrt{1 - u^2(t_1)/c^2}},$$

and so:

$$\frac{dt}{dt'} = \frac{1}{\sqrt{1 - u^2(t_1)/c^2}} = \gamma(u(t_1)). \qquad (12.43)$$

2. Next, use the velocity transformation relating u' and u. From equation (12.18),

$$\frac{du'}{dt'} = \frac{d}{dt'}\left[\frac{u - V}{1 - uV/c^2}\right]_{V=u(t_1)}.$$

Then by the chain rule of differentiation,

$$= \frac{du}{dt'}\frac{d}{du}\left[\frac{u - V}{1 - uV/c^2}\right]_{V=u(t_1)},$$

and differentiating a product,

$$= \frac{du}{dt'}\left[\frac{1}{(1 - uV/c^2)} + \frac{(-)(-V/c^2)(u - V)}{(1 - uV/c^2)^2}\right]_{V=u(t_1)},$$

putting on a common denominator,

$$= \frac{du}{dt'}\left[\frac{1}{(1 - uV/c^2)} + \frac{V/c^2(u - V)}{(1 - uV/c^2)^2}\right]_{V=u(t_1)},$$

$$= \frac{du}{dt'}\left[\frac{1 - V^2/c^2}{(1 - uV/c^2)^2}\right]_{V=u(t_1)},$$

and making the substitution $V = u(t_1)$,

$$= \frac{du}{dt'}\left(\frac{1}{1 - u^2/c^2}\right),$$

$$= \gamma^2(u)\frac{du}{dt'} = \gamma^2(u)\frac{du}{dt}\frac{dt}{dt'} = \gamma^3\frac{du}{dt}.$$

Therefore, summing it all up, we find

$$\frac{du'}{dt'} = \gamma^3\frac{du}{dt} \equiv \frac{d}{dt}[u\gamma(u)] \qquad (12.44)$$

is the Lorentz transformation of acceleration.

12.5 Apparent rotations and changes in shape

In S' a straight rod parallel to the x'-axis moves in the y'-direction. It may be assumed that it had zero initial displacement and velocity in the y'-direction.

(a) If the rod moves with constant velocity u in S', show that in S the rod is inclined to the x-axis at an angle

$$\theta = -\tan^{-1}(\gamma u V/c^2).$$

(b) If the rod moves with a constant acceleration a (in the y'-direction), show that to an observer in S the rod appears to be parabolic in shape, satisfying the equation of motion

$$y = 1/2 a\gamma^2 (t - Vx/c^2)^2.$$

(a) In S', we have $\dot{y}' = u$ and so $y' = ut' + C$. However, $y' = 0$ at $t' = 0$, therefore $C = 0$, and so $y' = ut'$.

In S, transform both sides of the equation for y':

$$y' = y \qquad \text{and} \qquad t' = \gamma(t - Vx/c^2),$$

hence

$$y = \gamma u(t - Vx/c^2),$$

which is a straight line with slope $\tan \theta = -\gamma u V/c^2$.

(b) We are given $\ddot{y}' = a$ $\dot{y}' = y' = 0$ at $t' = 0$, and $a \equiv$ constant. Therefore $\dot{y}' = at' + C_1$ and from the initial conditions we have $C_1 = 0$. Also, $y' = at'2/2 + C_2$ and the initial conditions yield $C_2 = 0$. Thus

$$y' = \frac{1}{2}at'2.$$

Lastly, Lorentz transform both variables:

$$y' = y \qquad \text{and} \qquad t' = \gamma(t - Vx/c^2),$$

and so

$$y = 1/2 a\gamma^2 (t - Vx/c^2)^2,$$

as required.

12.6 Exercises

12.1 In a frame S, particle A is at rest, while particle B is moving to the right with velocity u. Now consider a frame S' which moves to the right with velocity V relative to S. Find the value of u such that in S' the two particles are approaching each other with equal and opposite velocities, on the basis of: (a) a Lorentz transformation; and (b) a Galilean transformation.

12.2 A spaceship travels away from Earth with speed $c/2$. At some stage it launches a lifeboat at right angles to its direction of motion and with speed $c/3$, both measured in the spaceship's own inertial frame. What is the magnitude and direction of the velocity vector of the lifeboat, as observed from Earth?

12.3 A source of light at rest in frame S', emits a flash of light at an angle θ' to the x'-axis. Show that to an observer in S the flash of light makes an angle θ with the x-axis, where:

$$\cos\theta = \frac{c\cos\theta + V}{c + V\cos\theta},$$

where V is the speed of S' relative to S, in standard configuration.

The source emits light isotropically in its own frame of reference. Show that 50% of this light which is emitted into the forward hemisphere is concentrated into a cone of semi-angle $\cos^{-1}(V/c)$ in S. Also show that if V/c is close to unity, this angle is approximately γ^{-1}.

12.4 Show that equation (12.30) for stellar aberration may be rewritten in the form:

$$\tan(\theta'/2) = \left(\frac{c - V}{c + V}\right)^{1/2}\tan(\theta/2).$$

[Hint: the identity

$$\tan(\theta/2) = \frac{\sin\theta}{1 + \cos\theta},$$

may be helpful.]

12.5 A motorist accused of crossing on a red light, claims that it appeared to be green because of the Doppler effect. If this was true, what was the speed of the car? [Note: the relevant ratio of frequencies may be taken as $\nu(\text{green})/\nu(\text{red}) = 1.2$.]

12.6 A body moves with speed U in the x-direction in S'. What is its speed in S, assuming (a) a Galilean transformation; and (b) a Lorentz transformation?

If the rapidity α_u associated with any velocity u is defined by

$$\alpha_u = \tanh^{-1}(u/c),$$

show that the addition of velocities U and V by Lorentz transformation, can be expressed as

$$\alpha_W = \alpha_U + \alpha_V,$$

where W is the resultant velocity.
[Hint: you may assume the identity

$$\tanh(a + b) = \frac{\tanh a + \tanh b}{1 + \tanh a \tanh b}.]$$

12.7 A clock moves in a straight line with rapidity aT/c where T is the time registered by the clock and a is a constant. Show that a stationary clock which was synchronized with the moving clock at $T = 0$, now reads

$$t = (c/a)\sinh(aT/c),$$

and that the distance travelled by the moving clock away from the stationary one is given by

$$x = (c^2/a)[\cosh(aT/c) - 1].$$

Obtain the relationship between x and T in the limit of $c \to \infty$ and comment on the significance of the constant a in this limit.

13 Space-time geometry

In this chapter we shall introduce the concept of space-time as a four-dimensional vector space. We begin with the concept of an imaginary space axis which is proportional to the time, as being our fourth 'spatial' dimension. This may be seen as old-fashioned by some, but has the advantage of helping to supply some physical motivation for what otherwise may seem rather abstract mathematics. Then we make use of the idea—already encountered in Chapter 11—that the Lorentz transformation may be interpreted as a rotation of axes. After that, we consider some concepts like causality, world lines and the light cone, before going on to introduce four-vectors, and make use of the four-velocity and the four-acceleration to solve problems.

13.1 Vector spaces: background and motivation

We begin with the idea, probably familiar to most readers, that a vector space can have any number of dimensions. As we live in a three-dimensional Euclidean space we have a natural tendency to think of this as unique and special; but many of the concepts associated with vectors are not limited by the dimensionality of the space.[1] From our point of view here, the most important concepts that may be generalized (at least to begin with) are the basic definition of a vector and the idea of an inner or scalar product. Let us consider a hierarchy of vector spaces, beginning with two dimensions and going up to some general integer n dimensions.

We shall denote the number of dimensions by the symbol d and in each case we shall take as an example a position vector denoted by **s**. Our hierarchy of vector spaces is then as follows:

$d = 2$ A vector is an ordered pair of numbers. For example, a position vector can be written as $\mathbf{s} = (x, y)$, where x and y have their usual meanings as distances from the origin along mutually perpendicular coordinate axes. We define its scalar product (or inner product) as:

$$\mathbf{s} \cdot \mathbf{s} = (x^2 + y^2) \equiv s^2.$$

[1] The one obvious exception is the vector product, which has properties unique to the three-dimensional Euclidean space E_3.

$d = 3$ A vector is an ordered triad of numbers. The position vector is $\mathbf{s} = (x, y, z)$ and its scalar product with itself is

$$\mathbf{s} \cdot \mathbf{s} = (x^2 + y^2 + z^2) \equiv s^2.$$

$d = 4$ A vector is an ordered set of four numbers. The position vector is $\mathbf{s} = (x, y, z, w)$ where now we have four Cartesian coordinates, corresponding to the four rectangular coordinate axes. Its inner product is

$$\mathbf{s} \cdot \mathbf{s} = (x^2 + y^2 + z^2 + w^2) \equiv s^2.$$

$d = n$ A vector is an ordered set of n numbers, for any integer such that $2 \leq n \leq \infty$. The position vector may be written as $\mathbf{s} = (x_1, x_2, x_3, \ldots, x_n)$ and the inner product takes the form

$$\mathbf{s} \cdot \mathbf{s} = (x_1^2 + x_2^2 + x_3^2 + \cdots + x_n^2) = \sum_{i=1}^{n} x_i^2 \equiv x_i x_i \equiv s^2.$$

Note that a repeated index indicates that the index is to be summed. This allows us to dispense with the summation symbol and is known as the Einstein summation convention.

13.1.1 *Four-space: the old-fashioned version*

We have seen that the Lorentz transformations can be represented as a rotation of axes in a four-dimensional space with coordinates (x_1, x_2, x_3, x_4), known as Minkowski space. In terms of our earlier discussion, we interpret these coordinates as:

- $(x_1, x_2, x_3) \equiv (x, y, z)$: the usual space coordinates;
- $x_4 = ict$: an imaginary space coordinate, proportional to time.

Then Lorentz invariance of physical law is seen as invariance under rotations in four-dimensional space.

As a preamble, let us remind ourselves about rotations of axes in the real Euclidean 3-space. Referring to Figure 13.1, we note that the position of a point P is represented by the vector $\mathbf{r}$. The coordinates of $\mathbf{r}$ change according to which set of axes is the basis, but the vector itself is an entity in three-dimensional space and is itself unchanged. Only its *representation* changes, when we change basis. That is, $\mathbf{r}$ is at an angle θ to w, but at an angle $\theta - \phi$ to w'. But $r^2 = x^2 + w^2 = x'^2 + w'^2$ is constant, irrespective of rotation angle. This can be verified from the transformations:

$$x' = x \cos \phi - w \sin \phi$$

$$w' = w \cos \phi + x \sin \phi.$$

Now consider a rotation in Minkowski space, although we restrict ourselves for the sake of simplicity to only two axes. Taking x and w, with

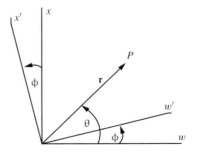

Fig. 13.1 Rotation of axes in two dimensions.

$w = \mathrm{i}ct$, as we saw in Chapter 11, the above transformation equations become:

$$x' = x\cos\phi - \mathrm{i}ct\sin\phi \implies x' = \gamma(x - Vt)$$
$$\mathrm{i}ct' = \mathrm{i}ct\cos\phi + x\sin\phi \implies t' = \gamma(t - Vx/c^2),$$

provided that

$$\cos\phi = \gamma \quad \text{and} \quad \sin\phi = \mathrm{i}\frac{V}{c}\gamma.$$

Thus invariance of the squared length $x^2 - c^2 t^2$ under rotation of axes corresponds to Lorentz invariance.

For the full four-dimensional space, we have the obvious generalization that Lorentz invariance corresponds to the fact that

$$-s^2 = x^2 + y^2 + z^2 - c^2 t^2, \tag{13.1}$$

is invariant under rotations of axes. The sign convention should be noted. Although we shall write it in different ways, we shall adhere at all times to this convention. We should also note that for the case $s = 0$, this relationship reduces to

$$x^2 + y^2 + z^2 - c^2 t^2 = 0, \tag{13.2}$$

which is the equation of a spherical wave front moving at the speed of light, c.

13.1.2 Four-space: the modern version

In the modern version of the formalism, we take the Cartesian coordinates of the four-dimensional Minkowski space to be labelled from 0 to 3. That is, the ordered set of four coordinates may be written as:

$$(x_0, x_1, x_2, x_3) \quad \text{or} \quad (x^0, x^1, x^2, x^3), \tag{13.3}$$

where the second version, with superscripts rather than subscripts, is equivalent to the first, but has a technical significance which will be explained later. The components x_1, x_2, x_3 are just the normal three-space components of the position vector, but $x_0 \equiv ct$ and is treated differently. We can introduce this difference by defining the inner product of the position vector with itself, thus:

$$\mathbf{s} \cdot \mathbf{s} = x_0^2 - x_1^2 - x_2^2 - x_3^2, \tag{13.4}$$

or, alternatively,

$$\mathbf{s} \cdot \mathbf{s} = x_0^2 - \mathbf{x} \cdot \mathbf{x}, \tag{13.5}$$

where the last term is just the usual three-dimensional scalar product. With this type of formalism, we do not need to have an imaginary coordinate axis, as the rules for manipulating imaginary quantities can be incorporated in the specification of the vector space. It should be noted that the expression for the squared-magnitude of the position vector is the same as that given in equation (13.1), including the choice of sign convention.

13.2 Lorentz-invariant interval between events in 4-space

In Euclidean 3-space, the squared interval between points, or the (differential distance)^2 is

$$ds^2 = dx^2 + dy^2 + dz^2, \qquad (13.6)$$

and is invariant under rotations and translations of the rectangular Cartesian coordinate axes.

In 4-space, an **event** is a point in space-time. By analogy with 3-space, the interval between events in 4-space is the 'distance' ds, such that

$$ds^2 = dx_0^2 - dx_1^2 - dx_2^2 - dx_3^2 = c^2 dt^2 - dx^2 - dy^2 - dz^2, \qquad (13.7)$$

is invariant under rotations and translations of rectangular coordinate axes.

But, as we have seen, rotations in 4-space correspond to Lorentz transformations in 3-space. Hence, we may say that:

ds^2 is Lorentz invariant.

This is true for all scalar products in 4-space and we shall make increasing use of this property in the remaining part of this book.

13.3 Causality

We may use the concept that the speed of light *in vacuo* provides an upper limit on the speed with which a signal can travel between two events, to establish whether or not any two events could be connected. In the interests of simplicity, we shall work with one space dimension $x_1 \equiv x$ and the time-like dimension $x_0 \equiv ct$. It is then a simple matter to generalize our conclusions to four dimensions.

Now, referring to Figure 13.2, let us consider events (1) and (2): their interval Δs satisfies the relationship:

$$\Delta s^2 = c^2 \Delta t^2 - \Delta x^2. \qquad (13.8)$$

Without loss of generality, we take Event 1 to be at $x = 0, t = 0$. Then, Event 2 can only be related to Event 1 if it is possible for a signal, travelling at the speed of light, to connect them. We illustrate three general possibilities in Figure 13.2, where Event 2 is at $(\Delta x, c\Delta t)$, its relationship to Event 1 depending on whether $\Delta s > 0, = 0$, or < 0.

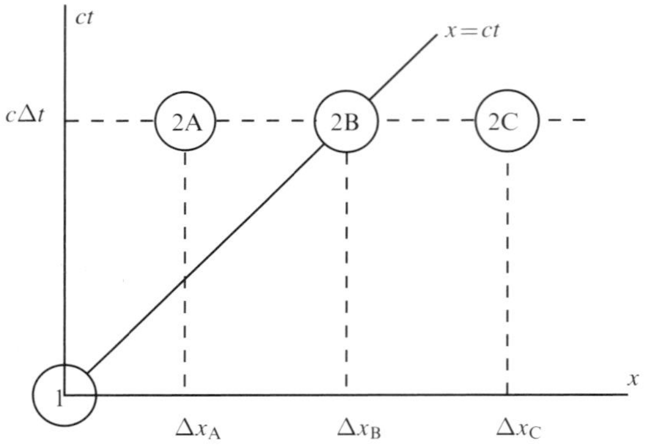

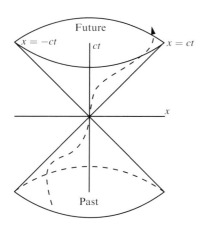

Fig. 13.2 Classification of space-time intervals: the intervals are between Event 1 at the origin and three possibilities for Event 2, viz., 2A, 2B and 2C.

We may summarize the three possibilities as follows:

Case A: timelike interval: $|\Delta x|_A < c\Delta t$ or $\Delta s^2 > 0$. Event 2 *can* be related to Event 1.

Case B: lightlike interval: $|\Delta x|_B = c\Delta t$ or $\Delta s^2 = 0$. Event 2 can *only* be related to Event 1 *by a light signal.*

Case C: spacelike interval: $|\Delta x|_C > c\Delta t$ or $\Delta s^2 < 0$. Event 2 *cannot* be related to Event 1.

Case B is the boundary between the other two cases and in general, for this case, if Event 1 is at the origin, Event 2 must lie on one of the lines $x = \pm ct$. The corresponding interval is also called a **null interval**.

13.3.1 *The light cone*

We have seen that an event is a point in 4-space. Now we introduce the term **world line**, which is the aggregate of points of any particle in four-dimensional space-time.

As the past and future of a particle must be connected, this implies that, in our simplified two-dimensional picture, the world line of a particle must lie within the lines $x = \pm ct$, as illustrated in Figure 13.3.

This concept is readily extended to the three-dimensional case:

$$\mathrm{d}s^2 = c^2\mathrm{d}t^2 - \mathrm{d}x^2 - \mathrm{d}y^2. \tag{13.9}$$

Here, the surface defined by $\mathrm{d}s = 0$ is called the **light cone**.

For the four-dimensional case, we have

$$\mathrm{d}s^2 = c^2\mathrm{d}t^2 - \mathrm{d}x^2 - \mathrm{d}y^2 - \mathrm{d}z^2, \tag{13.10}$$

and in this case, the **light cone** is a hypersurface in four-dimensional space.

Since a material particle always moves at speeds less than c, its world line lies within every light cone which has its vertex on it.

Fig. 13.3 The dotted line represents the world line of a material particle.

13.4 Four-vectors: spacelike, timelike and null

In this section we develop the idea of 4-vectors by making further use of analogies with the more familiar situation in Euclidean 3-space, where we have 3-vectors like, for instance, the position vector

$$\mathbf{x} \equiv (x_1, x_2, x_3) \equiv (x, y, z),$$

or the displacement

$$d\mathbf{x} \equiv (dx_1, dx_2, dx_3),$$

or the velocity

$$\mathbf{u} \equiv (u_1, u_2, u_3) \equiv \left(\frac{dx_1}{dt}, \frac{dx_2}{dt}, \frac{dx_3}{dt} \right),$$

and so on.

Similarly, in the Minkowski 4-space of *events*, we can define 4-vectors in terms of the coordinates, which are:

$$x^0 = ct; \qquad x^1 = x; \qquad x^2 = y; \qquad x^3 = z. \qquad (13.11)$$

Then, *by analogy*, we can define the position vector in 4-space as:

$$\mathbf{X} \equiv (ct, x, y, z) \equiv (ct, \mathbf{x}) \equiv X^\mu.$$

Note in particular our use of capital letters for 4-vectors and also Greek indices μ, ν, etc. such that $\mu = 0, 1, 2, 3$. Thus we may draw the comparison between our 4-space notation

$$X^\mu \equiv (X^0, X^1, X^2, X^3) \quad \text{where } \mu = 0, 1, 2, 3, \qquad (13.12)$$

and the corresponding 3-space notation

$$x_j \equiv (x_1, x_2, x_3) \quad \text{where } j = 1, 2, 3.$$

Similarly, we have the displacement vector,

$$d\mathbf{X} \equiv (dX^0, dX^1, dX^2, dX^3) = (cdt, d\mathbf{x}), \qquad (13.13)$$

and one can go on to define the 4-space analogue of any 3-space vector.

13.4.1 *Scalar product of 4-vectors*

These are defined by analogy with the scalar product in 3-space, but now we generalize the rules which we have already touched on in Section 13.1.2. For any two 4-vectors, $\mathbf{A}$ and $\mathbf{B}$, we have: $\mathbf{A} \equiv (A^0, A^1, A^2, A^3) \equiv (A^0, \mathbf{a})$ and $\mathbf{B} \equiv (B^0, B^1, B^2, B^3) \equiv (B^0, \mathbf{b})$. Their scalar product is defined as

$$\mathbf{A} \cdot \mathbf{B} = A^0 B^0 - A^1 B^1 - A^2 B^2 - A^3 B^3 \equiv A^0 B^0 - \mathbf{a} \cdot \mathbf{b}, \qquad (13.14)$$

where $\mathbf{a} \cdot \mathbf{b}$ is the usual three-dimensional scalar product and

$$A^2 = \mathbf{A} \cdot \mathbf{A} = (A^0)^2 - a^2, \tag{13.15}$$

with $a^2 = |\mathbf{a} \cdot \mathbf{a}|$, and with a similar result for $\mathbf{B}$.

As an example, we can take the displacement 4-vector and obtain its magnitude, thus:

$$ds^2 = d\mathbf{X} \cdot d\mathbf{X} \tag{13.16}$$

$$= (dX^0)^2 - |d\mathbf{x} \cdot d\mathbf{x}|. \tag{13.17}$$

Formally, we can write the 4-vector index as a superscript,

$$A^\mu \equiv (A^0, \mathbf{a}): \quad \text{the contravariant form;} \tag{13.18}$$

or as a subscript,

$$A_\mu \equiv (A^0, -\mathbf{a}): \quad \text{the covariant form.} \tag{13.19}$$

This procedure, in which raising or lowering the index changes the sign of the spacelike part of the 4-vector, gives us a compact method of writing down the rule for a scalar product in a very general way. That is,

$$A^2 = \mathbf{A} \cdot \mathbf{A} = A^\mu A_\mu = (A^0)^2 - a^2. \tag{13.20}$$

Obviously this procedure can be extended to any pair of 4-vectors.

Lastly, we note that in 3-space scalar products like $\mathbf{a} \cdot \mathbf{a} = a^2$ and $\mathbf{a} \cdot \mathbf{b}$ are invariant under rotation of axes. It follows that in 4-space, scalar products like $\mathbf{A} \cdot \mathbf{A} = A^2$ and $\mathbf{A} \cdot \mathbf{B}$ are also invariant under rotation of axes.

Therefore scalar products of 4-vectors are Lorentz invariant.

We have already noted this about ds^2—the squared interval given by

$$ds^2 = d\mathbf{X} \cdot d\mathbf{X} = (c^2 dt^2 - dx^2 - dy^2 - dz^2), \tag{13.21}$$

—and we shall make more use of this equivalence between Lorentz translations and rotations of axes later.

13.4.2 Classification of 4-vectors

In Section 13.3, we have previously classified the interval ds^2 as:

- $ds^2 > 0$: **timelike**;
- $ds^2 = 0$: **null (or lightlike)**;
- $ds^2 < 0$: **spacelike**.

Similarly, any arbitrary 4-vector **A** can be represented by a displacement vector in space-time and hence we classify *all* 4-vectors as:

- $A^2 = (A^0)^2 - a^2 > 0$: **timelike**;
- $A^2 = (A^0)^2 - a^2 = 0$: **null**;
- $A^2 = (A^0)^2 - a^2 < 0$: **spacelike**.

If, for example, the vector **A** represented the 4-space trajectory of a free particle, then **A** could only be timelike. If **A** were null, then it could only correspond to a pulse of light or the trajectory of a massless particle such as a photon or neutrino.[2] However, if **A** were to be spacelike, then it could not correspond to any particle trajectory. As it lies beyond the light cone with vertex on its starting point, it must be unphysical.

At this stage, it is convenient to formalize our definition of a world line, as given in Section 13.3.1, in terms of 4-vectors. We may say that the world line of a particle is the 4-space trajectory

$$\mathbf{X} = \mathbf{X}(\tau), \tag{13.22}$$

where τ is the time measured in the comoving frame of the particle, which corresponds to the Galilean trajectory of the particle $\mathbf{x} = \mathbf{x}(t)$ in Euclidean 3-space. For a material particle, $\mathbf{X}(\tau)$ must be timelike.

[2]At the time of writing, it appears that the neutrino may actually have a mass (*Physics World*, July 1998).

13.5 Proper time element $d\tau$

We know that ds^2 is an invariant, as is c^2. It is often convenient to work with a third invariant given by their ratio, viz.,

$$d\tau^2 = \frac{ds^2}{c^2} = dt^2 - \frac{dx^2 + dy^2 + dz^2}{c^2}. \tag{13.23}$$

The differential $d\tau$ is called the **proper time element** because, for consecutive events on a moving particle, it coincides with dt, which is the time differential measured by a clock fixed to the particle. This is because, in the rest frame of the particle,

$$dx = dy = dz = 0, \quad \text{so} \quad d\tau^2 = dt^2.$$

We can express the relation between $d\tau$ and dt in terms of the 3-velocity vector **u** of the particle:

$$d\tau^2 = dt^2 \left[1 - \frac{1}{c^2} \left(\frac{dx^2}{dt^2} + \frac{dy^2}{dt^2} + \frac{dz^2}{dt^2} \right) \right], \tag{13.24}$$

and so

$$\frac{d\tau^2}{dt^2} = 1 - \frac{u^2}{c^2},$$

hence

$$\frac{d\tau}{dt} = \sqrt{1 - \frac{u^2}{c^2}}. \tag{13.25}$$

Alternatively, we may rewrite this as

$$\frac{dt}{d\tau} = \left(1 - \frac{u^2}{c^2}\right)^{-1/2} = \gamma(u), \tag{13.26}$$

and this is a form which will be useful to us later on.

13.6 Four-velocity of a body

We shall denote the 4-velocity of a particle by **U** and define it by:

$$U^\mu = \frac{dX^\mu}{d\tau}, \tag{13.27}$$

where it should be noted that we have differentiated with respect to the proper time for the particle. Its relationship to the 3-velocity of the particle can be found as follows:

$$U^\mu = \frac{dX^\mu}{d\tau} = \frac{dX^\mu}{dt}\frac{dt}{d\tau} = \frac{dX^\mu}{dt}\gamma(u)$$

$$= \gamma(u)\frac{d}{dt}(ct, x, y, z)$$

$$= \gamma(u)\left(c, \frac{dx}{dt}, \frac{dy}{dt}, \frac{dz}{dt}\right). \tag{13.28}$$

Therefore the 4-velocity of a particle takes the form

$$\mathbf{U} = \gamma(u)(c, \mathbf{u}), \tag{13.29}$$

where **u** is its usual 3-velocity.

13.6.1 Invariant scalar product

We begin by noting that the scalar product $U^2 = \mathbf{U} \cdot \mathbf{U}$ is, like all scalar products, an invariant. Therefore, if we evaluate it in the rest frame of the particle, the result will be valid for all frames. Accordingly, from the definition of a scalar product given in equation (13.15), we have:

$$U^2 = \gamma^2(u)(c^2 - u^2), \tag{13.30}$$

which holds in all frames; and, setting $u = 0$, this reduces to

$$U^2 = c^2, \tag{13.31}$$

in the rest frame of the particle. Hence it follows that

$$U^2 = c^2, \qquad (13.32)$$

in all frames.

This particular result will be useful when tackling problems; but, more importantly, the reasoning which we have just used is an example of the kind of reasoning which we shall use again and again when working in 4-space.

13.7 Four-acceleration of a body

We denote the 4-acceleration of a particle by **A** and define it by

$$A^\mu = \frac{d^2 X^\mu}{d\tau^2} = \frac{dU^\mu}{d\tau}. \qquad (13.33)$$

Again, we do the differentiation with respect to the proper time.

As in the case of the 4-velocity, the relationship to the corresponding 3-vector—in this case the 3-acceleration **a**—is found as follows:

$$\mathbf{A} = \frac{d\mathbf{U}}{d\tau} = \gamma \frac{d\mathbf{U}}{dt} = \gamma \frac{d}{dt}(\gamma c, \gamma \mathbf{u}),$$

where we have used the chain rule of differentiation along with (13.26),

$$\mathbf{A} = \gamma \left(c \frac{d\gamma}{dt}, \frac{d\gamma}{dt}\mathbf{u} + \gamma \frac{d\mathbf{u}}{dt} \right)$$

$$= \gamma \left(c \frac{d\gamma}{dt}, \frac{d\gamma}{dt}\mathbf{u} + \gamma \mathbf{a} \right), \qquad (13.34)$$

and $\gamma \equiv \gamma(u)$, as before.

In the *instantaneous rest frame* of the particle, this result simplifies to

$$\mathbf{A} = (0, \mathbf{a}), \qquad (13.35)$$

as $u = 0$ in the rest frame. Thus the magnitude of the 4-acceleration of a particle $|\mathbf{A}| = 0$, if and only if, its *proper acceleration* vanishes. This result may be stated as:

**proper acceleration $\equiv$ the magnitude of the acceleration in the
rest frame.**

13.7.1 *Orthogonality of 4-velocity and 4-acceleration*

From equation (13.32) we have:

$$U^2 = \mathbf{U} \cdot \mathbf{U} = c^2,$$

and if we differentiate both sides of this relationship with respect to the proper time, we find

$$\frac{\mathrm{d}U^2}{\mathrm{d}\tau} = 2\mathbf{U} \cdot \frac{\mathrm{d}\mathbf{U}}{\mathrm{d}\tau} = 2\mathbf{U} \cdot \mathbf{A} = 0, \tag{13.36}$$

where we have introduced the 4-acceleration from (13.33) and the vanishing of the right-hand side comes from the differentiation of the constant c^2. Thus, we have the general relationship

$$\mathbf{U} \cdot \mathbf{A} = 0. \tag{13.37}$$

It should be noted that this is a general property of these two quantities and has nothing to do with the motion of any individual particle. This result should alert the reader to the fact that 4-velocity and 4-acceleration are very different in some ways from their 3-space analogues.

13.7.2 Example: *Express 4-acceleration in terms of 3-acceleration and 3-velocity*

Show that the acceleration 4-vector of a moving particle has components $(A^0, \mathcal{A})$ with

$$A^0 = \frac{\mathbf{u} \cdot \mathcal{A}}{c} \qquad \text{and} \qquad \mathcal{A} = \gamma(u)^2 \mathbf{a} + \gamma(u)^4 \frac{\mathbf{u}(\mathbf{u} \cdot \mathbf{a})}{c^2},$$

where the velocity and acceleration 3-vectors are $\mathbf{u}$ and $\mathbf{a}$.

We begin with the 4-acceleration, as defined by equation (13.33), viz.,

$$\mathbf{A} = \frac{\mathrm{d}\mathbf{U}}{\mathrm{d}\tau} \equiv (A^0, \mathcal{A}),$$

where $\mathcal{A}$ represents the 3-acceleration part of the 4-vector $\mathbf{A}$, and from equation (13.29) the 4-velocity is

$$\mathbf{U} = \gamma(u)(c, \mathbf{u}).$$

From the invariant

$$\mathbf{U} \cdot \mathbf{U} = c^2,$$

it follows that

$$\frac{\mathrm{d}}{\mathrm{d}\tau}(\mathbf{U} \cdot \mathbf{U}) = 0,$$

and hence

$$2\mathbf{U} \cdot \mathbf{A} = 0.$$

Ignoring the factor of 2, we have

$$\mathbf{U} \cdot \mathbf{A} = U^0 A^0 - \gamma \mathbf{u} \cdot \mathcal{A} = 0,$$

therefore

$$c\gamma A^0 = \gamma \mathbf{u} \cdot \mathcal{A}.$$

Now

$$A^0 = \frac{\mathbf{u} \cdot \mathcal{A}}{c},$$

and

$$\mathcal{A} = \frac{d}{d\tau}(\gamma(u)\mathbf{u}) = \frac{dt}{d\tau} \cdot \frac{d}{dt}\mathbf{u}\gamma(u) = \gamma(u)\left[\dot{\mathbf{u}}\gamma(u) + \mathbf{u}\frac{d}{dt}\gamma(u)\right].$$

Next we do the differentiations and substitute for the gamma factor, thus:

$$\gamma(u)\frac{d\gamma(u)}{dt} = \frac{1}{2}\frac{d}{dt}\gamma^2 = \frac{1}{2}\frac{d}{dt}\left(1 - \frac{u^2}{c^2}\right)^{-1}$$

$$= \frac{1}{2}\left(1 - \frac{u^2}{c^2}\right)^{-2}\frac{1}{c^2}\frac{d}{dt}u^2$$

$$= \frac{1}{2}\gamma^4\frac{1}{c^2}\frac{d}{dt}(\mathbf{u} \cdot \mathbf{u}) = \gamma^4\frac{\mathbf{u} \cdot \dot{\mathbf{u}}}{c^2},$$

therefore

$$\mathcal{A} = \gamma^2\dot{\mathbf{u}} + \gamma^4\mathbf{u}\left(\frac{\mathbf{u} \cdot \dot{\mathbf{u}}}{c^2}\right) = \gamma^2\mathbf{a} + \gamma^4\mathbf{u}\left(\frac{\mathbf{u} \cdot \mathbf{a}}{c^2}\right),$$

as required.

13.8 Transformation rules for 4-vectors

Under rotation of axes in 4-space, corresponding to the transformation of inertial frames $S \longrightarrow S'$, 4-vectors transform according to the same rules as the coordinates themselves. Thus, if the coordinates transform as:

$$(X^0)' = \gamma(X^0 - VX^1/c) \tag{13.38}$$

$$(X^1)' = \gamma(X^1 - VX^0/c) \tag{13.39}$$

$$(X^2)' = X^2 \tag{13.40}$$

$$(X^3)' = X^3. \tag{13.41}$$

Then, any 4-vector B^μ, transforms as:

$$(B^0)' = \gamma(B^0 - VB^1/c) \tag{13.42}$$

$$(B^1)' = \gamma(B^1 - VB^0/c) \tag{13.43}$$

$$(B^2)' = B^2 \tag{13.44}$$

$$(B^3) = B^3. \tag{13.45}$$

As we have seen, such rotations are equivalent to Lorentz transformations. We shall examine this correspondence in a little more detail when we reformulate special relativity in Section 13.10.4.

13.8.1 Example: *Obtain 3-velocity transformation laws from the rules for 4-vectors*

Show that the components of the 3-velocity **u** can be expressed in terms of the components of the corresponding 4-velocity **U**, thus

$$u_1 = c\frac{U^1}{U^0} \qquad u_2 = c\frac{U^2}{U^0} \qquad \text{and} \qquad u_3 = c\frac{U^3}{U^0}.$$

Write down the standard Lorentz transformations which relate the components of the 4-velocity **U′** in S' to those of the 4-velocity **U** in S. Hence obtain corresponding transformations for the components of the 3-velocity **u**.

From the definition of the 4-velocity, as given by equation (13.29), we have

$$\mathbf{U} = \gamma(u)(c, \mathbf{u}) = \gamma(u)(c, u_1, u_2, u_3) \equiv (U^0, U^1, U^2, U^3).$$

Hence

$$U^0 = \gamma c, \qquad U^1 = \gamma u_1, \qquad U^2 = \gamma u_2, \qquad U^3 = \gamma u_3,$$

and so

$$u_1 = c\frac{U^1}{U^0}, \qquad u_2 = c\frac{U^2}{U^0}, \qquad u_3 = c\frac{U^3}{U^0}.$$

The standard Lorentz transformations for **U**, as given by equations (13.42)–(13.45), are:

$$(U^0)' = \gamma(v)(U^0 - VU^1/c)$$

$$(U^1)' = \gamma(v)(U^1 - VU^0/c)$$

$$(U^2)' = U^2$$

$$(U^3)' = U^3.$$

Thus

$$u_1' = c\frac{(U^1)'}{(U^0)'} = c\frac{U^1 - VU^0/c}{U^0 - VU^1/c} = c\frac{(U^1/U^0 - V/c)}{(1 - VU^1/cU^0)} = \frac{u_1 - V}{1 - Vu_1/c^2}.$$

Similarly

$$u_2' = c\frac{(U^2)'}{(U^0)'} = \frac{cU^2}{\gamma(V)(U^0 - V/cU^1)} = \frac{u_2}{\gamma(V)(1 - Vu_1/c^2)},$$

and

$$u_3' = c\frac{(U^3)'}{(U^0)'} = \frac{cU^3}{\gamma(V)(U^0 - VU^1/c)} = \frac{u_3}{\gamma(V)(1 - Vu_1/c^2)}.$$

These are, of course, exactly the same results as we found in the previous chapter [equations (12.12)–(12.14)].

13.8.2 Example: *Obtain linear acceleration in the comoving frame*

A particle moves with velocity $u(t)$ along the x-axis in an inertial frame of reference. Prove that its acceleration relative to the instantaneously comoving inertial frame may be expressed as

$$\frac{\mathrm{d}}{\mathrm{d}t}[u\gamma(u)].$$

We have the 4-velocity

$$U = [c\gamma(u), u\gamma(u), 0, 0],$$

and the 4-acceleration

$$A = \frac{\mathrm{d}U}{\mathrm{d}\tau} = (A^0, A^1, 0, 0).$$

Here,

$$A^1 = \frac{\mathrm{d}U^1}{\mathrm{d}\tau} = \frac{\mathrm{d}}{\mathrm{d}\tau}[u\gamma(u)] = \gamma\frac{\mathrm{d}}{\mathrm{d}t}[u\gamma],$$

and

$$A^0 = \frac{\mathbf{u} \cdot \mathcal{A}}{c} = \frac{u_1 A^1}{c} = \frac{u\gamma}{c}\frac{\mathrm{d}}{\mathrm{d}t}[u\gamma].$$

We make a Lorentz transformation of these relationships to the instantaneously comoving frame with relative velocity u:

$$(A^1)' = \gamma(u)\left(A^1 - \frac{uA^0}{c}\right) = \gamma^2\left(1 - \frac{u^2}{c^2}\right)\frac{\mathrm{d}}{\mathrm{d}t}[u\gamma]$$

$$= \gamma^2\gamma^{-2}\frac{\mathrm{d}}{\mathrm{d}t}[u\gamma] = \frac{\mathrm{d}}{\mathrm{d}t}[u\gamma],$$

and

$$(A^0)' = \gamma(u)(A^0 - uA^1/c) = 0,$$

from the above equations for A^1 and A^0.

Let the 3-space part of the acceleration be denoted by

$$\mathcal{A} \equiv (A^1, A^2, A^3).$$

Then, for any inertial frame, the results of Section 13.7.2 reduce to

$$\mathcal{A} = \gamma^2 \mathbf{a} + \gamma^4 \frac{\mathbf{u}(\mathbf{u} \cdot \mathbf{a})}{c^2},$$

where $a = \ddot{x}$ is th]e usual 3-acceleration of the particle along the x-axis. In the comoving frame, $u = 0$, $\gamma = 1$ and $\mathcal{A} = a$. Hence, relative to the comoving frame

$$a = (A^1)' = d[u\gamma(u)]/dt.$$

13.8.3 Example: *Central acceleration in the comoving frame*

A particle moves at constant speed u on a circle of radius R relative to an inertial frame of reference. Calculate its acceleration relative to the instantaneously comoving inertial frame.

For simplicity, choose the x-axis parallel to the instantaneous velocity $\mathbf{U}$ and the y-axis in the plane of the motion. Then

$$\mathbf{u} = (u, 0, 0) \qquad \mathbf{a} = (0, u^2/R, 0), \qquad \text{and} \qquad \mathbf{u} \cdot \mathbf{a} = 0.$$

Hence the results of Section 13.7.2 reduce to:

$$A^0 = \frac{\mathbf{u} \cdot \mathcal{A}}{c} = \gamma^2 \frac{\mathbf{u} \cdot \mathbf{a}}{c} = 0,$$

and

$$\mathcal{A} = \gamma^2 \mathbf{a} + \gamma^4 \mathbf{u}(\mathbf{u} \cdot \mathbf{a})/c^2,$$
$$= \gamma^2 \mathbf{a} = \gamma^2(0, u^2/R, 0).$$

Therefore

$$\mathbf{A} = (A^0, \mathcal{A}) = (0, 0, \gamma^2 u^2/R, 0).$$

Now make the Lorentz transformation to the instantaneously comoving frame with relative velocity u. The new components of $\mathbf{A}$ are:

$$(A^0)' = \gamma(u)[A^0 - uA^1/c] = 0$$
$$(A^1)' = \gamma(u)[A^1 - uA^0/c] = 0$$
$$(A^2)' = A^2 = \gamma^2 u^2/R$$
$$(A^3)' = A^3 = 0.$$

Since in this frame $\mathcal{A}$ coincides with $\mathbf{a}$, the magnitude of the acceleration is

$$|\mathbf{a}| = u\gamma(u)^2/R.$$

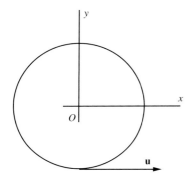

Fig. 13.4 A particle moving with constant velocity in a circle.

13.9 Usefulness of 4-vectors

We end our discussion of 4-vectors with a brief note on their usefulness and on the topic of Minkowski space. There is more to it than just an elegant formalism!

The basic postulate of special relativity is that physical laws take the same form in all inertial reference frames. This property is known as either **invariance of form** or, more simply, **form invariance**. Now, as we have seen in Section 11.7, in 4-space we have the equivalence:

Invariance under rotation of axes in 4-space

$$\equiv$$

invariance under Lorentz transformation.

This is a crucial property of the 4-space representation, and we may develop this aspect as follows. A *scalar* equation has the same form in all frames as both sides are invariant under rotations. That is, for two scalars *a* and *b*, the relationship

$$a = b,$$

holds for all frames. Although vectors are affected by a change of frame, *vector* relationships, such as,

$$f_i = g_i,$$

are also unaffected by rotations. Each side is transformed, but in the same way, thus:

$$f'_i = g'_i,$$

is **the same form of relation**, in a different coordinate frame.

This is true for a tensor of any rank. So if any physical law can be expressed as a 4-vector or 4-tensor equation, then it is invariant under imaginary rotation in 4-space; and naturally this means that it is Lorentz invariant.

13.10 Reconsideration of special relativity

In this section, we shall take a slightly more formal look at the 4-space formulation of special relativity and also introduce some ideas which we shall need for our brief introduction to general relativity in Chapter 15. The reader who is unfamiliar with the subject of tensors need not despair. Only simple ideas of matrices and vectors will be required.

13.10.1 *The metric tensor*

In Section 13.1 we discussed the inner product of vectors in *n* dimensions; and in Section 13.2 we introduced the squared interval $\mathrm{d}s^2$, for both Euclidean 3-space (13.6) and Minkowski 4-space (13.7). In fact these expressions are both special cases of the most general result in differential geometry for a path length. We can write down a general form for a path length at a point P in a vector space[3] of n dimensions as

$$ds^2 = g_{ij}dx_i dx_j, \qquad (13.46)$$

where both repeated indices i and j are summed from one up to n (or, equivalently, from zero up to $n-1$, if we prefer).

This expression for ds^2 is known as a **metric** and, although it may look quite formidable, we can readily simplify it by considering some familiar examples. However, first we should list some general points as follows:

1. The entity g_{ij} is a n-row $\times$ n-column array of coefficients.
2. The coefficients g_{ij} make up what is called the metric tensor.
3. The values of the coefficients may depend on the position of the point P.
4. The nature of the metric tensor may depend on the nature of the vector space and the choice of coordinate system.
5. The metric tensor is symmetric: $g_{ij} = g_{ji}$.

Those readers who are unfamiliar with this type of notation (or the very idea of tensors) may find it helpful to note that equation (13.46) can be rewritten as:

$$ds^2 = dx_i(g_{ij}dx_j).$$

The term in brackets can be thought of as being the multiplication of a matrix into a vector which generates a new vector. If we call this new vector $d\mathbf{y}$ then we may write it as

$$dy_i = g_{ij}dx_j, \qquad (13.47)$$

and hence

$$ds^2 = dx_i dy_i,$$

which is just the inner product of the two vectors. This step makes it easier to see the form that must be taken by g_{ij} for specific cases such as Euclidean 3-space and Minkowski 4-space. We shall consider these particular examples in the next section.

13.10.2 Example: *Metric tensors for Euclidean 3-space and Minkowski 4-space*

Let us begin by obtaining the form of the metric tensor appropriate to Euclidean 3-space. In effect we wish to choose g_{ij} such that equation (13.46) reduces to (13.6).

We first rewrite (13.6) so that the notation used is similar to that of (13.46), thus:

$$ds^2 = dx_1^2 + dx_2^2 + dx_3^2.$$

It is then easily verified that (13.46) reduces to this form provided first that $g_{ij} = 0$ for cases where $i \neq j$ (in other words, the array of coefficients has the form of a diagonal matrix), so that we have

$$ds^2 = g_{11}dx_1dx_1 + g_{22}dx_2dx_2 + g_{33}dx_3dx_3$$
$$= g_{11}dx_1^2 + g_{22}dx_2^2 + g_{33}dx_3^2,$$

and second, if

$$g_{11} = g_{22} = g_{33} = 1.$$

But this is just the definition of either the Kronecker delta δ_{ij} or the unit matrix. That is, we may now write the metric tensor for Euclidean 3-space as either

$$g_{ij} = \delta_{ij};$$

or as

$$g_{ij} = \text{diag}\,(1, 1, 1).$$

(The latter is just a compact notation for writing a diagonal matrix in terms of its only non-zero elements, viz., the diagonal elements.)

Now we turn to Minkowski 4-space and our first step is to change (13.46) into the notation introduced in Section 13.4 for 4-vectors. That is, we rewrite (13.46) as

$$ds^2 = g_{\mu\nu}dX^\mu dX^\nu, \tag{13.48}$$

where μ, ν take the values 0, 1, 2, and 3.

Next we argue just as we did for the previous example of Euclidean 3-space. If equation (13.48) is to reduce to (13.7) for 4-space: then, as before, $g_{\mu\nu}$ must be diagonal, so that (13.48) takes the intermediate form

$$ds^2 = g_{00}dX^0dX^0 + g_{11}dX^1dX^1 + g_{22}dX^2dX^2 + g_{33}dX^3dX^3,$$

and then the diagonal elements must satisfy the conditions

$$g_{00} = 1$$

$$g_{11} = g_{22} = g_{33} = -1;$$

or

$$g_{\mu\nu} = \text{diag}\,(1, -1, -1, -1). \tag{13.49}$$

It should be noted that this form of metric tensor, when combined with the expression (13.47), acts as a raising or lowering operator on the vector indices in 4-space. It is left as an exercise for the reader to verify that one may transform $X^\mu \longrightarrow X_\mu$, and *vice versa* in this way.

13.10.3 *Geodesics (timelike and null)*

In Galilean relativity, the trajectory of a free particle is a straight line, along which the particle moves with constant velocity. If we introduce special relativity, the same particle has a world line which is a straight line in Minkowski space. If one further wishes to introduce the concept of a curved space-time, as we shall in Chapter 15, then we shall be faced with the situation that a straight line is no longer the shortest distance between points. The general question then arises: what is the line which represents the shortest distance between two points in an arbitrarily curved space?

Fortunately the solution to this problem is known from geometry and the purpose of this section is to introduce the answer in the form of a **geodesic curve**. Most of us will be familiar with the fact that aircraft and ships on long journeys do not travel in straight lines, but rather along what are called great circle routes. This is because they are, in effect, constrained to move on the surface of a sphere. The concept is only surprising to us because we are accustomed to working with maps and charts which are produced by some form of projection of the surface of a sphere on to a plane. Obviously such charts are fine for short journeys where the curvature of the Earth can be neglected.

The geodesic curve can be defined formally by means of a variational principle. We shall not go into detail here, as a proper treatment would take us far beyond the scope of this book. However, the essential point to be recognized is that a geodesic is a curve joining two points in a vector space such that some associated property is minimized.

In using the mathematics of the geometry of vector spaces in n dimensions, we have to be cautious about the way in which we draw analogies between corresponding effects in different numbers of dimensions, although the associated mathematics usually can be generalized in quite a straightforward way. Suppose, for instance, we wish to determine whether a surface is flat. One way would be to use the Euclidean property that the angles of a triangle add up to 180 degrees. If one draws a triangle of the surface, then if this criterion is satisfied, it follows that the surface is flat.

If, instead, the triangle is drawn on the surface of a sphere, the three vertex angles must add up to more than 180 degrees. This may be seen from Figure 13.5, where a triangle is formed on the surface of the Earth by taking two lines of longitude and the equator. Obviously the lines of longitude each meet the equator at 90 degrees, so that the sum of the base angles is already 180 degrees. The angle at the polar vertex still has to be added on, and it could take any value up to 360 degrees. Similarly, one can construct a triangle on a different type of surface and find that the three vertex angles add up to less than 180 degrees. In this case, the surface is said to have *negative curvature*, whereas in the preceding case it has *positive curvature*. In either case, given the mathematical equation for the curve, one can find the radius of curvature (the usual measure of how curved a surface is, and in the case of the sphere just the normal radius), and other important features like the equation of the tangent vector at that point.

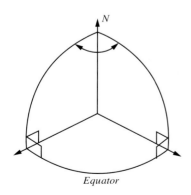

Fig. 13.5 A spherical triangle formed by two lines of longitude and the equator.

However, as we have said, one must be cautious about taking familiar ideas into an unfamiliar context. For instance, in plane Euclidean geometry it is well known that parallel lines never meet. But, if we take our analogue of a straight line on a flat surface to be a great circle on the surface of a sphere, and take the lines of longitude as examples of the latter, then it is apparent from Figure 13.5 that such lines, initially parallel near the equator, do in fact meet at each pole! Of course all is well if we restrict our attention to small, local areas and we may say that the surface of the sphere is *locally Euclidean*.

It is also of interest to note that, like other entities in Minkowski 4-space, geodesics can be classified as timelike, null or spacelike by referring to their tangent vector, **T**, say. The geometric characteristics of a geodesic curve at a point are the same as those of its tangent vector **T**, at the same point. Referring back to the discussion in Section 13.4.2, if its tangent vector satisfies $T^2 > 0$, a geodesic is timelike, and corresponds to a possible trajectory of a free particle in 3-space; and hence to a possible world line in 4-space. If, on the other hand, $T^2 = 0$, the geodesic is a null curve and such geodesics make up the surface of the light cone, and are therefore world lines for rays of light. The case where $T^2 < 0$ is of only technical significance. As spacelike geodesic curves lie outside the light cone, they cannot be world lines for any kind of particle and are therefore unphysical.

Lastly, although these statements are for a tangent vector at a point, it can be shown that the derivative of the tangent vector taken along the geodesic curve[4] is zero. In this sense the tangent vector is constant along the geodesic and this has the geometric interpretation that the geodesics are the 'straight lines' of the particular space under consideration. From our point of view here, its importance is that **the geodesic curve cannot change its type**. If it is timelike at one point, then it remains timelike at all points.

[4]Strictly, in tensor analysis, we should speak of the *covariant* derivative along the curve.

13.10.4 *Restatement of special relativity*

At this point it may be helpful to draw some threads together and make a formal statement of the Lorentz transformations as a tensor relationship in Minkowski 4-space. We have in fact given the basic relationships in Section 13.8, where they are seen as rotations of axes in 4-space. However, it is interesting at this stage to go back to our original formulations of the Lorentz transformations in Chapter 11, and make the connection with our 4-space formalism. It should be recalled that we are considering two frames S and S', in standard configuration, with relative speed V. If we introduce the factor $\beta = V/c$, and also the gamma factor which we have been using in this chapter, then equations (11.13) and (11.12) may be rewritten as:

$$t' = \gamma(t - \beta x/c),$$

and

$$x' = \gamma(x - \beta ct),$$

respectively. Note the change of order, as we are now going to treat time as the 'zero component' in Minkowski 4-space. Now, if we multiply the first equation through by a factor c, and rename $ct = x_0$, $ct' = x'_0$, and x, x' as x_1, x'_1, then the pair of equations may be written as:

$$x'_0 = \gamma(x_0 - \beta x_1)$$

$$x'_1 = \gamma(x_1 - \beta x_0).$$

Next, we can write this pair of equations as a single matrix equation, in terms of a two-dimensional matrix, thus:

$$\begin{pmatrix} x'_0 \\ x'_1 \end{pmatrix} = \begin{pmatrix} \gamma & -\beta\gamma \\ -\beta\gamma & \gamma \end{pmatrix} \begin{pmatrix} x_0 \\ x_1 \end{pmatrix}.$$

This may be readily verified by multiplying out the matrix and is equivalent to writing the equation in symbols as:

$$x'_i = a_{ij}x_j,$$

where the repeated or dummy index j is summed over the values 0 and 1, and the labelling index i is set equal to 0 and 1, in turn, in order to generate the two scalar equations.

We can extend this procedure to the full 4-vector notation by including the trivial transformations $x'_2 = x_2$ and $x'_3 = x_3$ as well. We do this by introducing a 4×4 matrix $L_{\alpha\beta}$, where as usual $\alpha, \beta = 0, 1, 2, 3$, which incorporates the above 2×2 matrix and whose coefficients are given by:

$$\begin{pmatrix} \gamma & -\beta\gamma & 0 & 0 \\ -\beta\gamma & \gamma & 0 & 0 \\ 0 & 0 & 1 & 0 \\ 0 & 0 & 0 & 1 \end{pmatrix}.$$

With this definition, the four-tensor form of the standard Lorentz transformations becomes

$$(X^\alpha)' = L_{\alpha\beta}X^\beta, \tag{13.50}$$

where we have restored the upper-case notation for full 4-vectors. The procedure for verifying that this relationship generates the requisite four equations which make up the standard set of Lorentz transformations is just a generalization of the above two-dimensional case. That is, we set the index α equal to 0, 1, 2 and 3 in turn, in order to generate the four equations, and for each value of α, the dummy index β is summed from 0 up to 3.

We conclude this section with a reformulation of the axiomatic basis of special relativity, with the aid of both hindsight and the existence of general relativity. In order to avoid any confusion with Einstein's original axioms, as given in Chapter 11, we shall refer to the revised axioms as postulates. We now set these out as follows:

Postulate 1 The space-time continuum is represented by a four-dimensional vector space with a flat metric.

Postulate 2 In this vector space, free particles travel along timelike geodesics while light rays travel along null geodesics.

These postulates could be stated with greater technical precision, but it will be sufficient for our purposes here to remark that the four-dimensional space referred to is the Minkowski space of events and is endowed with all the properties which we have discussed in the present chapter. With this in mind, Postulate 2 is equivalent to Einstein's second axiom, that the speed of light is the same for all inertial observers. Evidently Postulate 1 points us firmly in the direction of general relativity, when we consider the case where the metric is not flat but curved. We shall return to this point in Chapter 15.

13.11 Exercises

13.1 Three events E_1, E_2 and E_3 are observed to occur at space-time coordinates (x, t) of $(1, 2/c)$, $(4, 7/c)$ and $(6, 5/c)$, respectively, in some frame S. By considering the relevant space-time intervals, establish which pairs of events could be causally connected.

13.2 Two events E_1 and E_2 are observed as occurring at space-time coordinates (ct, x, y, z) of $(0, 0, 0, 0)$ and $(1, 2, 0, 0)$ respectively in some frame S. Find the speeds of frames in standard configuration with S, such that: (a) the two events are simultaneous; and (b) E_2 precedes E_1 by a time $1/c$.

13.3 Verify the invariance relation,

$$U^2 = c^2,$$

where U is the 4-velocity of a particle, by direct substitution for the gamma factor in equation (13.30).

13.4 Show that the Minkowski metric tensor acts as a raising or lowering operator on vector indices in 4-space; thus:

$$A_\mu = g_{\mu\nu} A^\nu,$$

for instance.

13.5 By direct differentiation of both sides of equation (13.48), obtain the general result

$$g_{\mu\nu} \dot{X}^\mu \dot{X}^\nu = c^2,$$

where the dot denotes differentiation with respect to proper time. Verify the equivalence of this result to that of equation (13.32).

13.6 A laboratory is situated on a ship which is moored where the Greenwich meridian crosses the equator. In spherical polar coordinates, its position at time $t = 0$ is given by $r = R$, $\theta = \pi/2$ and $\phi = 0$. Show that its world line takes the form

$$X^\mu = (\gamma c \tau, R, 0, \gamma R \omega \tau),$$

where ω is the rotational speed of the Earth, R is its radius, τ is the proper time, and

$$\gamma = (1 - R^2 \omega^2/c^2)^{-1}.$$

Further show that its 4-velocity is given by

$$U^\mu = (\gamma c, 0, 0, \gamma R \omega).$$

13.7 An ideal clock accelerates from a time t_i to a time t_f, as recorded in a frame S. Derive an expression for the proper time interval recorded by the clock between these two events, as a function of its speed $u(t)$ as measured in S.

13.8 By considering the transformation properties of the 4-acceleration, show that the components of the

3-acceleration **a** in S transform to an instantaneously comoving frame S', according to:

$$a_1' = \gamma^3 a_1 \qquad a_2' = \gamma^2 a_2 \qquad a_3' = \gamma^2 a_3,$$

where S and S' are instantaneously in standard configuration.

13.9 A spaceship accelerates with constant proper acceleration a_0 along the x-axis in S. If it started at the origin at $t = 0$, show that its world line takes the form

$$a_0 x^2 + 2c^2 - a_0 c^2 t^2 = 0.$$

13.10 By considering the behaviour of the solution in the preceding problem for very large values of x, show that a light signal emitted from a source at the origin at a time later than c/a_0 has a world line which never intersects the world line of the spaceship.

[Note: such an asymptote is known as the **event horizon** of the accelerated observer.]

Relativistic dynamics

14

As we have seen in the earlier parts of this book, Newton's laws are **postulates** about the physical world which are justified in practice for speeds that are much less than the speed of light. However, they are *not* Lorentz invariant: this shows up in their failure as $V \to c$. In order to set up an alternative to Newton's system of laws, we have to make new postulates in the light of *special relativity*.

14.1 Particle collisions

One objective of the present book is to extend our treatment of collisions to be compatible with special relativity. In view of the importance of high-energy particle collisions in modern physics, the need for a relativistic treatment of the subject should be quite evident. In fact, it turns out that this is not only a meritorious objective, but one that is also readily accomplished.

The important thing to bear in mind, at this juncture, is that previously we have used conservation laws in order to treat particle collisions in a purely Newtonian context. Now, in the context of special relativity, we might expect to have to regard the Newtonian conservation laws as low-speed approximations to something more accurate which we still have to formulate. Yet, conservation laws are by their nature statements of principle, so that we are also entitled to ask: how can they be modified or indeed why should they have to be changed?

The secret of the resolution of this paradox lies in the way in which the laws transform between inertial frames. Conservation of energy holds rigorously in both S and S', but the Galilean transformation from one frame to the other involves the addition or subtraction of kinetic energy. Analogous considerations hold for the Galilean transformation of momentum and angular momentum. Accordingly, we need to formulate the conservation laws in such a way that their transformation between frames is governed by the Lorentz transformations. An obvious approach would be to try to extend the conservation laws for 3-vectors to the case of 4-vectors, and this is what we shall do next.

14.2 Derivation of Lorentz-invariant 4-momentum conservation law

Consider collisions of particles, as we did in Chapter 9. In *Newtonian* mechanics, we have,

$$\sum_{\{states\}} \mathbf{p} = 0, \qquad (14.1)$$

where we are now introducing a more formal and succinct notation in which $\sum_{\{states\}}$ stands for a sum over particle states, and $\mathbf{p}$ is just the usual 3-momentum. Note that, with this convention, particle states *before* the collision count as positive, while particle states *after* the collision count as negative. Alternatively, another way of looking at this is that equation (9.2)—our previous statement of conservation of momentum for particle collisions—can be written in the same form as (14.1), by simply taking the terms on the right-hand side over to the left-hand side, and making the usual change of sign in the process.

At this stage we simply *guess* that a Lorentz-invariant theory results from applying this idea to 4-momentum $\mathbf{P}$. That is, we postulate that the law of conservation of momentum, as applied to particle collisions, should take the form

$$\sum_{\{states\}} \mathbf{P} = 0. \qquad (14.2)$$

Then, in order to take this approach further, we need a *definition* of 4-momentum for a particle. As a first step, we take the **proper mass** of a article to be m. Then we define the 4-momentum of the particle by *analogy* with 3-momentum. That is,

$$
\begin{aligned}
\mathbf{P} &= m\mathbf{U} & \text{(where } \mathbf{U} \equiv \text{4-velocity)} \\
&= m\gamma(u)[c, \mathbf{u}] & \text{(where } \mathbf{u} \equiv \text{3-velocity)} \\
&= [m\gamma(u)c, m\gamma(u)\mathbf{u}].
\end{aligned}
$$

Therefore we take the 4-momentum of a particle to be:

$$\mathbf{P} = [m\gamma(u)c, \mathbf{p}] \equiv (P^0, \mathbf{p}), \qquad (14.3)$$

where $\mathbf{p}$ is its relativistic 3-momentum.

Two new concepts now arise:[1] the relativistic momentum and the relativistic kinetic energy of a particle. We shall discuss these ideas in the next section.

[1] Some physicists would argue that there is also a third; viz., the concept of relativistic inertial mass. We shall discuss this further in Appendix C.

14.2.1 *Relativistic momentum*

From equation (14.3), we may identify the relativistic 3-momentum $\mathbf{p}$ as

$$\mathbf{p} = m\gamma(u)\mathbf{u}. \qquad (14.4)$$

Now, if overall conservation of 4-momentum requires:

$$\sum_{\{states\}} \mathbf{P} = \sum_{\{states\}} (m\gamma(u)c, \mathbf{p}) = 0, \qquad (14.5)$$

then the individual components of the vector must vanish separately. That is,

$$\sum_{\{states\}} m\gamma(u)c = 0; \qquad (14.6)$$

and

$$\sum_{\{states\}} \mathbf{p} = \sum_{\{states\}} m\gamma(u)\mathbf{u} = 0. \qquad (14.7)$$

Equation (14.7) is just a new version of Newton's conservation law for momentum, now modified by the inclusion of the γ-factor. For small u, $\gamma(u) \rightarrow 1$, and this reduces back to the familiar Newtonian form:

$$\sum_{\{states\}} m\mathbf{u} = 0, \qquad (14.8)$$

as one would require.

14.3 Kinetic energy and the mass–energy relation

Now let us return to equation (14.6): this appears to be analogous to the Newtonian equation of conservation of mass. That is, it reduces to

$$\sum_{\{states\}} m = 0, \qquad (14.9)$$

if we take the limit $\gamma \rightarrow 1$, for low speeds. Formally, we should stress the fact that this *Newtonian* statement is now to be reinterpreted as conservation of proper (or, rest) mass.

Next, we consider our new Lorentz-invariant statement of conservation of mass. This is obtained by cancelling the factor c—itself Lorentz invariant—from equation (14.6), to give:

$$\sum_{\{states\}} m\gamma = 0. \qquad (14.10)$$

Let us expand out the gamma factor in powers of u^2/c^2, where u is, of course, the speed of the particle, thus:

$$m\gamma(u) = m(1 - u^2/c^2)^{1/2} = m + \frac{1}{c^2}\left(\frac{1}{2}m\gamma u^2\right) + \cdots. \qquad (14.11)$$

If we interpret this as, in effect, an expansion for the inertial mass, then the *relativistic mass* of a *slowly moving* particle ($u \ll c$) exceeds its rest mass by $1/c^2 \times$ its kinetic energy. So the kinetic energy of a particle contributes to its mass in a way which is consistent with[2]

$$E = m\gamma(u)c^2. \tag{14.12}$$

This is the relativistic energy E of a particle, and we may use this result to rewrite equation (14.3) for the 4-momentum of the particle as

$$\mathbf{P} = (E/c, \mathbf{p}). \tag{14.13}$$

This is an important alternative form of the 4-momentum which is often used in applications.

14.4 Momentum–energy relationship

In this section we derive the relativistic analogue of the Galilean-invariant relation $E = p^2/2m$ in Newtonian mechanics. We follow that by showing that the resulting relationship is Lorentz invariant.

14.4.1 Derivation from definitions of energy and momentum

We start with the definition of the relativistic energy E, thus:

$$E = m\gamma(u)c^2,$$

and square both sides to get:

$$E^2 = m^2\gamma^2 c^4.$$

Next subtract m^2c^4 from each side

$$E^2 - m^2c^4 = m^2c^4(\gamma^2 - 1) = m^2c^4\left(\frac{1}{1 - u^2/c^2} - 1\right).$$

Then, putting the terms on the right-hand side on a common denominator and cancelling as appropriate, we obtain

$$E^2 - m^2c^4 = m^2c^4(\gamma^2 - 1) = m^2c^2\gamma^2u^2 = c^2p^2,$$

where the last step follows from the definition of the relativistic 3-momentum $\mathbf{p}$: see equation (14.5). Then, with some rearrangement, we have the usual form:

$$E^2 = c^2p^2 + m^2c^4. \tag{14.14}$$

14.4.2 *Verification of Lorentz invariance*

We may show that the above relationship is compatible with the Lorentz invariance of the scalar product $P^2 = \mathbf{P} \cdot \mathbf{P}$. Recall the definition of 4-momentum. From equation (14.3) we have

$$\mathbf{P} = (m\gamma(u)c, \mathbf{p}), \qquad (14.15)$$

and so the invariant P^2 takes the form:

$$P^2 = \mathbf{P} \cdot \mathbf{P} = m^2 \gamma^2 c^2 - p^2. \qquad (14.16)$$

Consider this relationship in the rest frame of the particle:

$$P^2 = m^2 c^2, \qquad (14.17)$$

which holds in all frames, so that we have

$$m^2 c^2 = m^2 \gamma^2 c^2 - p^2. \qquad (14.18)$$

Then, from the definition of the relativistic energy $E = m\gamma c^2$, we have $m\gamma = E/c^2$, so that equation (14.18) becomes

$$m^2 c^2 = \frac{E^2}{c^4} c^2 - p^2 = \frac{E^2}{c^2} - p^2, \qquad (14.19)$$

or, multiplying across by c^2, and rearranging:

$$E^2 = c^2 p^2 + m^2 c^4,$$

as expected.

14.4.3 Example: *Coalescence of two moving particles*

(a) Two particles of rest mass m_1 and m_2 which have velocities $\mathbf{u}_1$ and $\mathbf{u}_2$, as measured in some inertial reference frame, coalesce to form a single particle of mass M and velocity $\mathbf{u}$. Write down the total 4-momentum of the initial and final states and deduce that

$$\mathbf{u} = \frac{m_1 \gamma(u_1)\mathbf{u}_1 + m_2 \gamma(u_2)\mathbf{u}_2}{m_1 \gamma(u_1) + m_2 \gamma(u_2)}.$$

(b) What is the minimum mass M_{min} of a particle which can decay into two other particles of masses m_1 and m_2?

(a) Take m_1 and m_2 to be the rest masses of the initial particles.

- The initial state is $\mathbf{P}_I = (P_I^0, \mathbf{p}_I)$; where

$$P_I^0 = m_1 \gamma(u_1)c + m_2 \gamma(u_2)c$$

and

$$\mathbf{p}_I = m_1 \gamma(u_1)\mathbf{u}_1 + m_2 \gamma(u_2)\mathbf{u}_2.$$

- The final state is $\mathbf{P}_F = (P_F^0, \mathbf{p}_F)$; where

$$P_F^0 = M\gamma(u)c,$$

and

$$\mathbf{p}_F = M\gamma(u)\mathbf{u}.$$

- Conservation of 4-momentum implies $\mathbf{P}_I = \mathbf{P}_F$; and, in terms of the components of the vectors, this implies that

$$P_I^0 = P_F^0 \qquad \text{and} \qquad \mathbf{p}_I = \mathbf{p}_F.$$

Therefore, with some rearrangement,

$$\frac{\mathbf{p}_F}{P_F^0} = \frac{\mathbf{p}_I}{P_I^0},$$

and in turn this implies that

$$\frac{M\gamma(u)\mathbf{u}}{M\gamma(u)c} = \frac{m_1\gamma(u_1)\mathbf{u}_1 + m_2\gamma(u_2)\mathbf{u}_2}{m_1\gamma(u_1)c + m_2\gamma(u_2)c}.$$

Then, cancelling common factors as appropriate, we find that

$$\mathbf{u} = \frac{m_1\gamma(u_1)\mathbf{u}_1 + m_2\gamma(u_2)\mathbf{u}_2}{m_1\gamma(u_1) + m_2\gamma(u_2)},$$

as required.

(b) Conservation of energy for the reverse of process (a) is:

$$M\gamma(u)c^2 = m_1\gamma(u_1)c^2 + m_2\gamma(u_2)c^2.$$

In the CM frame, we have $u = 0$ and hence $\gamma(u) = 1$. Similarly, both $\gamma(u_1)$ and $\gamma(u_2)$ take their minimum values of unity when $u_1 = 0$ and $u_2 = 0$. Therefore the above expression reduces to:

$$M_{min} = m_1 + m_2.$$

14.5 Lorentz invariants

We have already seen that the geometric property that a scalar is invariant under rotations and translations of axes implies Lorentz invariance when we form the scalar from the inner product of two 4-vectors. Thus it follows that, if $\mathbf{P}$ is the 4-momentum of a moving particle, then the inner product of $\mathbf{P}$ with itself is Lorentz invariant. We may state this formally as:

$$\mathbf{P} \cdot \mathbf{P} = P^2 \text{ is a Lorentz invariant.} \tag{14.20}$$

The value of this invariant for any particular particle is found by considering it in the rest frame of the particle, where $u = 0$ and where

$$P^2 = m^2 c^2 = \text{its value in all frames.} \qquad (14.21)$$

The concept of a Lorentz invariant can be extended if we consider the elastic collision of two particles. If particle 1 and particle 2 have 4-momentum $\mathbf{P}_1$ and $\mathbf{P}_2$, respectively, then we may show that the inner product of the two 4-momenta is a Lorentz invariant. We do this as follows.

Conservation of 4-momentum during the collision implies that:

$$\mathbf{P}_1 + \mathbf{P}_2 = \mathbf{P}_1' + \mathbf{P}_2',$$

where the primed values are taken after the collision. Now square both sides of this equation to obtain

$$P_1^2 + 2\mathbf{P}_1 \cdot \mathbf{P}_2 + P_2^2 = (P_1')^2 + 2\mathbf{P}_1' \cdot \mathbf{P}_2' + (P_2')^2. \qquad (14.22)$$

But P_1^2 and P_2^2 are themselves both Lorentz invariants. Hence we may cancel $P_1^2 = (P_1')^2$ and $P_2^2 = (P_2')^2$ across both sides of the above equation to leave us with

$$\mathbf{P}_1 \cdot \mathbf{P}_2 = \mathbf{P}_1' \cdot \mathbf{P}_2', \qquad (14.23)$$

and hence

$$\mathbf{P}_1 \cdot \mathbf{P}_2 \text{ is a Lorentz invariant.}$$

This will turn out to be a very useful result when we consider some problems involving collisions.

14.6 Massless particles

In order to cater for massless particles like the neutrino and the photon, we need to have some way of working out their momentum. The momentum–energy relationship comes to our rescue here, and from equation (14.14) we have

$$E^2 = p^2 c^2 + m^2 c^4,$$

which is valid for any particle. Clearly for the special case of massless particles we just set $m = 0$ in this relationship, to obtain

$$E^2 = p^2 c^2,$$

and hence

$$p = E/c.$$

Formally, we therefore have the result that the magnitude of the relativistic 3-momentum of a massless particle is given by its relativistic energy E divided by the speed of light. If we take the direction of propagation of the massless particle to be given by the unit vector **n**, then we may write the 3-momentum vector of a massless particle as

$$\mathbf{p} = \frac{E}{c}\mathbf{n}, \tag{14.24}$$

and from equation (14.13) the corresponding 4-momentum is

$$\mathbf{P} = \left(\frac{E}{c}, \frac{E}{c}\mathbf{n}\right). \tag{14.25}$$

As an example, we could consider a photon of frequency ν. From quantum physics we have the relationship between the energy of a photon and its frequency in the form

$$E = h\nu,$$

where h is Planck's constant. In this case we can write the 4-momentum **Q**, say, of the photon as

$$\mathbf{Q} = \left(\frac{E}{c}, \frac{E}{c}\mathbf{n}\right) = \frac{h\nu}{c}(1, \mathbf{n}), \tag{14.26}$$

and this result will be of use, not only in problems, but also in the topic of Compton scattering which we shall discuss in the next section.

14.6.1 Example: *Compton scattering using conservation of 4-momentum*

A photon of momentum $h/\lambda = h\nu/c$, and energy $h\nu$, collides with a stationary electron of mass m, and is scattered at an angle θ, with new energy $h\nu'$. Show that the change of energy of the photon is related to the scattering angle by

$$\lambda' - \lambda = 2\lambda_c \sin^2 \theta/2,$$

where $\lambda_c = h/mc$ is known as the Compton wavelength.

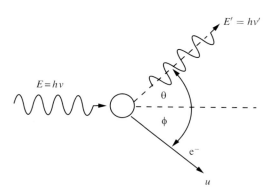

Fig. 14.1 Compton scattering of a photon by an electron in the LAB frame.

Let m be the mass of the electron and take the 4-momentum of the photon as $\mathbf{Q} = (E/c)(1, \mathbf{n})$. Also we have $E = h\nu$, where the frequency and wavelength of the photon are related by $\nu = c/\lambda$. We begin by establishing the initial and final 4-momenta, in the LAB frame, as follows.

- The total initial 4-momentum is $\mathbf{P}_I = \mathbf{Q}_I + \mathbf{P}_I^e$; where, for the photon,

$$\mathbf{Q}_I = (E/c, E/c, 0, 0),$$

and

$$\mathbf{P}_I^e = (mc, 0, 0, 0),$$

for the electron. Therefore, the total 4-momentum of the initial state is given by:

$$\mathbf{P}_I = (E/c + mc, E/c, 0, 0).$$

- The final total 4-momentum is $\mathbf{P}_F = \mathbf{Q}_F + \mathbf{P}_F^e$; where, for the scattered photon,

$$\mathbf{Q}_F = (E'/c, (E'/c)\cos\theta, (E'/c)\sin\theta, 0),$$

and

$$\mathbf{P}_F^e = (m\gamma(u)c, mu\gamma(u)\cos\phi, -mu\gamma(u)\sin\phi, 0),$$

for the recoiling electron. Therefore, the total 4-momentum for the final state is:

$$\mathbf{P}_F = (E'/c + m\gamma(u)c, (E'/c)\cos\theta + mu\gamma(u)\cos\phi,$$
$$(E'/c)\sin\theta - m\gamma(u)u\sin\phi, 0).$$

Next we invoke conservation of 4-momentum:

$$\mathbf{P}_I = \mathbf{P}_F.$$

We do this by equating individual components, thus:

$$\begin{aligned} P^0: & \quad E/c + mc = E'/c + m\gamma(u)c \\ p_x: & \quad E/c = (E'/c)\cos\theta + m\gamma(u)u\cos\phi \\ p_y: & \quad 0 = (E'/c)\sin\theta - m\gamma(u)u\sin\phi. \end{aligned}$$

Now, we wish to eliminate ϕ, and to do this we rearrange the equations for p_x and p_y, to have just a term involving ϕ on the right-hand side, square both sides of these equations, and then add the two resulting equations together:

$$E^2/c^2 - 2EE'/c^2\cos\theta + E'^2/c^2 = m^2\gamma^2u^2. \tag{14.27}$$

In order to eliminate $m^2\gamma^2 u^2$, we make use of relation (14.14) and apply this to the electron, thus:

$$E_e^2 = c^2 p_e^2 + m^2 c^4.$$

When we substitute from (14.12) for E and (14.4) for $\mathbf{p}$, this becomes

$$m^2\gamma^2 c^4 = m^2\gamma^2 u^2 c^2 + m^2 c^4. \tag{14.28}$$

Next, rearrange the equation for P^0 by taking E' to the left-hand side and square both sides:

$$m^2 c^2 + 2m(E - E') + E^2/c^2 - 2EE'/c^2 + E'^2/c^2 = m^2\gamma^2 c^2. \tag{14.29}$$

Now multiply equations (14.27) and (14.39) across by c^2, subtract one from the other and invoke (14.28) to find:

$$2m(E - E')c^2 = 2EE'(1 - \cos\theta).$$

Lastly, divide across by EE' then substitute $E/c = h/\lambda$, $E'/c = h/\lambda'$ and $(1 - \cos\theta) = 2\sin^2\theta/2$ to obtain,

$$\lambda' - \lambda = (2h/mc)\sin^2\theta/2,$$

as required.

14.6.2 *Compton scattering: alternative calculation using Lorentz invariants*

In order to underline the different treatment, we change our notation to that of Section 14.5, and write conservation of 4-momentum as

$$\mathbf{P}_1 + \mathbf{P}_2 = \mathbf{P}_1' + \mathbf{P}_2',$$

where particle 1 is the photon, particle 2 is the electron and the primes refer to the final state. This can be rewritten as

$$\mathbf{P}_2' = \mathbf{P}_1 + \mathbf{P}_2 - \mathbf{P}_1', \tag{14.30}$$

and we shall make use of it in this form presently. Now we summarize initial and final states in the LAB frame.

Initial state   From (14.26), we have

$$\mathbf{P}_1 = \frac{h\nu}{c}(1, \mathbf{n}),$$

where $\mathbf{n}$ is the unit vector in the direction of propagation of the incident photon. The target electron is at rest and so its 4-momentum is:

$$\mathbf{P}_2 = (m\gamma(u)c, 0) = (mc, 0).$$

Final state The scattered photon has frequency ν' and so its 4-momentum is

$$\mathbf{P}_1' = \frac{h\nu'}{c}(1, \mathbf{n}'),$$

where $\mathbf{n}'$ is the unit vector in the direction of propagation of the scattered photon. The 4-momentum of the recoiling electron is

$$\mathbf{P}_2' = (m\gamma(u')c, \mathbf{p}').$$

Now let us consider the Lorentz invariants. The 4-momenta of the photons are null vectors and it is easily verified therefore that

$$\mathbf{P}_1 \cdot \mathbf{P}_1 = \mathbf{P}_1' \cdot \mathbf{P}_1' = 0.$$

Also, from equation (14.17) for particles with mass, we have for the electron

$$\mathbf{P}_2 \cdot \mathbf{P}_2 = \mathbf{P}_2' \cdot \mathbf{P}_2' = m^2 c^2.$$

Having set up the problem, we next eliminate the final state of the electron by using the second of the above equalities and substituting from equation (14.31) for $\mathbf{P}_2'$, thus:

$$\mathbf{P}_2' \cdot \mathbf{P}_2' = m^2 c^2 = (\mathbf{P}_1 + \mathbf{P}_2 - \mathbf{P}_1') \cdot (\mathbf{P}_1 + \mathbf{P}_2 - \mathbf{P}_1');$$

or, multiplying out,

$$m^2 c^2 = \mathbf{P}_1 \cdot \mathbf{P}_1 + \mathbf{P}_2 \cdot \mathbf{P}_2 + \mathbf{P}_1' \cdot \mathbf{P}_1' + 2\mathbf{P}_1 \cdot \mathbf{P}_2 - 2\mathbf{P}_1 \cdot \mathbf{P}_1' - 2\mathbf{P}_2 \cdot \mathbf{P}_1'$$
$$= 0 + m^2 c^2 + 0 + 2\mathbf{P}_1 \cdot \mathbf{P}_2 - 2\mathbf{P}_1 \cdot \mathbf{P}_1' - 2\mathbf{P}_2 \cdot \mathbf{P}_1',$$

where we have made use of the invariants listed above. Cancelling and rearranging then gives:

$$\mathbf{P}_1 \cdot \mathbf{P}_2 - \mathbf{P}_2 \cdot \mathbf{P}_1' = \mathbf{P}_1 \cdot \mathbf{P}_1'.$$

Evaluating the remaining scalar products in terms of the given 4-momenta yields:

$$mh\nu - mh\nu' = \frac{h^2 \nu\nu'}{c^2}(1 - \mathbf{n} \cdot \mathbf{n}').$$

From Figure (14.1) it is easily seen that $\mathbf{n} \cdot \mathbf{n}' = \cos\theta$, and the result given in the previous section then follows.

14.7 Examples of high-energy collisions

In this section we shall consider three specific calculations where we use Lorentz invariants to solve problems involving collisions.

14.7.1 Example: *Photon absorbed by stationary proton*

A photon of energy E is absorbed by a stationary proton, resulting in a neutral pion and a recoiling proton. Show that the threshold energy for this process is: $E_{\min} = m(m + 2M)c^2/2M$, where m is the pion mass and M is the proton mass.

From equation (14.13) we have the 4-momentum as $\mathbf{P} = (m\gamma c, \mathbf{p}) = (E/c, \mathbf{p})$ and from (14.24) for the particular case of a photon $E = c|\mathbf{p}|$. Let us now consider matters in the CM and Lab frames of reference.

The CM frame   (Note that this is also the centre-of-momentum frame and that this means that the total 3-momentum is always zero.) As we are interested in the threshold energy for the reaction, we begin with the end result and ask: what is the minimum energy needed to achieve it? Evidently the total energy, after the collision, in this frame is given by:

$$E_{\mathrm{CM}} = \sum_\alpha m_\alpha c^2 \gamma(u_\alpha)$$

where the index α denotes a particle after the collision. The minimum value of the energy will be when all the particle velocities are zero and all the corresponding gamma factors are unity. Hence

$$E_{\mathrm{CM, min}} = \sum_\alpha m_\alpha c^2 = (m + M)c^2.$$

Then from the definition of the 4-momentum and the condition on the 3-momentum $\sum \mathbf{p} = 0$, we have the total 4-momentum in this frame is:

$$\mathbf{P} = \left(E_{\mathrm{CM}}/c, \sum \mathbf{p}\right) = (E_{\mathrm{CM}}/c, \mathbf{0}),$$

and hence

$$\mathbf{P} \cdot \mathbf{P} = (E_{\mathrm{CM}}/c)^2.$$

The LAB frame (or the target frame)   Here we have the proton at rest, such that $\mathbf{P}_{\mathrm{pr}} = (Mc, \mathbf{0})$, and for the incident photon we also have $\mathbf{P}_{\mathrm{ph}} = (E/c, \mathbf{p})$ such that $|\mathbf{p}| = E/c$. Therefore the total 4-momentum is:

$$\mathbf{P} = \left(\frac{E}{c} + Mc, \mathbf{p}\right),$$

and squaring both sides of this yields

$$\mathbf{P} \cdot \mathbf{P} = \left(\frac{E}{c} + Mc\right)^2 - p^2 = \left(\frac{E}{c} + Mc\right)^2 - E^2/c^2 = 2ME + M^2 c^2.$$

From the Lorentz invariance of P^2 we can then equate the two equations for $\mathbf{P} \cdot \mathbf{P}$:

$$\frac{E_{CM}^2}{c^2} = 2ME + M^2c^2,$$

and hence

$$E = \frac{E_{CM}^2 - M^2c^4}{2Mc^2}.$$

Now, from earlier, we have in the CM frame,

$$E_{CM,\,min} = (m + M)c^2,$$

and so

$$E_{min} = \left[\frac{(m + M)^2 - M^2}{2M} \right] c^2 = \frac{m(m + 2M)c^2}{2M},$$

for the LAB frame.

14.7.2 Example: *Two electrons scattering at right angles in the CM frame*

Two electrons of mass m and energy E collide elastically, with each moving off at right angles to their original directions, as determined in the CM frame. Show that in the LAB frame, the incident electron has initial energy $(2E^2 - m^2c^4)/mc^2$. Show also that the final energy of each electron is E^2/mc^2 in the LAB frame, and that the angle between their final directions of motion is $2\theta = 2\tan^{-1}(mc^2/E)$.

We begin by considering the situation in the CM frame.

The CM frame   (Figure 14.2) Each electron has energy E, before *and* after the collision, the total 3-momentum is zero, before *and* after; and so the total 4-momentum is:

$$\mathbf{P}_I = \left(\frac{2E}{c}, 0, 0, 0 \right),$$

and hence

$$(\mathbf{P}_I \cdot \mathbf{P}_I)_{CM} = \left(\frac{2E}{c} \right)^2.$$

The LAB frame   (Figure 14.3) Let the energy of the incident electron be E_I and the energy of *each* electron after the collision be E_F. Therefore the initial 4-momentum is:

$$\mathbf{P}_I = \left[\frac{E_I}{c} + mc, \left(\frac{E_I^2}{c^2} - m^2c^2 \right)^{1/2}, 0, 0 \right],$$

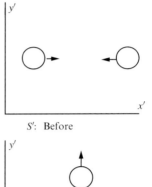

Fig. 14.2 Before and after the collision in the CM frame.

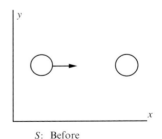

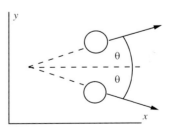

Fig. 14.3 Before and after the collision in the LAB frame.

where we have used the relationship

$$E^2 = p^2c^2 + m^2c^4,$$

rearranged to give us:

$$p = \left(\frac{E^2}{c^2} - m^2c^2\right)^{1/2}.$$

Similarly, the final 4-momentum is found to be:

$$\mathbf{P_F} = \left(\frac{2E_F}{c}, 2\left(\frac{E_F^2}{c^2} - m^2c^2\right)^{1/2} \cos\theta, 0, 0\right),$$

where each electron has a component of momentum resolved along the x-axis—hence the factor of 2—and the components along the y-axis cancel.

As usual, we make use of the fact that $\mathbf{P} \cdot \mathbf{P}$ is Lorentz invariant, thus:

$$(\mathbf{P_I} \cdot \mathbf{P_I})_{CM} = (\mathbf{P_I} \cdot \mathbf{P_I})_{LAB},$$

or

$$\left(\frac{2E}{c}\right)^2 = \left(\frac{E_I}{c} + mc\right)^2 - \left(\frac{E_I^2}{c^2} - m^2c^2\right) = 2m(E_I + mc^2),$$

and so

$$E_I = (2E^2 - m^2c^4)/mc^2.$$

Now, conservation of energy in the LAB frame takes the form:

$$2E_F = E_I + mc^2,$$

and hence

$$E_F = \frac{1}{2}(E_I + mc^2) = E^2/mc^2,$$

where we substituted for E_I from the preceding equation. Lastly we invoke conservation of 3-momentum in the LAB frame:

$$\mathbf{p_F} = \mathbf{p_I},$$

hence

$$2\left(\frac{E_F^2}{c^2} - m^2c^2\right)^{1/2} \cos\theta = \left(\frac{E_I^2}{c^2} - m^2c^2\right)^{1/2}.$$

Then substitute for E_F and for E_I:

$$\frac{2}{mc^3}(E^4 - m^4c^8)^{1/2} \cos\theta = \frac{2E}{mc^3}(E^2 - m^2c^4)^{1/2},$$

and

$$\cos\theta = \frac{E}{\sqrt{E^2 + m^2c^4}},$$

hence

$$\tan\theta = \frac{mc^2}{E},$$

which is the required result.

14.7.3 Example: *Proton makes head-on collision with photon*

Write down the 4-momentum of a photon of energy ε, travelling in the direction of the unit 3-vector **n** and show that this 4-momentum is a null vector. A proton of mass M, energy E and 3-momentum $(p, 0, 0)$ makes a *head-on* collision with a photon of energy ε, the recoil photon having an energy ε'. Show that

$$\varepsilon(E + cp) = \varepsilon'(E - cp + 2\varepsilon).$$

For a photon, $E = h\nu$ and $p = E/c$, therefore from equation (14.26),

$$\mathbf{P} = \frac{h\nu}{c}(1, \mathbf{n}),$$

and

$$P^2 \equiv \mathbf{P} \cdot \mathbf{P} = \frac{h^2\nu^2}{c^2}(1 - \mathbf{n} \cdot \mathbf{n}) = 0,$$

as $n^2 = 1$ by definition of a unit vector. Before the collision, we have (Figure 14.4):

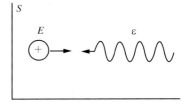

Fig. 14.4 Before the head-on collision between a proton and a photon.

- Proton 4-momentum: $\mathbf{Q} = (E/c, p, 0, 0)$.
- Photon 4-momentum: $\mathbf{P} = (\varepsilon/c)(1, -1, 0, 0)$.

After the collision, we have (14.5):

- Particle 4-momentum: $\mathbf{Q}' = (E'/c, p', 0, 0)$.
- Photon 4-momentum: $\mathbf{P}' = (\varepsilon'/c)(1, 1, 0, 0)$.

From the invariance of $\mathbf{P} \cdot \mathbf{Q}$

$$\frac{\varepsilon}{c}\left(\frac{E}{c} + p\right) = \frac{\varepsilon'}{c}\left(\frac{E'}{c} - p'\right),$$

and from conservation of energy:

$$\frac{E}{c} + \frac{\varepsilon}{c} = \frac{E'}{c} + \frac{\varepsilon'}{c},$$

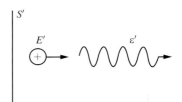

Fig. 14.5 After the head-on collision between the proton and the photon.

which may be rearranged as:

$$\frac{E'}{c} = \frac{E}{c} + \frac{\varepsilon}{c} - \frac{\varepsilon'}{c}.$$

Also, from conservation of momentum:

$$p - \frac{\varepsilon}{c} = p' + \frac{\varepsilon'}{c},$$

which may be rearranged as:

$$p' = p - \frac{\varepsilon}{c} - \frac{\varepsilon'}{c}.$$

Substitute from these two equations for p' and E' into the right-hand side of the equation arising from the Lorentz invariance of $\mathbf{P} \cdot \mathbf{Q}$, thus:

$$\frac{\varepsilon}{c}\left(\frac{E}{c} + p\right) = \frac{\varepsilon'}{c}\left(\frac{E}{c} + \frac{\varepsilon}{c} - \frac{\varepsilon'}{c} - p + \frac{\varepsilon}{c} + \frac{\varepsilon'}{c}\right).$$

Then, multiply both sides by c^2 and we have

$$\varepsilon(E + cp) = \varepsilon'(E - cp + 2\varepsilon),$$

as required.

14.8 *Example:* **Hard-sphere collisions revisited**

An electron moving at high speed strikes another electron, which is at rest, and is scattered through an angle θ, while the target electron moves off at an angle ϕ. Show that two angles satisfy the relationship

$$\tan \theta \tan \phi = \frac{1}{\gamma^2(u)},$$

where u is the speed of either electron in the CM frame. Also show that this reduces to the classical result for $c \to \infty$, and briefly state the effect of relativistic corrections on the joint scattering angle.

In order to interpret the picture of the collision shown in Figure 14.6, in terms of our usual two inertial frames, we may make the identifications $u = U/2$ and $V = u$, and the resulting equivalent picture is shown in Figure 14.7 for frames S (the LAB frame) and S' (the CM frame).

In S', we may resolve the velocity of the scattered particle into components thus:

$$u'_1 = u \cos \theta'$$
$$u'_2 = u \sin \theta',$$

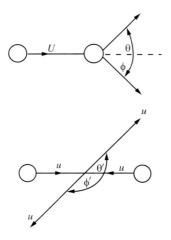

Fig. 14.6 The collision shown in both LAB and CM frames.

and so

$$\tan \theta = \frac{u_2'}{u_1'}.$$

In S, the velocity transformations give the components of the velocity of the scattered particle as

$$u_1 = \frac{u_1' + V}{1 + Vu_1'/c^2} = \frac{u\cos\theta' + u}{1 + Vu\cos\theta'/c^2}$$

$$u_2 = \frac{u_2'}{\gamma(1 + Vu_1'/c^2)} = \frac{u\sin\theta'}{\gamma(1 + Vu\cos\theta'/c^2)},$$

and so

$$\tan\theta = \frac{u_2}{u_1} = \frac{u\sin\theta'}{\gamma(1 + Vu\cos\theta'/c^2)} \cdot \frac{(1 + Vu\cos\theta'/c^2)}{u(1 + \cos\theta')},$$

therefore

$$\tan\theta = \frac{\sin\theta'}{\gamma(u)(1 + \cos\theta')}.$$

Similarly,

$$\tan\phi = \frac{\sin\theta'}{\gamma(u)(1 - \cos\theta')},$$

hence

$$\tan\theta\tan\phi = \frac{\sin^2\theta'}{\gamma^2(u)(1 - \cos^2\theta')}$$

$$= \frac{\sin^2\theta'}{\gamma^2(u)\sin^2\theta'} = \frac{1}{\gamma^2(u)}.$$

For $c \to \infty$, $\gamma(u) \to 1$, and so, in the limit,

$$\tan\theta\tan\phi = 1.$$

From this it follows that

$$\frac{\sin\theta\sin\phi}{\cos\theta\cos\phi} = 1$$

which implies that

$$\cos\theta\cos\phi - \sin\theta\sin\phi = 0,$$

and hence

$$\cos(\theta + \phi) = 0,$$

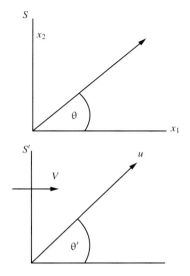

Fig. 14.7 Equivalent picture in our usual inertial frames in standard configuration.

therefore

$$\theta + \phi = \pi/2,$$

as required. Evidently, the effect of relativistic corrections is to reduce the joint scattering angle to values less than $\pi/2$.

14.9 Rockets revisited

In this section we return to our variable mass problem as posed by the rocket problem in Section 8.3, but now we assume that the rocket speeds are large enough that the dynamics must correspond to special relativity. However, we begin by considering the case of a rocket which is propelled by the emission of photons.

14.9.1 *The photon rocket*

A 'photon rocket' uses electromagnetic radiation as its method of propulsion. The rocket, initially at rest and having mass M_I emits photons in one direction until it is moving in the opposite direction with speed V, its mass now being M_F. Write down the initial and final 4-momenta of the rocket and emitted photons and deduce that

$$M_I/M_F = [(c+V)/(c-V)]^{1/2}.$$

From equation (14.3) we have the 4-momentum $\mathbf{P} = (P^0, \mathbf{p})$ of a moving body and specifically for a photon we have $P^0 = |\mathbf{p}|$, which follows from equation (14.25).

Take the x-axis to be the direction of motion: then the 4-momentum of the nth photon emitted *backwards* is

$$\mathbf{P}_{\text{photon}} = (|\mathbf{p}_n|, -|\mathbf{p}_n|, 0, 0).$$

Now the initial 4-momentum (rocket and photons) is given by

$$\mathbf{P}_I = (M_I c, \mathbf{0});$$

while the final 4-momentum (rocket and photons) is just

$$\mathbf{P}_F = \left(M_F c \gamma(V) + \sum_n |\mathbf{p}_n|, M_F V \gamma(V) - \sum_n |\mathbf{p}_n|, 0, 0 \right).$$

Then we invoke conservation of 4-momentum, viz.,

$$P^0: \quad M_F c \gamma(V) + \sum_n |\mathbf{p}_n| = M_I c;$$

and

$$p: \quad M_F V \gamma(V) - \sum_n |\mathbf{p}_n| = 0,$$

which gives

$$\sum_n |\mathbf{p}_n| = M_F V \gamma(V).$$

Substituting back into the equation for the zero-order component:

$$M_F(c + V)\gamma(V) = M_I c,$$

and so

$$M_I = M_F \sqrt{\frac{c + V}{c - V}},$$

as required.

14.9.2 Example: *Conventional rockets*

A rocket accelerates from rest in a straight line, until it attains a speed u. During this time, it ejects gases at a constant speed V_G, relative to the rocket. If there are no external forces acting, show that the differential equation for its speed as a function of its mass is

$$m\frac{du}{dm} + V_G\left(1 - \frac{u^2}{c^2}\right) = 0,$$

where m and u are the mass and the speed of the rocket at any time.

Our method here is to consider energy and momentum balances in the instantaneous rest frame of the rocket, and then to transform these balance equations to its initial rest frame. As usual, we work with inertial frames S and S' in standard configuration, and these are chosen to be:

- $S \equiv$ initial rest frame of rocket;
- $S' \equiv$ instantaneous rest frame of rocket at time t.

We note that the rocket mass is m in both frames.

In S: The rocket has mass m and speed u at time t, and has mass $m + dm$ and speed $u + du$ at time $t + dt$.

In S': The speed of rocket is zero (by definition) but there is a change du' in time interval dt.

We use the velocity transformations of Chapter 12 in order to obtain an expression for du'. At time t, we have

$$u' = \frac{u - V}{1 - uV/c^2},$$

which gives us $u' = 0$ when $u = V$, which is consistent!
    Now, at time $t + dt$,

$$du' = \frac{u + du - u}{1 - u(u + du)/c^2} = \frac{du}{1 - u^2/c^2} + 0(du)^2,$$

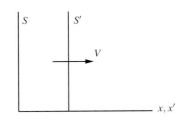

Fig. 14.8 The initial and current instantaneous rest frames of the rocket.

where we neglect the second order of infinitesimals. Next we make use of the conservation laws in S':

Conservation of energy in S' When applied at times t and $t + dt$ this takes the form

$$m\gamma(u')c^2 = (m + dm)\gamma(u' + du')c^2 + dE_G,$$

where dE_G is energy of the emitted gas in S'. By definition, $u' = 0$ therefore $\gamma(u') = 1$. Also

$$dE_G = dm_G\gamma(V_G)c^2.$$

(Note the relativistic relation for the emitted gas which is moving relative to S': in this case, unlike the analysis in Section 8.3, $dm_G \neq -dm$ and we no longer have conservation of mass!) Hence conservation of energy becomes

$$mc^2 = (m + dm)\gamma(du')c^2 + dE_G.$$

Cancelling terms as appropriate, and retaining first order in infinitesimals, this becomes

$$\frac{dE_G}{c^2} = -dm.$$

Conservation of momentum in S' Again, we consider the situation at the successive instants t and $t + dt$ in order this time to invoke conservation of momentum, thus:

$$0 = (m + dm)\gamma(du')du' - \left(\frac{dE_G}{c^2}\right)V_G.$$

We may write this to first order in infinitesimals as

$$mdu' - \left(\frac{dE_G}{c^2}\right)V_G = 0,$$

and, substituting for dE_G/c^2 from the preceding equation,

$$mdu' + dmV_G = 0.$$

Then, from the velocity transformation for du',

$$m\frac{du}{1 - u^2/c^2} + dmV_G = 0,$$

and hence the required result follows after some rearrangement.

14.9.3 Example: *Comparison of conventional and photon rockets*

Prove that the ratio of the initial mass M_I to the final mass M_F of the rocket is given by

$$\frac{M_I}{M_F} = \left(\frac{c + U}{c - U}\right)^{c/2V_G}.$$

Note that this reduces to the result for the 'photon rocket' when we set $V_G = c$. Why was the case of the photon rocket relatively easy? Why, specifically, is it easier to treat the ordinary rocket in the classical limit, without relativistic effects?

From the previous example, with some rearrangement, we have:

$$m\frac{du}{dm} + V_G\left(1 - \frac{u^2}{c^2}\right) = 0.$$

Separate the variables to obtain

$$\frac{dm}{m} = -\frac{1}{V_G}\frac{du}{(1 - u^2/c^2)} = -\frac{c}{V_G}\frac{d\beta}{(1 - \beta^2)},$$

where $\beta = u/c$ and $du = c\,d\beta$. We use partial fractions for the right-hand side, and integrate both sides to find:

$$\begin{aligned}
\ln M_F - \ln M_I &= -\frac{c}{V_G}\int\frac{d\beta}{(1 - \beta)(1 + \beta)} \\
&= -\frac{c}{2V_G}\int\left[\frac{d\beta}{1 + \beta} + \frac{d\beta}{1 - \beta}\right] \\
&= -\frac{c}{2V_G}[\ln(1 + \beta) - \ln(1 - \beta)],
\end{aligned}$$

therefore

$$\ln M_I - \ln M_F = -[\ln(1 + \beta)^{c/2V_G} - \ln(1 - \beta)^{c/2V_G}],$$

or

$$M_I/M_F = \left(\frac{1 + \beta}{1 - \beta}\right)^{c/2V_G} = \left(\frac{c + u}{c - u}\right)^{c/2V_G}.$$

In the case of the photon rocket, we have $V_G = c$ which is the same in both S and S'. In other words, Einstein's second axiom of special relativity! The simplifying factor in the classical case is that forces are invariant under Galilean transformation and so the equation of motion is the same in both S and S'.

14.10 Relativistic treatment of forces

We conclude this chapter with a brief discussion of the way in which the concept of force is accommodated in the relativistic 4-space formulation of mechanics; and follow this by giving the appropriate generalization of Newton's second law to a form involving 4-vectors. To begin with, we consider some of the conceptual difficulties which underlie this process.

At first sight, the appropriate strategy may seem to be obvious. We have, after, all been quite successful in introducing 4-space quantities purely by analogy with the more familiar operations in 3-space. In this way we have introduced the 4-space forms of the velocity, momentum and acceleration for a particle. Surely the next step is to write down a 4-space analogue of N2 in such a way that it reduces to the usual form of N2 when we take the limit of small velocities?

In fact the answer to the question is 'yes'. That is indeed what we shall do; but first we must consider how such a procedure will differ from steps taken earlier. In doing so, we need to remind ourselves that we have operational definitions for quantities like velocity and acceleration of a particle, in that these are quantities which we can measure. That is to say, the very meanings of quantities like displacement, velocity and acceleration are bound up with their measurement. They are, in an obvious sense, primary quantities in mechanics; and other quantities like angular momentum derive their fundamental status directly from them. In contrast, concepts like force and torque, although they have some intuitive basis, rely directly on Newton's second law for their formal definition and hence for their measurement in terms of the primary quantities. Accordingly our concept of force, which we wish to generalize to 4-vector form, relies on a hypothesis about nature, which, as we are about to correct it, cannot be true!

Evidently this poses a problem; but we can get some guidance on how to deal with it by going back to the beginning of this book and reflecting on some of our earlier ideas. For instance, we have from an early stage put more emphasis on the conservation laws of mass, energy, and momentum (including, of course, angular momentum) than on Newton's laws. This is not just because the use of conservation laws gives a more complete picture of the basic physics; and, in addition, often leads to easier methods of solving problems; but also because we believe that they are indeed more fundamental than Newton's second law. They are in fact **null statements**, and if they are found to have been violated even once, then we have to rethink the whole of physics.

Of course, as we have seen from Chapter 2 onwards, N2 is entirely compatible with the conservation of energy and momentum and, moreover, implies the existence of these laws. So this gives us some more guidance. We now wish to postulate a general law which not only defines 4-force, is analogous to N2 and reduces to N2 for low speeds but is also compatible with the Lorentz-invariant conservation laws based on the introduction of 4-momentum and relativistic energy.

$d\tau$ in equation (13.26), we can make the change of time variable in the differentiation as

$$\frac{d}{d\tau} = \frac{dt}{d\tau}\frac{d}{dt} = \gamma(u)\frac{d}{dt},$$

so that

$$K_\alpha = \gamma(u)\frac{d}{dt}[m\gamma(u)c, p_i] = \gamma(u)\left[\frac{1}{c}\frac{dE}{dt}, k_i\right], \qquad (14.37)$$

where we have substituted $m\gamma = E/c^2$ and $k_i = dp_i/dt$. Therefore, we have the 4-force as

$$K_\alpha = \gamma(u)\left[\frac{1}{c}\frac{dE}{dt}, k_i\right], \qquad (14.38)$$

and we see that the power dE/dt—that is, the rate at which the force transfers energy to the particle—is the partner of k_i in the 4-force.

This is as far as we shall take this topic, but it should be noted that this generalization of N2 to a 4-vector form (or perhaps this choice, with the benefit of hindsight, of a Minkowski equation of motion which reduces to the Newtonian form, in the classical limit?) has some interesting aspects. First, just as the classical concepts of the energy and the momentum are linked together in the 4-momentum, so we now find here that the classical force and its rate of doing work are linked in the Minkowski 4-force. Second, the acceleration of a particle is not necessarily in the same direction as the force acting on the particle. Both these aspects are taken further in two exercises at the end of this chapter.

14.11 Exercises

14.1 A stationary atom of proper mass m is struck by a photon of frequency ν, and recoils. If the atom absorbs the photon, what are the speed and mass of the recoiling particle?

14.2 A tachyon is a hypothetical particle which moves faster than light. Show that if the energy of such a particle were to be measurable, then its mass would have to be imaginary.

14.3 A charged particle moves with velocity $\mathbf{u}$ under the influence of a uniform magnetic field $\mathbf{B}$. Show that if the initial direction of the velocity is at right angles to the magnetic field, the orbit of the particle will be a circle; and find its radius.
[Hint: equation (1.17) for the Lorentz force will be needed here.]

14.4 Show that the spatial part of the Lorentz 4-force K^α, as given by equation (14.33), reduces to the Lorentz 3-force, as given by equation (1.17), for time-independent potentials and in the limit of small particle velocities.

14.5 By considering the inner product of the 4-velocity U^α with the Minkowski 4-force K^α, as given by equation (13.36), show that the timelike component of the 4-force can be expressed in terms of the rate of doing work $\mathbf{F} \cdot \mathbf{u}$, where $\mathbf{F}$ is the 3-force and $\mathbf{u}$ is the 3-velocity of a test particle. Hence show that the relativistic energy of the particle is given by $E = m\gamma(u)c^2$, where all the symbols have their usual meanings.

14.6 Show that, according to the relativistic equation of motion (14.35), the force $\mathbf{k}$ and the corresponding acceleration $\mathbf{a}$ are no longer in the same direction unless: either $\mathbf{a}$ is perpendicular to $\mathbf{u}$; or $\mathbf{a}$ is parallel to $\mathbf{u}$. Show that the inertial coefficients in the two cases are $m\gamma(u)$ and $m\gamma^3(u)$ respectively.

15 Towards general relativity

It may be helpful if we begin by pointing out that general relativity could equally well be referred to as Einstein's relativistic theory of gravitation. In other words, just as we have seen that special relativity replaces Galilean relativity, so we now turn our attention to the theory which superseded Newton's theory of gravitation. We have also seen that special relativity reduces to Galilean relativity, provided that the velocities of the bodies involved are very small compared to the speed of light. In the same way, we should note that general relativity reduces to special relativity, provided that we restrict our attention to situations where the effects of the gravitational field may be neglected. We shall discuss this aspect further at the appropriate point. Now we begin with the basic axiom of general relativity. This is nothing more than the principle of equivalence, which we have already met in Chapter 1.

15.1 Principle of equivalence

The statement given in Section 1.4.3 is just one version of the principle of equivalence and we shall presently discuss others. However, it is important to be clear about the difference between a statement of principle on the one hand and an experimental result on the other.

In this case we begin with an experimental result—originally due to Galileo—but since confirmed by many others. This result states that the constant of proportionality in the expression for the acceleration of a body due to an applied force is (in suitable units) the same as the constant of proportionality in the expression for the acceleration of the same body due to gravity, *to within experimental error*. That is, putting this experimental result in the form of an equation, we have:

$$m_I = m_G \pm \delta m,$$

where we use the notation of Chapter 1 for inertial and gravitational mass, and δm is the experimental error involved in measuring the mass of the body.

In the Newtonian picture of the universe, if instances were found where this result did not hold, the basic theory would not be affected: it does not depend on there being an equivalence. However, if we elevate the experimental result to a principle, then we must write down a different equation, thus:

$$m_I \equiv m_G.$$

That is to say, we are asserting that the two properties are exactly the same thing. In this case, a single contrary instance would be enough to destroy the principle.

The situation is exactly analogous to the relationship between the Michelson–Morley experiment and special relativity. In that case, the null result of the Michelson–Morley experiment was elevated by Einstein to the status of a principle and made the basic axiom of special relativity. In the case of general relativity, Einstein elevated another null result—the failure to detect any difference between the inertial and the gravitational mass of a body—to a principle; and, in one form or another, to be the basic axiom of his theory of gravitation.

15.1.1 *Einstein's lift experiments*

In Section 7.1, we saw that the gravitational and inertial forces acting on a body could cancel out in an accelerating frame of reference. This possibility led Einstein to suggest that gravity was in some sense an inertial force. That is, he postulated that the gravitational force was an effect which—like centrifugal force—arose from the use of a non-inertial frame. In a paper which was published in 1914 he discussed a series of thought experiments which are sometimes referred to as the 'lift experiments'. As these experiments are no longer thought experiments, having actually been carried out, we shall think of an experimenter in a rocket ship, rather than in a lift. The essential feature is that the experimenter can only observe the behaviour of a test mass inside the rocket ship. We consider four cases, as illustrated in Figures 15.1–15.4. In each of these cases, we take our inertial frame S to have its origin at the centre of the Earth and its orientation determined by (say) the fixed stars. When the rocket ship is accelerating its frame is denoted by S'_a, in the notation of Chapter 7, and when it is moving at constant speed with respect to S it is our usual second frame S'.

We begin with our rocket ship before take-off, when it is still at rest on the Earth, as shown in Figure 15.1. The experiment consists of releasing the test mass and allowing it to fall to the floor. The experimenter measures its acceleration and finds to be g, the local acceleration due to gravity. No surprises there! The result is just the same in both S and S', as initially both these frames are the same.

Now the rocket ship takes off. At some later time it is a long way from Earth, and is moving with a constant speed relative to the Earth. Under these circumstances, when the observer releases the test mass, it remains where it was released. In both S and S' the test mass has no net forces acting on it. In S it moves uniformly with S', while in S' it is of course at rest.

Next we consider the case where the rocket motors have been switched on, and it accelerates at a uniform rate a. It is still a long way from the

Fig. 15.1 Rocket ship at rest on Earth.

Fig. 15.2 Rocket ship remote from Earth and at rest or in a state of uniform motion.

Fig. 15.3 Rocket ship remote from Earth and accelerating at a rate a.

Fig. 15.4 Rocket ship in free fall towards Earth.

Earth, so that Earth's gravity can still be neglected; and to an observer in S, its floor accelerates up towards the test mass when the experimenter releases it. However, to the observer in S'_a, the mass appears to accelerate towards the floor, as if under the influence of a force of magnitude ma.

Lastly, we imagine that the rocket is now in free fall back to Earth, and is close enough for the Earth's gravitational field to affect it. The situation is depicted in Figure 15.4. To an observer in the Earth's frame S, the experimenter, rocket and test mass all appear to be falling to Earth with an acceleration g. However, to the experimenter in the rocket ship, the test mass remains motionless and he concludes that no forces are acting on it.

The overall conclusion from these experiments must be that the experimenter in the rocket ship is unable to tell whether forces acting on the test mass are due to gravity or to some inertial force. The observer in S can tell, but he or she is in a privileged position, being completely outside the non-inertial frame which is under consideration. Considerations of this kind led Einstein to the idea that perhaps gravity was also an inertial force. Of course in this case the privileged inertial observer has to be outside the universe! This in itself must raise questions about the validity of the concept of an inertial frame and we shall reexamine this question once we have discussed Einstein's postulate of the equivalence of gravity and acceleration.

15.1.2 *Gravitation as an inertial force*

Einstein's conclusions were formulated as the principle of equivalence. The formulation is different from that given in Section 1.4.3, and in order to draw a distinction, we shall refer to the form which we are about to give as PE2 and state it as follows:

PE2: The principle of equivalence states that a frame undergoing constant acceleration is locally indistinguishable from a frame at rest or in uniform motion in a gravitational field.

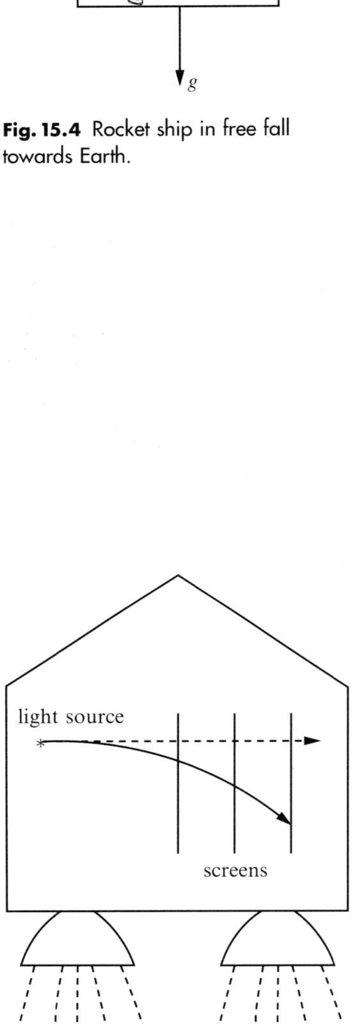

This is known as the *weak* form of the principle of equivalence. We should note the two qualifications, both of which have to do with non-uniformity. We restrict the acceleration to *constant acceleration* and the region to a *local region* over which the gravitational effects have no spatial inhomogeneity.

Einstein took this idea further and suggested that it should also apply to the electromagnetic field. Again it is helpful to consider an experiment, but this time we do an optical experiment. The situation is illustrated in Figure 15.5. A ray of light is shone parallel to the floor of the laboratory in the rocket ship (this is the local definition of the horizontal!) and passes through a series of translucent screens where its passage may be noted by a bright dot on each of the screens. If the rocket ship is in uniform motion (that is, moving with constant velocity in S) the light ray is parallel to the floor and is not affected in any way by the speed of the rocket. However, the situation will be different if the rocket accelerates. Taking the screens to be equispaced and the acceleration to

Fig. 15.5 Optical experiments on a rocket ship: the horizontal dotted line shows a light ray when the ship is in uniform motion; the continuous line shows a light ray when the ship is accelerating.

be constant, the time taken for the light to reach each screen will increase in arithmetic progression 1, 2, 3, ..., whereas the vertical displacements of the screens due to the acceleration will be in geometrical progression 1, 4, 9 .... Thus the light spots on the screens will now lie on a parabola, just like the trajectory of a projectile in a uniform gravitational field. Hence, if the principle of equivalence is to apply to the electromagnetic field, it follows that a light ray should be bent by a gravitational field.

This may seem a rather surprising conclusion. After all, in terrestrial classical physics a light ray is our practical method of establishing a straight line. However, calculation shows that the effect of gravity on a light ray would be too small to observe under normal terrestrial conditions and one has to look for an astronomical effect, where the light from a distant star would be deflected by the gravitational field of the Sun. In order to observe such a deflection, it is necessary to have an eclipse of the Sun. In 1919 a British astronomical expedition to Africa observed a solar eclipse and were able to verify Einstein's quantitative prediction of the amount by which the light rays were deflected.[1]

15.1.3 *The inertial frame revisited*

It was a view held by philosophers such as Bishop Berkeley and later Ernest Mach (1893) that one can attribute no meaning to the concept of motion, as such, but only to relative motion. This was essentially the same idea as that of Galileo (and Newton), but taken to a deeper level. The argument runs that, in an otherwise empty universe, there is simply no way of telling that a body is in motion. The very idea is meaningless.

A body can only be in motion relative to something else. In an otherwise empty space, there is nothing to which the motion can be referred. If we put one other body, however small, in the space then the concept of motion relative to that other body becomes possible.

Mach took the view that it was the presence of the mass of the universe which validated the concept of motion, and that this mass therefore constituted the ultimate inertial frame. As the mass of the universe is mainly concentrated in the fixed stars, then we may regard the fixed stars as constituting the ultimate inertial frame.

Some support for this idea comes from the use of Foucault's pendulum, as discussed in Section 7.2.6. If such a pendulum is set swinging at the North Pole, then on the Newtonian view of absolute space, the pendulum swings in a plane which is fixed, while the Earth rotates underneath it. At the same time, the Earth is rotating relative to the fixed stars. The time taken for the Earth to rotate through 360 degrees can be obtained by comparing a fiduciary mark on its surface with: (a) the fixed plane of rotation of the pendulum; and (b) the fixed stars. The two times are known to be the same, within experimental error.

This result lends support to the hypothesis that Newton's absolute space can be replaced by an inertial frame defined with respect to the fixed stars. For instance, in laboratory experiments we normally take our coordinate system (however implicitly) as being fixed to the Earth. Then, providing the duration and scale of the experiment are such that

[1] The idea that the light from stars could be deflected by the gravitational field of the Sun was not actually new. Previously the corpuscular theory of light was a rival to the wave theory; and, on the assumption that light consisted of a stream of particles, nothing could be more natural than the idea that such particles would be affected by a gravity field. On this basis, predictions of the deflection of the light from a star go back to the beginning of the 19th century.

the stars apparently remain fixed, our empirical frame of reference should be a perfectly adequate inertial frame.

This pragmatic concept can be formalized in the shape of a **local inertial frame**. While it is not possible to shield a body from the gravitational field, we can eliminate gravitational effects by restricting our attention to a frame which is in *free fall* in a region where the spatial variation of the gravitational field is too small to be measured. In addition, in order to exclude any effects from inertial forces the frame must not be rotating. Thus, under these restrictions, we have an inertial frame in which special relativity is **locally** valid. That is, its validity holds only for the local region where we may neglect spatial variations of the gravitational field.

This concept leads to a further restatement of the principle of equivalence, which we shall denote by PE3, thus:

PE3: No local experiment can distinguish between the free fall of a body in a gravitational field and the uniform motion of the same body in the absence of a gravitational field.

This statement is the basis of general relativity and is very profound. Yet it is no more than a statement of the results of the lift experiments. Its form should be compared with the similar statements which constitute the corollary to the principle of Galilean relativity in Section 1.8 and the corollary to the principle of special relativity in Section 11.2.

15.2 Curved space

In Section 15.1.2, we saw that the principle of equivalence leads us to the conclusion that a light ray will be deflected by a gravitational field. As we have also seen, this effect has been confirmed experimentally. Since our only practical method of defining straight lines is by the use of light rays, we are therefore led inevitably to the conclusion that, in the presence of a gravitational field, the shortest distance between two points is a curved line and so space itself must be curved. As special relativity has taught us that space and time are inextricably linked, this must further imply that the four-dimensional space-time continuum is intrinsically curved in the presence of a gravitational field. We shall see that this inevitably holds consequences for time as well as space. First we shall discuss some other consequences.

15.2.1 *The necessity for a curved metric*

At this stage it is helpful to return to the ideas of Section 13.10, where we introduced the metric of a vector space and also the subject of geodesic curves. Suppose that, just as in Section 13.10.3, we investigate the geometric properties of the space-time continuum by measuring the vertex angles of a triangle, in order to find out whether or not they show the Euclidean property of adding up to 180 degrees. Obviously such an experiment would require us to work on astronomical scales and the

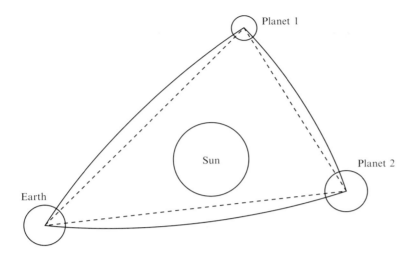

Fig. 15.6 Triangulation of space in the solar system.

planets provide an obvious choice for distant objects with which to define a triangle.

In principle we could use standard surveying techniques. (Needless to say, this must be a *Gedanken* experiment, as we shall require in addition to the observer on Earth, one observer to be on Planet 1 and another to be on Planet 2.) Since the rays of light will be bent by the gravitational field of the Sun, the situation will be as shown in Figure 15.6. The dotted lines are the Euclidean straight lines which would be traced out by the light rays in the absence of the Sun. The continuous lines are the paths taken by the light rays in the presence of the Sun and it should be noted that their shape is determined by the fact that the rays will be deflected towards the Sun. As we noted earlier, when we discussed the geometry of spherical triangles in Section 13.10.3, a triangle with vertex angles which add up to greater than 180 degrees is on a surface which possesses positive curvature. In the present case, we may draw an analogous conclusion that the triangle defined by taking the three planets as its vertices has angles which add up to greater than 180 degrees, and hence is embedded in a space with positive curvature.

In these circumstances, the light rays take the shortest distances between points by travelling along the geodesics of the curved three-dimensional (3D) space. In the curved $4D \equiv (1 + 3)D$ space,[2] their world lines are the null geodesics, in accordance with the principles of special relativity. It should be noted that in this context the concept of Euclidean straight lines is literally without meaning.

The geodesics of both Euclidean 3D space and Minkowski $(1 + 3)D$ space are straight lines and it may be shown that their corresponding metric tensors are flat. From differential geometry, it is known that the curvature of a vector space resides in its metric. We shall expand on this topic a little when we discuss curved vector spaces. But for the moment, we should note that the implication of the principle of equivalence is that the space-time continuum should be represented by a vector space with a non-flat metric.

[2]The right-hand side of the identity is known as the **signature** of the space. We shall give a more general statement of this in Section 15.2.3.

15.2.2 *Equivalence of a non-inertial frame and curved space*

Consider now experiments which are done, like the lift experiments, in a non-inertial frame, but this time in a rotating frame rather than in a linearly accelerating one. For this purpose, we may envisage a laboratory on a rotating turntable. We denote its reference frame by S'_{rot} relative to the local inertial frame S, which is fixed to the Earth. In this case, the experiment of dropping a mass to the floor in S'_{rot} will just be essentially the same thing as doing the identical experiment in S. However, an observer in the laboratory frame S'_{rot} will note that a ball set on the floor at the centre of the turntable will tend to roll out radially to the wall of the laboratory.

To an observer in frame S, this is just an example of the effects of centrifugal force. But, to a poorly informed observer in S'_{rot}, it would appear that the ball was experiencing a repulsive central force. In particular, it would seem to be a gravitational force which repelled a body rather than attracting it. The obvious question now arising is whether the inertial force is associated with a curvature of space. This is indeed the case, but it is rather more difficult to demonstrate than in the lift experiments, where an accelerating lift gave rise to a parabolic ray of light in the non-inertial frame of the lift. In the rotating frame, in order to find out what is going on, we have to invoke special relativity, rather than just rely on classical mechanics.

We now wish to test the local space of the turntable, to find out whether it is Euclidean or not. As in the previous cases, we construct a triangle and measure its angles. The situation is as illustrated in Figure 15.7, where we look down on the rotating turntable from above.

The first point to note is that both observers—in S' and S'_{rot}, respectively—agree that the diameter of the turntable is unaffected by the rotation.

Moreover, if the turntable is at rest in S, they also agree on the distances AB, BC and CA. In order to do this, they agree on a measuring rod (i.e. the unit of length) and both then agree on the number of units

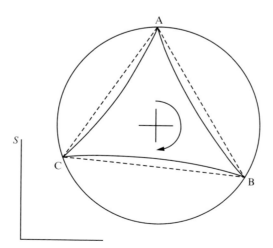

Fig. 15.7 Triangulation of space on a rotating turntable.

making up the length AB, and so on. In addition, if the frame S is locally Euclidean, they also agree that the line segments AB, BC and CA are straight lines.

However, if the turntable is rotating, the measuring rod belonging to the observer in S'_{rot} will suffer a Lorentz–Fitzgerald contraction, the magnitude of the reduction in its length depending on its position and orientation in S'_{rot}. If this observer tries to measure the length of the line segment AB, say, then his problem is that his measuring rod becomes shorter as he moves from the point A to the centre of the line segment and then starts to increase in length as he moves out again towards the point B. Obviously this is a variational problem of how to measure the line of stationary length between the points A and B in S'_{rot}. If the observer moves the measuring rod radially towards the centre of the turntable, then he is minimizing the effect of the Fitzgerald contraction (to move outwards would make matters worse!), so that we may expect the geodesic in this problem to be curved towards the centre, as shown in the figure.

If we now consider the use of optical methods, then we may conclude that, as the 'gravity' observed in the rotating frame is repulsive, the ray of light would be bent away from the centre (the opposite of the case where the rays of light were bent by the gravitational field of the Sun), as illustrated. However, perhaps the best evidence for a non-Euclidean space in the rotating frame comes from the measurement of π from the circular turntable. When this is rotating, the observer on the turntable will find that his measuring rod is suffering the maximum contraction as he measures the circumference, while being unaffected (at least as regards its length) when he measures the diameter. Accordingly, he finds the diameter to have the same number of units as the observer in S, but that the circumference has a greater number of units. Accordingly, in S'_{rot} the ratio of the circumference to diameter of a circle, that is π, is larger than in S.

15.2.3 Signature of a vector space

Consider a vector space with dimension $d = n$, as discussed in Section 13.1. The position vector $\mathbf{s}$ is an ordered set of the n numbers $x_1, x_2, \ldots, x_n$. Suppose that the inner product discussed in Section 13.1 is now found to take the form

$$\mathbf{s} \cdot \mathbf{s} = (-x_1^2 - x_2^2 - x_3^2 - \cdots - x_m^2 + x_{m+1}^2 + x_{m+2}^2 + \cdots + x_n^2), \quad (15.1)$$

where an ordered subset containing m vectors makes negative contributions to the inner product and an ordered subset of $p = n - m$ vectors makes only positive contributions to the inner product. Then the two numbers m and p are said to be the signature of the space, thus:

$$\text{signature} \equiv (m + p)\text{D}.$$

Obvious examples are:

- signature of Eucledean 3-space: $(0 + 3)D \equiv 3D$
- signature of Minkowski 4-space: $(1 + 3)D$.

These are the only examples which will be of interest here, but it is worth mentioning that even when we include curvature, the space-time continuum will still be represented by a vector space with signature $(1 + 3)D$. This notation merits this brief discussion here as, not only is it used in works on general relativity, but it is also becoming current in other areas of mathematical physics.

15.3 Curved vector spaces

Many of the basic ideas of general relativity are really quite simple, but the mathematical aspects of the theory can be quite intimidating. In this section we shall give the most abbreviated of accounts of the relevant geometry of abstract vector spaces. Our emphasis will be on explaining the significance and meaning of various important quantities, rather than on giving derivations or detailed mathematical treatments. Yet, nevertheless, we shall hope to achieve sufficient degree of completeness, in order to make the general idea intelligible.

When Einstein began to formulate the ideas that led to general relativity, he found that the necessary mathematics was already waiting for him. Riemann had generalized the work of Gauss and others on curved surfaces in three-dimensional space to spaces with any number of dimensions. Indeed, Riemann himself had speculated that curved trajectories, for example, of planets round the Sun, might be due to some underlying curvature of three-dimensional space. Although this did not work, it is now accepted that the same idea, when applied to $(1 + 3)D$ **space-time** does work.

What we have to keep in mind is that we are dealing at all times with the analytical geometry of abstract vector spaces. What then concerns us is the degree of correspondence between the premises and predictions associated with such a space and that of the real world in which we live. For example, the Euclidean 3D space is in good correspondence with geometric operations in the classroom or laboratory. If we go to a $(1 + 3)D$ space, and it is left uncurved (i.e. Minkowski space), then we shall expect a good correspondence with those aspects of the physical world which are described by special relativity. On the other hand, if we work with a $(1 + 3)D$ space and curvature is admitted, then so is gravity and we have a good correspondence with those aspects of the physical world which are described by general relativity.

15.3.1 Geometry of space curves in 3D

If we are to formulate a theory relating gravity to the curvature of space then, as we shall certainly wish to know the magnitude of the gravitational force in given circumstances, this implies the need to quantify the degree of curvature of space. In order to introduce the topic of curvature, we shall begin with simple line segments, before going on to general vector spaces.

The idea of curvature is based on the circle and may be generalized to other types of curve. We know that a circle is characterized by its radius. If we consider an infinitesimal segment of any other curve, then to an approximation it may be regarded as an arc of a circle and the radius of this circle is taken to be the **radius of curvature** of the curve at that point.

The **curvature** of the curve at a point is then defined as the **reciprocal of the radius of curvature** at that point.

To take a simple but important example, a straight line has an infinite radius of curvature and hence possesses zero curvature. If we extend these ideas to two-dimensional surfaces, then we can form elementary arcs of circles in two mutually perpendicular coordinate directions. If the curvature is zero in both directions, then the surface is said to be flat.

Further extension to an arbitrary number of dimensions brings with it impossible problems of visualization, but we can still speak of an n-dimensional hypersurface as being flat when, by means of analytical tests, it is found to have zero curvature.

Now we turn to the topic of these analytical tests. Let us consider an arbitrary curve $\mathbf{x} = \mathbf{x}(t)$ in Euclidean 3D space. Here t could be the time and $\mathbf{x}(t)$ a particle trajectory. As the parameter t varies over a given interval, we generate the curve $\mathbf{x(t)}$. The situation is as illustrated in Figure 15.8.

If the variable s represents arc length along the curve, then it can be shown that the unit tangent vector $\mathbf{u}$ at the point P is given by

$$\mathbf{u} = \frac{\mathrm{d}\mathbf{x}}{\mathrm{d}s}. \tag{15.2}$$

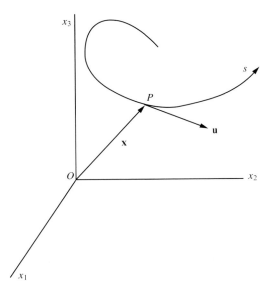

Fig. 15.8 Geometry of a curve in 3D space.

It can be further shown that if we differentiate again, and introduce the radius of curvature ρ, we may write

$$\frac{d\mathbf{u}}{ds} = \frac{d^2\mathbf{x}}{ds^2} = \frac{\mathbf{n}}{\rho}, \tag{15.3}$$

where $\mathbf{n}$ is the unit vector perpendicular to the tangent at the point P. Equally one can write the expression for the radius of curvature at P as:

$$\frac{1}{\rho} = \sqrt{\left(\frac{d^2 x_1}{ds^2}\right)^2 + \left(\frac{d^2 x_2}{ds^2}\right)^2 + \left(\frac{d^2 x_3}{ds^2}\right)^2}. \tag{15.4}$$

We shall not be making quantitive use of these expressions, but it is worth noting that the **amount of curvature** depends on the magnitude of second derivatives of the curve with respect to distance along it. This essential point should be borne in mind when we consider the curvature tensor in the next section.

15.3.2 *Curvature in (1 + 3)D*

We have seen in the preceding section that the determination of the curvature of a surface involves the variation of the surface coordinates x_i with distance s along the surface (in some particular coordinate direction). Accordingly, for multi-dimensional spaces, it may be seen as inherently plausible that we should consider the expression for the metric, as given in Section 13.10.2. In particular, as we are only concerned here with $(1 + 3)$D space, we shall remind ourselves of the particular form of the metric as given by (13.48), viz.,

$$ds^2 = g_{\mu\nu}dX^\mu dX^\nu, \tag{15.5}$$

where we are using the 4-space notation introduced in Section 13.4 and each Greek index runs over 0, 1, 2, 3.

As the $\{X^\mu\}$ are just the coordinates of the space, the structure or nature of the space must be contained in the metric tensor $g_{\mu\nu}$. We have seen in Section 13.10.2 that the metric tensor takes the simple forms $g_{\mu\nu} = \text{diag}(1, 1, 1)$ for Euclidean 3D space and $g_{\mu\nu} = \text{diag}(-1, 1, 1, 1)$ for Minkowski $(1 + 3)$D space. We may guess that as these are **flat** spaces, with only constant terms in the metric, all differentials would vanish. In contrast, curvature would be associated with a metric which depends on position (i.e. on the coordinates), and we may emphasize this by rewriting (15.5) as

$$ds^2 = g_{\mu\nu}(\mathbf{X})dX^\mu dX^\nu, \tag{15.6}$$

as a necessary condition for the associated vector space to possess non-zero curvature.

What we need to do now is to introduce a rather formidable looking entity called the **Riemann tensor**; or, alternatively, the **curvature tensor**. We shall do this in easy stages; as follows.

First we wish to obtain an analogue of equation (15.3) for the curvature of a $(1 + 3)$D space. There are two considerations to be borne in mind. First, just as vector differentiation is more complicated than scalar differentiation, so too is tensor differentiation more complicated than vector differentiation. The need to preserve certain transformation properties constrains the final result. Second, it is not actually the metric tensor which we differentiate, but something else[3] which depends on it. The result is the introduction of the so-called Christoffel symbol of the second kind, which we shall write in modern notation:[4]

$$\Gamma^{\mu}_{\alpha\beta} = \frac{1}{2} g^{\mu\nu} \left(\frac{\partial g_{\nu\alpha}}{\partial X^{\beta}} + \frac{\partial g_{\nu\beta}}{\partial X^{\alpha}} - \frac{\partial g_{\alpha\beta}}{\partial X^{\nu}} \right). \qquad (15.7)$$

Given that the metric tensor is symmetric (in its indices), or that

$$g^{\mu\nu} = g^{\nu\mu},$$

it follows, by construction, that the Christoffel symbol is symmetric in its lower indices, or:

$$\Gamma^{\mu}_{\alpha\beta} = \Gamma^{\mu}_{\beta\alpha}, \qquad (15.8)$$

and the reader should verify that point. It is also instructive to note that as each index can take four values, the Christoffel symbol has $4 \times 4 \times 4$ elements. However, the above symmetry means that if we represent the $\alpha\beta$ elements for any fixed μ by a 4×4 matrix, then the elements above the diagonal are equal to those below, reducing the $\alpha\beta$ array from 16 to 10 *independent* elements and, taking into account all values of μ, the full array becomes $4 \times 10 = 40$ independent elements.

Now we are ready to state the defining form of the Riemann tensor which contains all the information needed about the intrinsic curvature of a vector space. Again, its form arises from the operations of (covariant) differentiation of tensors and it is defined in terms of Christoffel symbols as:[5]

$$R^{\alpha}_{\mu\rho\sigma} \equiv \frac{\partial}{\partial X^{\rho}} \Gamma^{\alpha}_{\sigma\mu} - \frac{\partial}{\partial X^{\sigma}} \Gamma^{\alpha}_{\rho\mu} + \Gamma^{\alpha}_{\rho\beta}\Gamma^{\beta}_{\sigma\mu} - \Gamma^{\alpha}_{\sigma\beta}\Gamma^{\beta}_{\rho\mu}. \qquad (15.9)$$

The reader should note the inner products in the latter two terms on the right-hand side due to the repeated index β. Indeed, it is always sensible with tensor expressions to do the detailed 'book keeping' and check that labelling indices balance across an equation and that dummy indices tie up in pairs.

The curvature tensor is a formidable object but we can domesticate it to some extent. First, there is the question of taxonomy. A scalar is a tensor of rank zero, a vector is a tensor of rank one, a matrix has rank two and, by a process of induction, the Riemann curvature tensor is of rank four.

As each of its indices takes one of four values, this means that the Riemann tensor has $4 \times 4 \times 4 \times 4 = 256$ elements. There are various symmetries and these can be used, just as for the Christoffel symbol

[3]The Langrangian.

[4]The traditional notation in differential geometry is based on curly brackets but capital gamma now seems to be used in general relativity, and is also known as the 'metric connection'.

[5]A more compact notation is usually employed for the partial differentials but as we will not be carrying out any extensive manipulations there would be no point in risking confusion by a change of notation.

above, to establish the number of independent elements as just 20. Even so, it may be seen that one feature of this theory which makes life difficult is just the sheer number of terms!

15.3.3 *The Einstein tensor*

Further simplification can be made, our ultimate goal being the Einstein tensor. Again, we proceed in stages. First we introduce the **Ricci tensor**, which is defined in terms of the Riemann tensor (as given by (15.9)); thus:

$$R_{\mu\sigma} = R^{\alpha}_{\mu\alpha\sigma}. \tag{15.10}$$

That is, we set the upper index of the Riemann tensor equal to the middle lower one, and sum over the repeated indices. This operation is known as a **contraction** of tensor indices. It can be shown from the properties of the metric and Riemann tensors that the Ricci tensor is symmetric, or:

$$R_{\mu\sigma} = R_{\sigma\mu}. \tag{15.11}$$

We can then use this tensor to derive the **curvature scalar** R, which is defined as

$$R = g^{\mu\sigma} R_{\mu\sigma}; \tag{15.12}$$

and from these two quantities we define the **Einstein tensor** $G_{\mu\nu}$, thus:

$$G_{\mu\nu} = R_{\mu\nu} - \frac{1}{2} g_{\mu\nu} R, \tag{15.13}$$

where the choice of the letter G for the tensor stands for **gravity**.

The Einstein tensor is also symmetrical, thus

$$G_{\mu\nu} = G_{\nu\mu}. \tag{15.14}$$

At this stage we should perhaps remind ourselves that the Einstein tensor has been obtained by various combinations of contractions of the Riemann curvature tensor and in turn that this latter quantity depends on partial derivatives of the metric tensor with respect to the basic coordinates of the $(1 + 3)$D space which we are considering. Our next step is to obtain a governing equation which we can solve for the Einstein tensor $G_{\mu\nu}$.

15.4 The field equations of general relativity

In arriving at possible field equations for general relativity, it is reasonable to look for a formulation which reduces to the Newtonian result in the limit of weak fields. As we shall see, the Einstein field equations, dealing as they do with curvature, involve second derivatives of the metric. Thus, although they are very much more complicated than the Poisson–Laplace equations, they are of a similar type, with the metric playing the part of the potential.

15.4.1 Einstein field equations for gravity

In Section 1.3.1, we introduced the Newtonian field equations for gravity. We may recall that in equation (1.39) we derived a Poisson equation in the form

$$\nabla^2 U = 4\pi G \rho(\mathbf{x}), \tag{15.15}$$

where U is the gravitational potential, G is the gravitational constant and $\rho(\mathbf{x})$ is the local mass density. For the case where space is empty (the vacuum), this reduces to Laplace's equation, thus:

$$\nabla^2 U = 0. \tag{15.16}$$

These equations fix the form of the Newtonian gravitational potential and our goal now is to find analogous governing equations which determine the form of the Einstein tensor.

As we have said, it is a reasonable requirement that such equations should reduce to the Newtonian forms in the limiting case of weak fields, and so our arguments here will be based on an analogy between the two cases. We shall sketch the barest outlines of such an argument here. The underlying idea is that we should concentrate on non-uniform gravitational effects. That is, where the gravitational potential is not constant, but varies with spatial position.

We begin with the Newtonian case. The experiment that we shall consider is where two identical particles of unit mass, are in free fall under gravity. Initally their trajectories are parallel and separated by a small vector $\boldsymbol{\lambda}$. If they enter a region where the gravitational field is non-uniform on a scale $\lambda = |\boldsymbol{\lambda}|$, then their trajectories will either converge or diverge, and $\boldsymbol{\lambda}$ will depend on time. The situation is as illustrated in Figure 15.9.

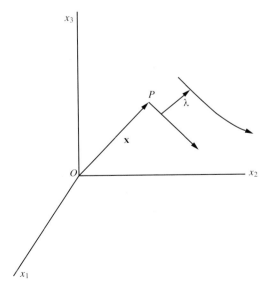

Fig. 15.9 Deviation of particle trajectories due to non-uniform gravitational potential.

Under these circumstances, it is easily shown that the equation of motion for λ takes the form:

$$\ddot{\lambda}_i = -\lambda_j \frac{\partial}{\partial x_i} \frac{\partial}{\partial x_j} U(\mathbf{x}), \tag{15.17}$$

where the double dot signifies double differentation with respect to time. Or, introducing a tensor K_{ij}, such that

$$K_{ij} = \frac{\partial}{\partial x_i} \frac{\partial}{\partial x_j} U(\mathbf{x}), \tag{15.18}$$

then the equation for the Newtonian deviation of trajectories can be rewritten as:

$$\ddot{\lambda}_i + K_{ij}\lambda_j = 0. \tag{15.19}$$

From equation (15.15) we have the general condition on the trace of K_{ij}, that

$$K_{ii} = 4\pi G\rho(\mathbf{x}), \tag{15.20}$$

Or, in the case of empty space, from (15.16) we have

$$K_{ii} = 0, \tag{15.21}$$

the Newtonian vacuum field equation.

Now we extend the idea to general relativity and to $(1 + 3)$D space. Here particles move along world lines which are timelike geodesies. Accordingly, gravitational non-uniformity corresponds to curvature and hence implies geodesic deviation for two identical test particles on initially parallel trajectories. Hence an analysis similar to the Newtonian one leads to the result

$$\frac{\mathrm{D}^2\lambda^\alpha}{\mathrm{D}\tau^2} + \dot{X}^\mu \dot{X}^\nu R^\alpha_{\mu\nu\sigma}\lambda^\sigma = 0, \tag{15.22}$$

where a dot here denotes differentiation with respect to proper time τ. The capital D denotes an absolute differential but we need not be concerned about that here. We shall merely take this to be the equivalent of $\ddot{\lambda}^\alpha$ in the geodesic case and this will be our basis for comparison with equation (15.19) for the Newtonian case.

In order to facilitate such a comparison, we introduce the tensor

$$K^\alpha_\sigma = \dot{X}^\mu \dot{X}^\nu R^\alpha_{\mu\nu\sigma}, \tag{15.23}$$

so that equation (15.22) for the geodesic deviation may we written as

$$\frac{\mathrm{D}^2\lambda^\alpha}{\mathrm{D}\tau^2} + K^\alpha_\sigma\lambda^\sigma = 0. \tag{15.24}$$

Now, if we pursue our analogy, the Newtonian vacuum field equations arose from the trace of K_{ij} vanishing. Accordingly, we postulate that the

Einstein vacuum field equations arise from the analogous step:

$$K_{\alpha\alpha} = \dot{X}^\mu \dot{X}^\nu R^\alpha_{\mu\nu\alpha} = 0. \tag{15.25}$$

As this condition should hold for arbitrary $\dot{\mathbf{X}}$, it further implies that[6]

$$R^\alpha_{\mu\alpha\nu} = 0, \tag{15.26}$$

which implies the vanishing of the Ricci tensor, viz.,

$$R_{\mu\nu} = 0, \tag{15.27}$$

and in turn the vanishing of the Einstein tensor

$$G_{\mu\nu} = 0. \tag{15.28}$$

Equation (15.28) is the general statement of the vacuum field equations of general relativity, where the form of the Einstein tensor $G_{\mu\nu}$ is given by equation (15.13).

15.4.2 The full field equations of general relativity

In order to complete our analogy between the Newtonian theory of gravitation and general relativity, we wish to find the full field equations which would correspond to putting a source term on the right-hand side of (15.28).

In the Newtonian case, the source in Poisson's equation (1.39) involves the mass density in 3D space. From special relativity, we have the mass-energy equivalence, so that we might conjecture that an energy density would be relevant. Further, and also from special relativity, we know that energy is inextricably linked with momentum, in the 4-momentum P^μ. Accordingly we introduce a **density** of 4-momentum through the volume integral

$$P^\mu = \int_V T^{\mu\nu} \, dV_\nu, \tag{15.29}$$

where it should be borne in mind that this is a volume in 4-space, and that the energy–momentum density $T^{\mu\nu}$ is a tensor, which is also known as the **matter tensor**.

Accordingly we can, with some plausibility, take our source for the field equations, when space is not empty, to be proportional to the matter tensor; and equation (15.28) now becomes generalized to

$$G_{\mu\nu} = \kappa T_{\mu\nu}, \tag{15.30}$$

where κ is some constant of proportionality. Alternatively, if we express the Einstein tensor in terms of the Ricci tensor, then these equations take what is perhaps their most usual form, viz.,

$$R_{\mu\nu} - \tfrac{1}{2} R g_{\mu\nu} = \kappa T_{\mu\nu}, \tag{15.31}$$

as the full field equations of general relativity.

[6]We use the antisymmetry of the Riemann tensor in its last pair of indices here.

The constant κ is known as the **coupling** constant. It can be shown that equation (15.31) reduces to equation (1.39) for the limit of weak fields, provided that

$$\kappa = 8\pi G/c^4. \qquad (15.32)$$

15.5 Solving the field equations

Einstein's basic idea in general relativity was to determine the metric of the space-time continuum by means of assumptions about the distribution and velocity of matter. In this brief treatment, we shall only be able to discuss this process for the vacuum.

15.5.1 *The Schwarzschild metric*

Formally we may pose the problem of solving the vacuum field equation (15.28) as an initial value problem; just as we did with N2 in Section 1.1.2. Repeating (15.28) here for convenience, viz.,

$$G_{\mu\nu} = 0,$$

we should remind ourselves that the Einstein tensor depends on the metric tensor, up to and including its second derivatives. Accordingly, we wish to solve equation (15.28) for the metric tensor, given its values, and those of its first derivatives at some initial time.

In practice, solution of the Einstein field equations involves many technicalities and we shall not pursue them here. However, in principle, we could employ a standard technique, as follows.

The Einstein tensor contains second-order derivatives of the metric. By repeated differentiation, we can obtain derivatives of all orders, and these can be expressed in terms of the metric tensor $g_{\alpha\beta}$ and its first derivatives at the initial time. In this way we may make use of Taylor series to develop a general expression for the metric as a power series in time.

The first non-trivial solution of the Einstein field equations, for the simplest case of spherical symmetry, was obtained by Schwarzschild in 1916. He obtained the following expression for the invariant line element:

$$c^2 \, d\tau^2 = \left(1 - \frac{2\Lambda}{r}\right) dt^2 - \left(1 - \frac{2\Lambda}{r}\right)^{-1} dr^2 - r^2 d\theta^2 - r^2 \sin^2\theta \, d\phi^2,$$

$$(15.33)$$

where (r, θ, ϕ) are the usual spherical polar coordinates and Λ is a constant. This is the Schwarzschild line element and comparison of this expression with that given in equation (15.5) allows one to deduce the corresponding metric tensor.

In order to assess a theory of gravitation, one needs a test particle of mass M, say. It can be shown that Λ is simply the mass of the test particle expressed in relativistic units; or

$$\Lambda = GM/c^2, \tag{15.34}$$

and is sometimes called the **geometric mass**. With the speed of light squared appearing in the denominator, it is clear that Λ will normally be a small quantity.

We can use the smallness of Λ to assess the effect of curved space on the radial distance between events. If we take these to be r_A and r_B, at constant t, θ and ϕ, equation (15.33) implies for the purely spatial part

$$dS^2 = \frac{dr^2}{(1 - 2\Lambda/r)}. \tag{15.35}$$

Then, taking the square root of both sides, and integrating, we find the radial distance S_{AB} between r_A and r_B to be:

$$S_{AB} = \int_{r_A}^{r_B} \frac{dr}{\sqrt{(1 - 2\Lambda/r)}} = \int_{r_A}^{r_B} dr \left[1 + \frac{\Lambda}{r} + O(\Lambda^2) \right], \tag{15.36}$$

from which

$$S_{AB} \simeq (r_B - r_A) + \Lambda \ln(r_B/r_A), \tag{15.37}$$

where we have made use of the fact that Λ is small.

This shows us the effect of a curved metric on the apparently simple question of the radial distance between two points. However, if we take the limit as $r \to \infty$, in equation (15.33) for the Schwarzschild metric, then the expression for the line element reduces to

$$c^2 d\tau^2 = dt^2 - dr^2 - r^2(d\theta^2 + \sin^2 \theta \, d\phi^2), \tag{15.38}$$

and we have recovered the flat Minkowski metric, as specialized to spherical polar coordinates. Accordingly, the spherically symmetric solution of the Einstein field equation for the vacuum is **asymptotically flat**.

15.5.2 *The matter tensor*

So far we have been considering the effect of a curved metric. That is, we have been concerned with the second half of the classic summary of general relativity, viz.,

... space tells matter how to move.

Now we should consider the first half of that statement, viz.,

Matter tells space how to curve ...

and this brings us to the full field equations of general relativity in the form of equation (15.31). This is analogous, in classical field theory, to going from the homogeneous Laplace's equation to the inhomogeneous Poisson equation.

The problem now becomes two-fold (to put it at its simplest level!): (a) choosing a form for the matter tensor $T^{\mu\nu}$; (b) solving the field equations for the metric tensor. For convenience, we repeat equation (15.30) here, thus:

$$G^{\mu\nu} = \kappa T^{\mu\nu}.$$

The first step in the problem is to choose an appropriate form or model for the matter tensor $T^{\mu\nu}$. This really takes us into a major area of research, extending on into cosmology and how one models the universe. We shall merely give two very simple models here, so that the reader is left with at least some idea of what might be involved.

First let us consider a **dust**. This is an assembly of particles which have straight, mutually parallel world lines. If the velocity of each dust particle is **V**, then the corresponding matter tensor is just

$$T^{\mu\nu} = \rho V^\mu V^\nu, \tag{15.39}$$

where ρ is the (proper) mass density.

Second, for a **stationary gas**, the matter tensor is given by

$$T^{\mu\nu} = \text{diag}(\rho, p, p, p), \tag{15.40}$$

where ρ and p are the mass density and the gas pressure.

We could go on to include the effects of electromagnetic radiation, motion of a gas and even more complicated distributions and states of matter. But we have made as much of a point as is appropriate to this short account and we shall stop there!

15.6 The predictions of general relativity

Tests of general relativity can, to some extent, be regarded more as tests of the general postulate, embodied in the principle of equivalence, rather than of the precise form of the field equations. We have already mentioned the experimental confirmation of the bending of a light beam in a gravitational field in Section 15.1.2. After the first spectacular support for Einstein's ideas, later measurements have varied to quite an extent, indicating that there is considerable observational error, but at least the spread of measurements is still consistent with Einstein's prediction.

In this section we shall conclude with a brief examination of some of the other consequences of general relativity.

15.6.1 *Precession of the planetary orbits*

In Chapter 3, we saw that the orbit of a planet (say) was obtained by eliminating the time as a variable so that one was left with an orbital equation which could be solved for the locus of points corresponding to a permitted trajectory. It is easily shown that if the orbital equation (3.56) is specialized to the case of an attractive inverse-square law, it takes the form

$$\frac{\mathrm{d}^2 u}{\mathrm{d}\phi^2} + u = \frac{k}{h^2}, \tag{15.41}$$

where (r, ϕ) are the usual polar coordinates, $u = 1/r$ is a convenient transformation (as discussed in Chapter 3), k is a constant and h is the angular momentum (per unit mass) of the planet. The corresponding orbit is an ellipse with the Sun at one focus. The motion, as we saw in Section 3.2.1, is planar, and angular momentum is conserved.

In general relativity, the planet moves along a timelike geodesic, and its orbit is a projection on a surface corresponding to $t = $ constant. As the gravitational field of the Sun is spherically symmetric the Schwarzchild metric, along with its associated geodesics, is an appropriate choice. It can be shown that the orbit in 3D space remains planar and angular momentum is still conserved. However, the detailed calculation would take us beyond the scope of this book, so we merely quote the result for the orbital equation, which now becomes:

$$\frac{\mathrm{d}^2 u}{\mathrm{d}\phi^2} + u = \frac{\Lambda}{h^2} + 3\Lambda u^2, \tag{15.42}$$

where Λ is the 'geometric mass' which appears in the Schwarzschild expression for the invariant interval as given by equation (15.33). For small Λ, this equation may be solved as a perturbation of the Newtonian case, and the orbit is still found to be approximately elliptical. However, the ellipse is no longer closed, and its major axis rotates, so that the value of ϕ corresponding to its position of closest approach increases with each orbit.

In the solar system, the effect is largest for Mercury. Even allowing for many-body effects due to other planets, it was a long-standing puzzle in astronomy, known as the 'advance of the perihelion of Mercury', that lacked an explanation. Einstein's theory supplied the explanation, as it predicted this precession to within observational error.

15.6.2 *The gravitational redshift*

Let us return to the 'lift experiments', and consider another optical experiment. This time we consider the case where the rocket ship is in free fall towards Earth, and the light ray is perpendicular to the floor of the laboratory.

Let us assume that we have a light source, with proper frequency ν_0, situated on the floor of the rocket and shining on the ceiling, a height H above it. When a light pulse is emitted at time t, an observer on Earth will measure a Doppler-shifted frequency which is given by equation (12.34). As there is no transverse motion in this case, we simply call the radial velocity V, which we take to be the speed of the rocket relative to Earth. Then, with some rearrangement, (12.34) gives:

$$\nu = \nu_0 (1 + V/c)^{-1} (1 - V^2/c^2)^{1/2} \simeq \nu_0 (1 - V/c) + O(V^2/c^2), \tag{15.43}$$

where the second step results from the use of the binomial series for both brackets, and we shall assume that terms of order (V^2/c^2), and higher, may be neglected.

At a later time $t + \delta t$, the light pulse reaches the ceiling of the rocket ship laboratory and during the interval δt, the speed of the freely falling rocket ship has increased to

$$V(t + \delta t) = V(t) + g\delta t. \tag{15.44}$$

The interval of time δt is just the time taken for the light pulse to travel the distance H from floor to ceiling and is

$$\delta t = H/c. \tag{15.45}$$

Hence, from (15.44) and (15.45) it follows that at time $t + \delta t$, equation (5.43) yields for the Doppler-shifter frequencies observed in the reference frame of the Earth,

$$\nu + \delta\nu = \nu_0\left[1 - (V + gH/c^2)\right] + O(V^2/c^2). \tag{15.46}$$

Subtracting (15.43) from (15.46) then gives the change of frequency observed on Earth as

$$\delta\nu = -\nu_0 gH/c. \tag{15.47}$$

Thus the frequency observed in the reference frame on Earth is reduced, corresponding to a **redshift**. As the amount of the redshift depends on the Earth's gravitational acceleration g, this is a purely gravitational effect.

15.6.3 *Black holes*

We conclude this brief tour of general relativity with a look at one of the most fascinating aspects of the subject. Namely, that gravitational collapse of a star may lead to the formation of a singularity in space-time. Because no light can escape from such a singularity, it is popularly known as a 'black hole'.

We begin by examining the statement that 'Matter tells space how to curve, and space tells matter how to move', in a little more detail. As is so often the case in this field, it is helpful to proceed by analogy, and we shall consider the game of billiards. This is a game of Newtonian hard-sphere collisions, but if we view it from above, without the benefit of stereoscopic vision, it appears to involve circular discs moving on a flat surface. Needless to say everything we see under these restricted circumstances is still according to the Newtonian rules!

Now suppose that we see some mysterious events, as illustrated in Figure 15.10, where the motion of a disc appears to be influenced by the proximity of another disc. This would violate our theory that only 'hard-sphere' collisions were involved. As we actually know that the objects we are watching are three-dimensional (even although we can't see that) we might be tempted to postulate the existence of either a bump or a dip in the surface of the table affecting the motion of the cue ball.

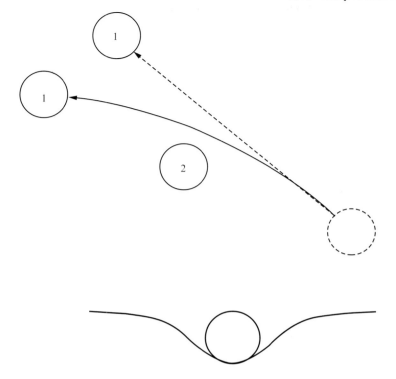

Fig. 15.10 Mysterious events in two-dimensional billiards. The dotted line shows the trajectory of disk 1, if disk 2 is not present.

Fig. 15.11 Billiards played on a thin rubber sheet: we postulate that disk 2 is really the 2D projection of sphere 2, which deforms the sheet that it rests on.

From such an explanation, plus the fact that the mysterious interaction only occurred when another ball was nearby, it would be but a short step to the theory that the second ball was causing either the hollow or the bump. That is, we might end up with the hypothesis that this game of billiards was being played on a tensioned sheet of rubber. Then the second ball would cause a curvature of the space, as indicated in Figure 15.11.

If we wanted to take this analogy further, and model the orbital motion of a planet around the Sun, say, then the second ball would need to be much larger, and more massive than the cue ball, so that the effects of the latter on the rubber sheet could be neglected. Of course, frictional forces would ultimately bring the cue ball to rest and hence spoil the analogy!

This is a qualitatively correct picture of what general relativity has to tell us, but in order to form some impression of the concept of a 'black hole' and its physical properties, we shall have to return to the Schwarzschild metric, as given by equation (15.33). This is based on an assumption of spherical symmetry, and it may be regarded as an appropriate vacuum field solution for a single spherical star.

For simplicity, we restrict ourselves to the static case $t = $ constant and further concentrate solely on the radial direction by setting $\phi = \theta = $ constant. Thus equation (15.33) reduces to

$$c^2 \, d\tau^2 = \frac{-dr^2}{(1 - 2\Lambda/r)}. \tag{15.48}$$

Also, if we take r to be the X^1 coordinate, comparison with the defining expression in equation (13.48) then gives the relevant element of the metric tensor as

$$g^{11} = \frac{-1}{1 - 2\Lambda/r}. \qquad (15.49)$$

We have already noted that this reduces to the flat Minkowski metric in the limit as $r \to \infty$. Now let us consider possible singularities for small values of the radial coordinate. By inspection it is clear that there are two interesting cases, viz:

1. $r \to 0$ and $g^{11} \to 0$;
2. $r \to 2\Lambda$ and $g^{11} \to \infty$.

In fact the singularity at $r = 2\Lambda$ can be eliminated by an appropriate coordinate transformation and is termed **removable**. In contrast, the apparently benign behaviour of the metric tensor as $r \to 0$ is misleading. We are actually interested in the **curvature** of the metric tensor; and, as we saw earlier, obtaining this involves double differentiation, so that from (15.49) we have:

$$\text{curvature} \sim \Lambda/r^3.$$

Thus, there is a singularity in the curvature, and hence in the gravitational attraction, at the origin. This cannot be transformed away and is known as an **essential singularity**.

All of this does not mean that the behaviour of the metric at $r = 2\Lambda$ is uninteresting. Far from it! The spherical surface at $r = 2\Lambda$ can be shown to be an **event horizon**. On this surface, radially outgoing photons stay where they are, while all others are attracted to the singularity at the origin. For $r < 2\Lambda$, **all** photons are attracted to the singularity.

In this context, the parameter 2Λ is known as the Schwarzschild radius and its value can be calculated from equation (15.34). In the case of the Sun this works out to be slightly less than 3 km, while in the case of the Earth it is just under 1 cm. The question then arises: why should we be interested in the behaviour of the metric—singular or otherwise—at distances many orders of magnitude less than the physical extent of typical astronomical bodies?

The answer lies in the theoretical possibility that a sufficiently massive spherical star will undergo gravitational collapse. That is, internal gravitational attraction will overcome the internal pressure, and the star will contract, until all its matter is contained in the singularity at its centre. To an observer of this process, it would only be possible to see the surface of the star until it reached the Schwarzschild radius. Once the radius of the star is less than 2Λ, no light can escape from the surface to the rest of the universe. The star therefore becomes a black hole in that it emits no radiation. If anything—radiation or matter—strays too close, it will be irreversibly attracted into the black hole.

15.7 Exercises

15.1 A circular turntable of diameter D rotates with angular speed ω, in an inertial frame S. Show that an observer in the frame of the turntable S'_{rot} measures the ratio of the circumference to the diameter of the turntable as $\pi' = \pi / \sqrt{1 - D^2\omega^2/2c^2}$, where π is the ratio measured in S, when the turntable is stationary.

15.2 Estimate the distance that a photon 'falls' as it travels horizontally over a distance of 30 m, say, in a laboratory which is itself in free fall near the surface of the Earth.

15.3 Estimate the angle that a photon is deflected through as it passes the Sun. [Hint: you will need to work out the Sun's gravitational acceleration in the vicinity of its surface.]

15.4 Work out the radius of curvature at the point $(0, 0)$ of the curve

$$y^2 - 2ay + x^2 = 0.$$

[Note: for the case of a simple curve connecting only x and y, equation (15.4) for the radius of curvature reduces to:

$$\rho = \frac{(1 + y_1^2)^{3/2}}{y_2}, \quad \text{where} \quad y_1 = \frac{dy}{dx} \quad \text{and} \quad y_2 = \frac{d^2y}{dx^2}.]$$

15.5 The position vector $\mathbf{x} = (x_1, x_2, x_3)$ of a moving particle is given in parametric form by the relationships:

$$x_1 = a \cos t, \qquad x_2 = a \sin t, \qquad x_3 = ct.$$

Verify that the particle's trajectory is a right circular helix and obtain its radius of curvature. [Note: Equation (15.4) for the radius of curvature may be usefully recast in this situation as:

$$\rho = \frac{|\dot{\mathbf{x}}|^3}{|\dot{\mathbf{x}} \times \ddot{\mathbf{x}}|}.]$$

15.6 Work out the numerical value of the Schwarzschild radius Λ for the Earth.

15.7 Show that the components of the metric tensor corresponding to the Schwarzschild metric are given by:

$$g_{\mu\nu} = \text{diag}\{(1 - \Lambda/r), (1 - \Lambda/r)^{-1}, -1, -1\}.$$

15.8 Considering the equatorial laboratory on the rotating Earth, as discussed in Exercise 13.6 of Chapter 13, show that an inertial observer will measure the laboratory clock as running slow by a modified gamma factor γ^* such that

$$\gamma^* = \frac{1}{\sqrt{1 - (R^2\omega^2/c^2 + \Lambda/R)}},$$

where R is the radius, and ω is the rotational speed, of the Earth; and Λ is the disposable constant in the Schwarzschild metric. [Hint: impose the invariance requirement

$$g_{\mu\nu} \dot{X}^\mu \dot{X}^\nu = c^2,$$

taking $g_{\mu\nu}$ to be the tensor of the Schwarzschild metric.]

15.9 It is proposed that a perpetual motion machine may be constructed as follows. Energy $E = mc^2$ is used to create a particle of mass m, which is allowed to fall through a height H, so that it gains energy mgH. The available energy $E = mc^2 + mgH$ is converted into a photon which propagates back up to the starting point, where it can release its energy with, apparently, a net gain of amount mgH. What is the flaw in this proposal?

15.10 Show that a photon can only just escape from the surface of a star if the radius R of the star is given by the critical value:

$$R = 2GM/c^2,$$

where G is the gravitational constant, M is the mass of the star and c is the speed of light. Comment on the physical significance of this result. [Hint: This is purely a matter of Newtonian mechanics! Start by working out the escape velocity for a particle of mass m at the surface of a gravitational mass. Then assume that the particle is a photon and the escape velocity must equal the speed of light.]

Appendices

Integration using a dummy variable

This is a useful technique for tidying up integrals, by incorporating the initial conditions in the limits. It is also essential for the symbolic manipulations which will be a central part of later, more advanced treatments of the subject. Of course that would take us beyond the scope of the present book; but we do find a use for it in Section 1.1.2, and it is helpful to become familiar with the idea at an early stage in one's education.

We are already familiar with the simple **indefinite** integral, for example:

$$I = \int x\,dx = \frac{x^2}{2} + C,$$

where C is the constant of integration. Similarly, we might encounter a simple **definite** integral, for example

$$\int_a^b x\,dx = \left[\frac{x^2}{2}\right]_a^b = \frac{b^2}{2} - \frac{a^2}{2}.$$

Note that a and b are the limits of integration and that the integral results in a numerical value (given, that is, the numerical values of a and b); and there is no unknown constant of integration remaining to be found.

Now, we can write I as if it were a definite integral, thus:

$$I = \int_a^x x'\,dx' = \left[\frac{x'^2}{2}\right]_a^x = \frac{x^2}{2} - \frac{a^2}{2}.$$

Comparison of the two results shows that, in this instance, the constant of integration is $C = -a^2/2$. In this context, x' is known as a **dummy variable** and x is the **labelling variable**.

We can extend this procedure as follows. Let us consider a well-behaved function $f(x)$, where x is a continuous real variable on the interval $0 \leq x \leq 1$. (The specific choice of limits on the interval is purely for definiteness and simplicity.) Then the integral of $f(x)$ over this interval defines a new function $g(x)$, such that

$$g(x) = \int_0^x f(x')\,dx',$$

and the dummy variable x' is just like the variable x on the interval $0 \leq x' \leq x \leq 1$.

The procedure can be extended to higher orders of integration, such as

$$h(x) = \int_0^x g(x')\, \mathrm{d}x' = \int_0^x \int_0^{x'} f(x'')\, \mathrm{d}x''\, \mathrm{d}x'$$
$$= \int_0^x \int_0^{x'} f(x'')\, \mathrm{d}x'\, \mathrm{d}x'',$$

where x'' is yet another dummy variable, such that $0 \leq x'' \leq x' \leq x \leq 1$; and so on, to any desired order.

Solid angles

B

In Chapter 10, we make use of the concept of a solid angle, in order to define the scattering cross-section. For those who are not familiar with this idea, we present a brief discussion here. Basically a solid angle is just the three-dimensional analogue of the usual two-dimensional angle; and we introduce it by means of a simple generalisation of the latter concept.

An ordinary angle is related to the circle and an elementary angle $d\theta$ is defined by reference to a segment of a circle with arc length dl. Then, for a circle of radius r, we define the angle $d\theta$ by means of the relationship:

$$d\theta = \frac{\text{arc length}}{\text{radius}} = \frac{dl}{r}.$$

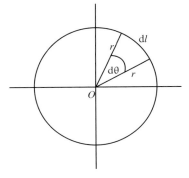

Fig. B.1 Definition of an ordinary angle.

In other words, we say that the angle $d\theta$ is the angle *subtended* at the centre of the circle by the arc of length dl.

The angle $d\theta$, defined this way, is measured in **radians**. The angle subtended by the whole circumference (i.e. the arc length $l = 2\pi r$) is just

$$\theta = \frac{2\pi r}{r} = 2\pi \text{ radians.}$$

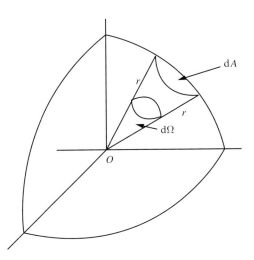

Fig. B.2 Definition of an elementary solid angle.

If we now generalize to three dimensions, then the **solid angle** is related to the **sphere**. An elementary solid angle $d\Omega$ is defined by reference to an elementary cone with base dA which is an element of the surface area of a sphere of radius r. We define the elementary solid angle $d\Omega$ as the apex angle of such a cone, by means of the relationship

$$d\Omega = \frac{dA}{r^2}.$$

In other words, we say that the solid angle $d\Omega$ is the angle subtended at the centre of a sphere of radius r by the element of surface area dA.

The solid angle defined in this way is measured in **steradians**. The angle subtended by the whole surface of the sphere (i.e. the surface $A = 4\pi r^2$) is just

$$\Omega = \frac{4\pi r^2}{r^2} = 4\pi \text{ steradians.}$$

Relativistic inertial mass

In both Chapters 11 and 14 we remarked (in footnotes) that we would not use the concept of relativistic mass in this book. It is certainly seen by some physicists as old-fashioned; but, before we briefly discuss it, we should perhaps emphasize that its use or otherwise is really just a matter of taste and indeed the usage is still widespread.

If we consider equation (14.4) for the relativistic 3-momentum $\mathbf{p}$, with the rest (or proper) mass now denoted by m_0 we may write this as

$$\mathbf{p} = m_0 \gamma(u) \mathbf{u},$$

where the other symbols have their usual meanings. Then, is we introduce the relativistic mass by the definition:

$$m(u) \equiv m_0 \gamma(u),$$

the equation for $\mathbf{p}$ may be written in the form:

$$\mathbf{p} = m(u) \mathbf{u}.$$

thus the relativistic 3-momentum has exactly the same form as the Newtonian case and we interpret $m(u)$ as a relativistic mass which increases with the speed of the particle, tending towards infinity as the speed of the particle tends towards the speed of light.

One advantage of using the relativistic inertial mass is that equation (14.12) for the relativistic energy of a particle becomes:

$$E = mc^2.$$

In other words, with this convention, Einstein's most famous equation now takes its most familiar form.

D The Penrose– Terrell rotation

[1]R. Penrose, *Proc. Camb. Phil. Soc.* **55**, 137–139 (1958).
[2]J. Terrell, *Phys. Rev.* **116**, 1041–1045 (1959).

We have seen that the introduction of special relativity, and the associated Lorentz transformations, leads to predictions that moving clocks run slow and that moving bodies contract in their direction of motion. We have also noted that the phenomenon of time dilation has received ample confirmation by experiment but that the practical difficulty of accelerating macroscopic bodies up to speeds near the speed of light stands in the way of any direct confirmation of the Fitzgerald contraction. Despite this, it had always been assumed that the Fitzgerald contraction could in principle be seen, or at least photographed. The fact that it could not was first pointed out by Penrose[1] and, independently, by Terrell,[2] more than half a century after Einstein's original paper in 1905. Apart from its inherent interest, this topic is worth considering as it helps us to take a more critical and formal look at what we mean by terms like 'seeing' and 'observing'.

The operations of seeing or of taking a photograph are the same in principle, because in both cases we register in our brains an image which is composed of photons, all of which arrived on the retina or the photographic emulsion at much the same time. For sake of simplicity, we shall ignore practical considerations such as persistence of vision or the 'speed' of the film and just refer to the photons forming such an image as arriving simultaneously. However, if they arrived simultaneously on the photographic plate then, due to the finite speed of light, they would have left the various parts of an extended body at a variety of different times. If there is relative motion between the body and the observer, then clearly this must have implications for how the observer views the body.

At this point it is convenient to distinguish between a 'world picture' and a 'world map', as follows:

World picture. This is what an observer sees or photographs. It is a two-dimensional projection of events that occurred earlier, their time of occurrence being in proportion to their distance from the observer.

World map. This is an instantaneous three-dimensional map of the observer's spatial reference frame. It is obtained by freezing space-time at some instant throughout the three-dimensional space.

A typical example of a world picture is when one looks at the sky at night. In contrast, a world map can only be determined by measurement. When we state conclusions from special relativity, we are doing so in the context of a world map. What Penrose and Terrell considered was: what is the effect of examining such conclusions in the context of the world picture, where the finite speed of light has to be taken into account. In the interests of completeness, we shall give a very brief account of this work here.

Both of these authors made use of the idea of aberration, as discussed in the present book in Section 12.2.5, and considered two observers momentarily at the same point in space, with O' moving at speed V relative to O and comoving with the object being viewed. Taking the latter to be a sphere, one may imagine the projection of its image on a common screen by both O and O'. The screen is assumed to be flat, and the angular positions of the image points of a single object point on the sphere will satisfy the aberration formula, as given by equation (12.30). In fact, we shall use the form derived in Exercise 12.4, thus:

$$\tan(\theta'/2) = \left(\frac{c-V}{c+V}\right)^{1/2} \tan(\theta/2).$$

Then for small angles, one may conclude that the images on both screens will be the same (i.e. a circular outline for a sphere) and differ only in their size, the ratio or magnification M being given by

$$M = \left(\frac{c-V}{c+V}\right)^{1/2}.$$

Terrel also examined the case of a moving rod and concluded that it would appear to rotate in the frame of reference of O by an amount $\theta - \theta'$. He also concluded that, as the light reaching O from its rear end would arrive later than that from its front end, the image seen would suffer an elongation which just cancelled out the Fitzgerald contraction. Similarly, although a moving sphere would contract in its direction of motion, exactly the same 'time lag' effect would cancel out this effect and it would not, as a result, *appear* to be flattened. This is a particularly apposite example since, as Terrell points out, Einstein made a specific prediction in his 1905 paper that a moving sphere would appear flattened, becoming a disc at $V = c$.

E Solutions to exercises

Chapter 1

1.1 From equation (1.17):

$$\mathbf{F} = q\mathbf{E} + q(\mathbf{u} \times \mathbf{B}) = 0$$

hence

$$-\mathbf{E} = \mathbf{u} \times \mathbf{B} = -\mathbf{B} \times \mathbf{u},$$

along the y-axis. Thus

$$E = uB$$

and so

$$u = E/B.$$

1.2 Equation (1.20):

$$\phi = q\left(\frac{1}{x_+} - \frac{1}{x_-}\right)$$

Referring to Figure 1.1, we may use the cosine law for the triangle to write:

$$x_+^2 = (a/2)^2 + x^2 - 2(a/2)x\cos\theta$$

with a similar result for x_-. Take $a \ll x$ such that terms of order $(a/x)^2$ can be neglected:

$$x_+ = \left[x^2 - ax\cos\theta + O\left(a/x^2\right)\right]^{1/2} \simeq x\left[1 - \left(\frac{a}{2x}\right)\cos\theta\right]$$

$$\simeq x - \frac{1}{2}a\cos\theta,$$

with an analogous result for x_-.

Hence,

$$\phi \simeq q\left(\frac{1}{x - \frac{1}{2}a\cos\theta} - \frac{1}{x + \frac{1}{2}a\cos\theta}\right),$$

and putting both terms on a common denominator,

$$\phi \simeq qa\cos\theta \left[\frac{1}{x^2 - \frac{1}{4}a^2\cos^2\theta} \right] \simeq \frac{qa\cos\theta}{x^2}.$$

1.3 Take the geometry to be as in Figure 1.2, but with the ring centred on point O. The potential at the test point P due to an elementary length dl of the ring is given by equation (1.23) in the form:

$$dU = -\frac{g\lambda dl}{\omega}.$$

This is the same for all elements dl and so we get the total potential by adding up all the elements to give the circumference:

$$U = -\frac{2\pi aG\lambda}{\omega} = -\frac{2\pi aG\lambda}{(a^2 + x^2)^{1/2}},$$

from the geometry of Figure 1.2.

1.4 Given:

$$\rho = \rho_0(1 - r^2/2R^2),$$

equation (1.45) $\Rightarrow \dfrac{\partial}{\partial r}\left(\dfrac{r^2}{\rho}\dfrac{\partial p}{\partial r} \right) = -4\pi Gr^2\rho = -4\pi G\rho_0(r^2 - r^4/2R^2).$

Integrate twice with respect to r:

$$p(r) = -4\pi G\rho_0^2\left(\frac{r^2}{6} - \frac{1}{15}\frac{r^4}{R^2} + \frac{1}{120}\frac{r^6}{R^4} \right) + C_1\rho_0\left(-\frac{1}{r} - \frac{r}{2R^2} \right) + C_2.$$

Boundary conditions: $p(r)$ is finite at $r = 0$ therefore $C_1 = 0$, and, $p(r) = 0$ at $r = R$ thus

$$C_2 = \frac{13}{30}\pi G\rho_0^2 R^2.$$

Pressure at centre $= p(0) = C_2$. Thus:

$$p(0) = \frac{13}{30}\pi G\rho_0^2 R^2,$$

as required.

1.5 Earth radius (mean) $R = 6371\,\text{km}$. Equation (1.51) gives:

$$g = \frac{GM}{r^2},$$

and:

$$GM = 3.986 \times 10^{14}\,\text{m}^3\,\text{s}^{-2}.$$

Hence:

$$g = \frac{3.986 \times 10^{14}}{(6.371 \times 10^6)^2} = \frac{3.986}{6.371^2} \times 10^2 = 9.8202 \, \text{m s}^{-2}.$$

Take $R \pm 5 \, \text{km}$ as the variation.

$$g_- = 10^2 \times \frac{3.986}{6.376^2} = 9.8048;$$

$$g_+ = 10^2 \times \frac{3.986}{6.366^2} = 9.8357,$$

and the variation is about 0.32% .

1.6 This is a discussion problem and no solution is given.

1.7 The discussion given in Section 11.8 provides a solution for this problem.

1.8 For the first part we use conservation of energy:

$$KE + PE = \text{constant},$$

which implies

$$\frac{1}{2}mu^2 + 0 = 0 + mgh,$$

and cancelling m and rearranging:

$$u^2 = 2gh,$$

thus:

$$h = u^2/2g.$$

For the second part, we use N2 and, measuring y as positive upwards,

$$\ddot{y} = -g.$$

Integrating twice, with respect to time, and invoking boundary conditions

$$t = 0, \qquad y = 0, \qquad \dot{y} = u,$$

to fix values of the constants of integration, we find:

$$y = ut - gt^2/2,$$

for the height at any time t. The time taken to reach the highest point y_{max} is denoted by t_{m} and found by substituting in the above equation, thus:

$$y_{\text{max}} = ut_{\text{m}} - gt_{\text{m}}^2/2.$$

From the first part we have

$$y_{max} = u^2/2g,$$

and rearranging the equation for t_m, and substituting for y_{max}, we find

$$(t_m - u/g)^2 = 0 \Rightarrow t_m = u/g,$$

and, since the time taken to return is the same,

$$t_{flight} = 2t_m = 2u/g.$$

In S,

$$y = ut - \frac{1}{2}gt^2, \qquad t' = t.$$

In S'

$$y' = y, \qquad x' = x - Vt, \qquad t' = t.$$

Hence:

$$y' = ut - \frac{1}{2}gt^2; \qquad x' = -Vt.$$

Substitute $t = -x/V$ from the second equation into the first:

$$y' = -\left(\frac{u}{V}\right)x' - \frac{1}{2}\frac{g}{V^2}(x')^2.$$

1.9 In S:

$$E = \frac{1}{2}m\dot{x}^2; \qquad p = m\dot{x}.$$

Substitute $\dot{x} = p/m$ in the equation for the energy:

$$E = \frac{1}{2}m\left(\frac{p}{m}\right)^2 = p^2/2m.$$

In S':

$$E' = \frac{1}{2}m\dot{x}'^2 = \frac{1}{2}m(\dot{x} - V)^2 = \frac{1}{2}m\dot{x}^2 - m\dot{x}V^2$$

$$p' = m\dot{x}'.$$

Assume the relationship holds: i.e.,

$$E' = \frac{p'^2}{2m} = \frac{1}{2m}(m\dot{x} - mV)^2 = \frac{1}{2}(\dot{x}^2 - 2\dot{x}V + V^2)$$

$$= \frac{1}{2}m\dot{x}^2 - m\dot{x}V + \frac{1}{2}mV^2,$$

in agreement with the Galilean transformation of the kinetic energy.

1.10 Galilean transformations:

$$x' = x - Vt, \qquad u' = u - V, \qquad t = t'.$$

Transform the given equation from $S \rightarrow S'$.

$$u = u' + V.$$

Take the equation term by term:

$$f(x, t) \rightarrow f'(x', t) \qquad \text{given}$$

$$\frac{\partial u}{\partial x} = \frac{\partial}{\partial x}(u' + V) = \frac{\partial u'}{\partial x} \qquad \text{as } V = \text{const}$$

$$= \frac{\partial x'}{\partial x} \cdot \frac{\partial u'}{\partial x'} + \frac{\partial t'}{\partial x} \cdot \frac{\partial u'}{\partial t'} = \frac{\partial u'}{\partial x'} \qquad \text{as} \qquad \frac{\partial x'}{\partial x} = 1 \quad \text{and} \quad \frac{\partial t'}{\partial x} = 0$$

$$\frac{\partial u}{\partial t} = \frac{\partial}{\partial t}(u' + V) = \frac{\partial u'}{\partial t} \qquad \text{as } V = \text{const}$$

$$= \frac{\partial x'}{\partial t} \cdot \frac{\partial u'}{\partial x'} + \frac{\partial t'}{\partial t} \cdot \frac{\partial u'}{\partial t'}$$

$$= -V \frac{\partial u'}{\partial x'} + \frac{\partial u'}{\partial t'} \qquad \text{as} \qquad \frac{\partial x'}{\partial t} = -V \quad \text{and} \quad \frac{\partial t'}{\partial t} = 1.$$

The given equation becomes

$$\frac{\partial u'}{\partial t'} - V \frac{\partial u'}{\partial x'} + (u' + V) \frac{\partial u'}{\partial x'} = f' \quad \text{or} \quad \frac{\partial u'}{\partial t} + u' \frac{\partial u'}{\partial x'} = f',$$

as required.

1.11 Adopting the hint given in the question, we consider events from the point of view of an observer in frame S' fixed to the buoy. In this frame, the boat travels away from the origin for 15 minutes and then returns to it in 15 minutes. Hence, in S', the bridge moved 1 km in 30 minutes with a speed of 2 km/hr. Thus, in S, the **water** flows with speed $V = 2$ km/hr

Chapter 2

2.1 Take the buoyancy force F_B to act vertically upwards and gravitational force acting downwards. Accordingly N2 gives:

$$F_B - mg = -ma,$$

or

$$F_B = m(g - a).$$

If we remove δm from the mass of the balloon, then we require:

$$F_{\text{B}} - (m - \delta m)g = (m - \delta m)a$$

or

$$F_{\text{B}} = (m - \delta m)(a + g).$$

Equating the two expressions for F_{B} :

$$m(g - a) = (m - \delta m)(a + g).$$

Cancelling and rearranging:

$$\delta m = \frac{2ma}{a + g}.$$

2.2 If the body is moving with constant speed down the slope, then N2 yields:

$$\text{frictional force} = \text{gravitational force}$$

or

$$\mu mg \cos \alpha = mg \sin \alpha$$

hence

$$\mu = \frac{\sin \alpha}{\cos \alpha} = \tan \alpha,$$

where μ is the coefficient of friction and $mg \cos \alpha$ is the normal reaction of the body on the sloping surface. Now tip the slope to $\beta > \alpha$. N2 becomes:

$$ma = mg \sin \beta - \mu mg \cos \beta.$$

Substitute for the coefficient of friction μ and cancel factors of m:

$$a = g \sin \beta - g \frac{\cos \beta \sin \alpha}{\cos \beta}$$

$$= g \frac{\sin \beta \cos \alpha}{\cos \alpha} - g \frac{\cos \beta \sin \alpha}{\cos \beta} = g \frac{\sin(\beta - \alpha)}{\cos \alpha}.$$

2.3 Show the kinetic energy changes according to:

$$\frac{\mathrm{d}T}{\mathrm{d}t} = \mathbf{F} \cdot \mathbf{u}.$$

Equation (2.5) gives:

$$W_{12} = \int_1^2 \mathbf{F} \cdot \mathrm{d}\mathbf{x} = \int_1^2 \frac{\mathrm{d}}{\mathrm{d}t}(m\mathbf{u}) \cdot \mathrm{d}\mathbf{x}$$

$$= \frac{1}{2} \int_1^2 m \frac{\mathrm{d}}{\mathrm{d}t}(u^2)\,\mathrm{d}t = \int_1^2 \frac{\mathrm{d}T}{\mathrm{d}t}\,\mathrm{d}t.$$

therefore

$$\int_1^2 \mathbf{F} \cdot d\mathbf{x} = \int_1^2 \left(\frac{dT}{dt}\right) dt.$$

This equation cannot depend on the arbitrary end-points 1 and 2. Thus:

$$\mathbf{F} \cdot d\mathbf{x} = \left(\frac{dT}{dt}\right) dt.$$

Divide across by dt and take the limit as $dt \to 0$: hence

$$\mathbf{F} \cdot \frac{d\mathbf{x}}{dt} = \left(\frac{dT}{dt}\right),$$

and so

$$\mathbf{F} \cdot \mathbf{u} = \frac{dT}{dt}.$$

2.4 Measure the coordinate x along the barrel of the gun, taking $x = L$ at the muzzle. Assume speed of bullet $u = 0$ at $x = 0$. From N2

$$F = ma$$

where $a = du/dt$. Also, by the usual transformation:

$$u\frac{du}{dx} = a,$$

and integrating both sides along the gun barrel

$$\int_0^{u(L)} u \, du = \int_0^L a \, dx,$$

hence

$$\left[\frac{u^2}{2}\right]_0^{u(L)} = \left[ax\right]_0^L = aL,$$

therefore

$$a = \frac{u^2(L)}{2L},$$

given that the pressure of the gases on the bullet is constant.
   Hence

$$F = \frac{mu^2(L)}{2L} = \frac{T(L)}{L},$$

where $T(L)$ is the kinetic energy of the bullet as it leaves the muzzle.
Now

$$\frac{dT}{dt} = \mathbf{F} \cdot \mathbf{u} = \frac{Tu(L)}{L},$$

where we substitute for F from the previous result.
$[T = 1.3 \, \text{kJ}, \ dT/dt = 0.4 \, \text{MW}.]$

2.5 Given

$$f(r) = \kappa/r^2.$$

From equation (2.10):

$$U(r) = -\int_\infty^r \frac{\kappa}{r'^2}\, \mathrm{d}r' = \int_r^\infty \frac{\kappa}{r'^2}\, \mathrm{d}r'$$

$$= -\left[\frac{\kappa}{r'}\right]_r^\infty = -\left[0 - \frac{\kappa}{r}\right] = \frac{\kappa}{r}.$$

2.6 Conservation of energy:

$$\frac{1}{2}u^2 + U(x) = E: \quad \text{total energy.}$$

Given

$$f(x) = \lambda x^{-3},$$

from (2.10),

$$U(x) = -\lambda \int_{x_0}^x x'^{-3}\, \mathrm{d}x' = -\lambda \left[\frac{-x'^{-2}}{-2}\right]_{x_0}^x$$

$$= \frac{\lambda}{2}[x'^{-2}]_\infty^x = \frac{\lambda}{2}[x^{-2} - 0]$$

$$= \frac{\lambda x^{-2}}{2}.$$

Given $x = a$, $u = 0$ therefore $U(a) = E = \lambda a^{-2}/2$. From Section 2.2.3 (for a particle of unit mass)

$$t_1 - t_2 = \int_a^0 \frac{\mathrm{d}x}{[\lambda a^{-2} - \lambda x^{-2}]^{1/2}}$$

Let $x = ay$; $\mathrm{d}x = a\,\mathrm{d}y$: therefore:

$$t_1 - t_2 = \int_1^0 \frac{\lambda^{-1/2}a\,\mathrm{d}y}{\sqrt{a^{-2}[1 - y^{-2}]}} = \frac{a^2}{\sqrt{\lambda}} \int_1^0 \frac{\mathrm{d}y}{[1 - y^{-2}]^{1/2}}.$$

Set $y = 1/u$, $\mathrm{d}y = -U^{-2}\,\mathrm{d}u$, hence:

$$t_1 - t_2 = \frac{a^2}{\sqrt{\lambda}} \int_0^1 \frac{\mathrm{d}u}{u^2[1 - u^2]^{1/2}}.$$

2.7 Two spheres of mass M_1 and M_2.

Before impact, speeds: V_{1i} and V_{2i}.

After impact, speeds: V_{1f} and V_{2f}.

Conservation of momentum:

$$M_1 V_{1f} + M_2 V_{2f} = M_1 V_{1i} + M_2 V_{2i}.$$

Conservation of energy:

$$M_1 V_{1f}^2 + M_2 V_{2f}^2 = M_1 V_{1i}^2 + M_2 V_{2i}^2.$$

Rewrite as:

$$M_2(V_{2f} - V_{2i}) = M_1(V_{1i} - V_{1f})$$

$$M_2(V_{2f}^2 - V_{2i}^2) = M_1(V_{1i}^2 - V_{1f}^2).$$

By dividing one into another and using

$$(a^2 - b^2) = (a + b)(a - b),$$

$$V_{2f} + V_{2i} = V_{1i} + V_{1f},$$

or

$$V_{2f} - V_{1f} = -(V_{2i} - V_{1i}).$$

$M_2 \to \infty$ makes $V_{2f} = V_{2i} = 0$, therefore the equation gives

$$-V_{1f} = V_{1i}.$$

2.8 Solution can be obtained by adapting the discussion given in Section 1.8.3.

2.9 The child's top slows down because there is a torque acting on it, in accordance with the analogous version of N2 for angular motion. The torque is due to the friction between the rotating 'point' and the surface upon which it rotates.

2.10 N2, for angular motion, is given by (2.19) as

$$\frac{d\mathbf{l}}{dt} = \mathbf{\Gamma}.$$

The torque is given by

$$\mathbf{\Gamma} = \mathbf{x} \times \mathbf{F}$$

and for $\mathbf{F}$ parallel to $\mathbf{x}$ the vector product vanishes and

$$\frac{d\mathbf{l}}{dt} = 0.$$

2.11 From equation (2.27)

$$(M_1 + M_2)\frac{d\bar{\mathbf{v}}}{dt} = \mathbf{F}_1 + \mathbf{F}_2$$

where $\bar{\mathbf{v}}$ is the velocity of the centre of mass. For two-dimensional motion, write $\mathbf{v} \equiv (\bar{u}, \bar{v})$ in the horizontal and vertical directions respectively and the only force acting is that of gravity in the vertical direction. Hence the equation of motion becomes:

$$(M_1 + M_2)\frac{d\bar{v}}{dt} = M_1 g + M_2 g$$

$$(M_1 + M_2)\frac{d\bar{u}}{dt} = 0.$$

Evidently these equations are now the same as for the elementary single-particle case, provided that the particle has mass $(M_1 + M_2)$ and moves with velocity $(\bar{u}, \bar{v}, 0)$. Thus it is easily verified that the centre of mass describes a parabola.

2.12 From equation (2.27):

$$(M_1 + M_2)\ddot{R} = F_1 + F_2.$$

Now, although F_1 and F_2—the forces due to gravity on the two masses—act vertically downwards, because of the pulley they act in opposite directions on the string. Hence, the equation of motion is:

$$(M_1 + M_2)\ddot{R} = -M_1 g + M_2 g$$

and

$$\ddot{R} = \frac{M_2 - M_1}{M_1 + M_2} g.$$

2.13 From equation (2.25), the centre of mass is given by:

$$R = \frac{M_1 x_1 + M_2 x_2}{M_1 + M_2} = \frac{M_2(x_2 + M_1 x_1/M_2)}{M_2(1 + M_1/M_2)}.$$

Cancelling, and taking the limit,

$$\lim_{M_2 \to \infty} \frac{(x_2 + M_1 x_1/M_2)}{(1 + M_1/M_2)} = x_2.$$

Relative coordinates become:

$$r_2 = x_2 - R = 0;$$

$$r_1 = x_1 - R = x_1 - x_2 \equiv \text{distance between the particles.}$$

2.14 From the example considered in Section 2.4.5, we use conservation of energy in the form:

$$V(r) + \frac{1}{2}\mu \dot{r}^2 = C,$$

where C is a constant and μ is the reduced mass, as given by equation (2.44),

$$\mu = \frac{M_1 M_2}{M_1 + M_2}.$$

Also, from equation (1.23),

$$V(r) = -\frac{GM_1 M_2}{r}.$$

Substituting for μ and $V(r)$, conservation of energy becomes

$$\frac{1}{2}\frac{M_1 M_2}{M_1 + M_2}u^2 - G\frac{M_1 M_2}{r} = C,$$

and the required result follows.

Chapter 3

3.1 Read the solution to Exercise 2.10 in conjunction with the discussion of a central force in Section 3.1.

3.2 The first part is covered in Section 3.2.4. For the second part, we are given:

$$dA = \frac{1}{2}|\mathbf{x} \times d\mathbf{x}|.$$

In plane polars,

$$\mathbf{x} = r\mathbf{e}_r \quad \text{and} \quad d\mathbf{x} = r\,d\phi\,\mathbf{e}_\phi,$$

hence

$$dA = \frac{1}{2}r^2\,d\phi,$$

with the vector sense being in the direction of unit vector $\mathbf{k}$. Thus

$$\frac{dA}{dt} = \frac{1}{2}r^2\frac{d\phi}{dt} = \frac{1}{2}r^2\dot\phi.$$

This can be written in terms of the angular momentum, and from the first part of the exercise:

$$\frac{dA}{dt} = \frac{l}{2M} \equiv \text{constant.}$$

3.3 From equation (3.20), the effective potential is:

$$U^*(r) = U(r) + \frac{l^2}{2mr^2},$$

The fictitious force $f^*(r)$ is given by:

$$f^*(r) = -\frac{\partial U^*}{\partial r} = -\frac{\partial U}{\partial r} - \frac{\partial}{\partial r}\left(\frac{l^2}{2mr^2}\right)$$

$$= f(r) - \frac{l^2}{2mr^3}(-2) = f(r) + \frac{l^2}{mr^3},$$

as required.

3.4
$$U^* = -\frac{\kappa l^{-\lambda r}}{r} + \frac{l^2}{2mr^2}.$$

Consider two extreme cases:

$r \to 0$:

$$e^{-\lambda r} \;\to\; 1 - \lambda r \;\Rightarrow\; U^* \to -\frac{\kappa(1-\lambda r)}{r} + \frac{l^2}{2mr^2}.$$

Thus

$$U^* \;\to\; \kappa\lambda - \frac{\kappa}{r} + \frac{l^2}{2mr^2}.$$

As $r \to 0$, the last term will dominate and the potential is repulsive: see Figure 3.4. $r \to \infty$: $e^{-\lambda r}$ falls off faster than any power law.
 Hence

$$U^* r \to \frac{l^2}{2mr^2}$$

and the potential is repulsive at large r as well.

3.5 (a) Repulsive force, vanishing as $r \to \infty$. Hence $f(r) > 0$ for all r.

$$U(r) = -\int_{r_0}^{r} \mathbf{F}(\mathbf{r}') \cdot d\mathbf{r}' = -\int_{r_0}^{r} f(r')\, dr' = \int_{r}^{r_0} f(r')\, dr'.$$

Hence

$$U(r) = \int_{r}^{\infty} f(r')\, dr',$$

as $U(\infty) = 0$ in this case.
 Since $U^*(r) > 0$ for all r, and

$$\dot{r}^2 = \frac{2}{M}[E - U^*(r)],$$

we must have $E > 0$ for an orbit to exist. We can only have an unbounded orbit as in Figure E.1.

(b) Attractive force, vanishing as $r \to 0$. Hence $f(r) < 0$ and $f(0) = 0$. In this case,

$$U(r) = -\int_{r_0}^{r} f(r')\, dr' = -\int_{0}^{r} f(r')\, dr' \geq 0 \quad \text{for all } r.$$

Thus can have bounded orbits as shown in Figure E.2.

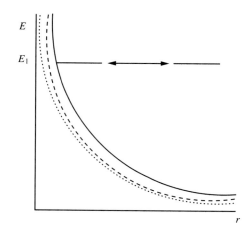

Fig. E.1 Potentials in the solution to Exercise 3.5(a).

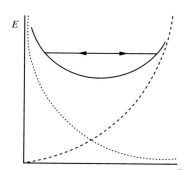

Fig. E.2 Potentials for the solution of Exercise 3.5(b).

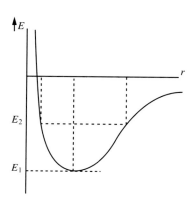

Fig. E.3 Effective potential for Exercise 3.6.

3.6 See Figure E.3.

3.7 We use equation (3.44) to find the appropriate $f(r)$ for each case.

(a)
$$r = ae^{-b\phi};$$

$$\frac{\mathrm{d}r}{\mathrm{d}\phi} = ae^{-b\phi} = -br$$

$$\frac{\mathrm{d}^2 r}{\mathrm{d}\phi^2} = ae^{-b\phi} = b^2 r.$$

Hence, from equation (3.44)

$$f(r) = \frac{mh^2}{r^4}\left[b^2 r - \frac{2}{r}\cdot b^2 r^2 - r\right] = -\frac{mh^2}{r^3}(b^2 + 1).$$

(b)
$$r = a + b\cos\phi$$

$$\frac{\mathrm{d}r}{\mathrm{d}\phi} = -b\sin\phi$$

$$\frac{\mathrm{d}^2 r}{\mathrm{d}\phi^2} = -b\cos\phi.$$

Hence, from equation (3.44),

$$f(r) = \frac{mh^2}{r^4}\left[-b\cos\phi - \frac{2}{r}(-b\sin\phi)^2 - r\right]$$

$$= mh^2\left[2\frac{(a^2 - b^2)}{r^5} - \frac{3a}{r^4}\right].$$

(c)
$$r = a/(1 + b\cos\phi)$$

$$\frac{dr}{d\phi} = \frac{ab\sin\phi}{(1 + b\cos\phi)^2};$$

$$\frac{d^2r}{d\phi^2} = \frac{ab\cos\phi}{(1 + b\cos\phi)^2} + \frac{2ab^2\sin^2\phi}{(1 + b\cos\phi)^3},$$

and substituting into (3.44), followed by some manipulation, gives

$$f(r) = -\left(\frac{mh^2}{a}\right)\frac{1}{r^2}.$$

[Note: we used the given orbital equation to eliminate the constant b, by rewriting it as:

$$(1 + b\cos\phi) = a/r.]$$

3.8 This is just example (b) of Section 3.6.1, in slight disguise.

3.9 From equation (3.56) we have:

$$u''(\phi) + u(\phi) = -\frac{1}{mh^2u^2}f(1/u),$$

where the dash denotes differentiation with respect to ϕ. Given

$$f(r) = -(ar^{-2} + br^{-3}) = -(au^2 + bu^3),$$

the equation of motion becomes:

$$u''(\phi) + (1 - b/mh^2)u/p = \frac{a}{mh^2},$$

or

$$u''(\phi) + \lambda^2 u(\phi) = \frac{a}{mh^2},$$

where

$$\lambda^2 = (1 - b/mh^2).$$

Taking $b < mh^2$ such that λ is real, the general solution of the equation of motion can be written as the sum of a particular integral and a complementary function, thus:

$$u(\phi) = \frac{a}{mh^2\lambda^2} + A\cos(\lambda\phi - b).$$

Initial conditions:

$$\dot{r}(0) = 0 \quad \text{and} \quad r(0) = R.$$

Now

$$\dot{r} = -h\frac{\mathrm{d}U}{\mathrm{d}\phi}$$

(see Section 3.7.4), and substituting from the expression for $u(\phi)$,

$$\dot{r} = -h\lambda A \sin(\lambda\phi - B).$$

From initial conditions $\dot{r}(0) = 0$ hence $B = 0$. Thus

$$\dot{r}(\phi) = -h\lambda A \sin \lambda\phi = 0$$

when $\phi = n\bar{u}/\lambda$ for $n = 0, 1, 2, \ldots$.

3.10 Let r, ϕ be the coordinates of the particle on the plane, with r measured from the hole. Let z be the distance of the second particle beneath the plane. Hence

$$s = r + z.$$

Let T be the tension in the string. From equations (3.51) and (3.52), the equations of motion of the particle on the table are:

$$m[\ddot{r} - r\dot{\phi}^2] = -T;$$

$$m[r\ddot{\phi} + 2\dot{r}\dot{\phi}] = 0.$$

For the particle below the table, N2 gives:

$$m\ddot{z} = mg - T,$$

or, using $s = r + z$, the last equation of motion may be written

$$m\ddot{r} = T - mg,$$

where $\dot{s} = \dot{r} + \dot{z} = 0$ for an inextensible string. Eliminating tension T, the radial equation of motion becomes:

$$m[\ddot{r} - r\dot{\phi}^2] = -m\ddot{r} - mg$$

or

$$2m\ddot{r} - mr\dot{\phi}^2 + mg = 0.$$

As usual, with central forces, the first integral of N2 for the $\mathbf{e}_\phi$ direction gives us conservation of angular momentum as

$$mr^2\dot{\phi} = l$$

and substituting for $\dot{\phi}$ in the radial equation of motion yields

$$2m\ddot{r} - \frac{l^2}{mr^3} + mg = 0.$$

If we multiply through by $\dot{r}$ and integrate with respect to time, we obtain the energy equation as

$$\frac{1}{2}(2m)\dot{r}^2 + \frac{l^2}{2mr^2} + mgr = C,$$

where C is a constant. If we take the constant to be

$$C = E + mgs,$$

so that E is the total energy of the two-particle system when $z = 0$ and the lower mass is still at the level of the plane, then

$$\frac{1}{2}(2m)\dot{r}^2 + \frac{l^2}{2mr^2} + mg(r - s) = E.$$

This is the equation of conservation of energy for a particle of mass $2m$ moving under the influence of an effective potential:

$$U^*(r) = \frac{l^2}{2mr^2} + mg(r - s).$$

For circular motion of the mass on the plane with $r = s/2$, we have $\dot{r} = \ddot{r} = 0$, and the equation of motion reduces to:

$$\frac{l^2}{m(s/2)^3} = mg,$$

and the required result follows.

3.11 Note that $f(r)$ is given as the magnitude of the force, and as the force is stated to be attractive, we write equation (3.55) as:

$$m\ddot{r} - \frac{mh^2}{r^3} = -f(r),$$

where as usual we have substituted for $\dot{\phi}$ in terms of h, the angular momentum per unit mass.

For a circular orbit, $r = R$, $\ddot{r} = 0$ and above equation yields

$$h^2 = \frac{1}{m}R^3 f(R).$$

For a small deviation from the circular orbit, $r = R + d$ and $\ddot{r} = \ddot{d}$ and the equation of motion becomes:

$$m\left(\ddot{d} - \frac{h^2}{(R+d)^3}\right) = -f(R + d).$$

Expand out in powers of d/R (taken to be small), using the binomial theorem on the left-hand side and Taylor series on the right-hand side, to obtain

$$\ddot{d} + \left[\frac{3h^2}{R^4} + \frac{1}{m}f'(R)\right]d = \frac{h^2}{R^3} - \frac{1}{m}f(R) = 0$$

and invoking the equation for circular motion for the last step,

$$\ddot{d} + \frac{1}{m}\left[\frac{3f(R)}{R} + f'(R)\right]d = 0,$$

where we have again invoked the equation for circular motion. For stability, we need this equation to reduce to SHM with

$$d = A\cos(\lambda\phi + B)$$

where

$$\lambda^2 = \frac{1}{m}\left[\frac{3f(R)}{R} + f'(R)\right].$$

Evidently we must have $\lambda^2 > 0$ and so the criterion for stability becomes:

$$\frac{Rf'(R)}{R} > -3.$$

3.12 From equation (1.22) we can write an expression for the gravitational force between the Sun and a planet as

$$\mathbf{F} = -\frac{GM_pM_s}{x^2}\,\hat{\mathbf{x}},$$

where M_p is the mass of the planet and M_s is the mass of the Sun. Also, with the definition of a central force in equation (3.1)

$$\mathbf{F}(\mathbf{x}) \equiv \mathbf{f}(\mathbf{r})\mathbf{e}_r \equiv \mathbf{f}(|\mathbf{x}|)\hat{\mathbf{x}},$$

we can identify

$$f(r) = -\frac{GM_pM_s}{r^2}$$

and so $\kappa = GM_pM_s$.

Kepler's laws are listed in Section 3.8, and their proofs are given respectively in Sections 3.7.3, 3.2 and 3.8.1.

Chapter 4

4.1 From the equation for the angular frequency of the simple pendulum given in Section 4.2.2,

$$\omega = \sqrt{\frac{g}{l}} = \frac{2\pi}{T}.$$

Thus:

$$T = 2\pi\sqrt{\frac{l}{g}} = 2\pi\sqrt{\frac{0.49}{9.81}} = 1.40\,\text{s}.$$

Take S to be the reference frame of the pendulum and S' to be the frame moving with the strip of paper at speed V (say), and in standard configuration relative to S.

The coordinates of the pendulum bob are as follows: In S:

$$y = A\cos(\omega t + \epsilon), \quad \text{from equation (4.4), and } x = 0 \text{ (say)}.$$

In S':

$$y' = A\cos(\omega t + \epsilon) \quad \text{and} \quad x' = -Vt.$$

Substitute for $t = -x'/V$ from the second equation into the first:

$$y' = A\cos(-\omega x'/V + \epsilon) = A\cos(\omega x'/V - \epsilon),$$

as the cosine is an even function of its argument.

4.2 For equilibrium $\rho Ah = $ mass of cylinder where Ah is the volume of water displaced and ρ is the material density of the cylinder.

Displace vertically through a distance z: the buoyancy force is increased by $\rho g Az$ and N2 gives:

$$\rho Ah \frac{d^2z}{dt^2} = -g\rho Az.$$

Thus the equation of motion becomes:

$$\frac{d^2z}{dt^2} + \frac{g}{h}z = 0.$$

This is SHM with period:

$$T = 2\pi\sqrt{h/g}.$$

4.3 Referring to Figure E.4, at point P a particle of mass m experiences a gravitational force

$$F = \frac{GMmr}{a^3},$$

where M is the mass of the Earth. The component along AB is $f = F\cos\theta = (x/a)F$.

Now $r = \sqrt{a^2 + x^2}$, and so

$$F = \frac{GMm\sqrt{a^2 + x^2}}{a^3} = \frac{GMm}{a^2} + 0\left(\frac{x^2}{a^2}\right).$$

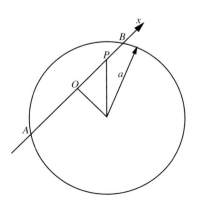

Fig. E.4 Figure for the solution of Exercise 4.3.

Hence

$$f = \frac{x}{a} F = \frac{GMm}{a^3} x.$$

From N2

$$m\ddot{x} = -\frac{GMmx}{a^3},$$

which is of the form

$$\ddot{x} + n^2 x = 0,$$

if

$$n^2 = \frac{GM}{a^3} = g/a,$$

and hence this is SHM with period

$$T = 2\pi\sqrt{a/g}.$$

4.4 Sufficient information to solve this is given at the end of Section 2.2.5.

4.5 Referring to Figure E.5, when the string is in the stretched position, all the tensions are the same:

$$T = c(a - a_0)$$

where c is a constant.

From N2, the exact equations of motion are:

$$m\ddot{x}_1 = -T_1 \sin\theta_1 + T_2 \sin\theta_2$$

$$m\ddot{x}_2 = -T_2 \sin\theta_2 - T_3 \sin\theta_3.$$

For small displacements from equilibrium, we have small angles θ_1, θ_2 and θ_3. Therefore

$$\sin\theta_1 = \frac{x_1}{\sqrt{a^2 + x_1^2}} \simeq \frac{x_1}{a}$$

and similarly

$$\sin\theta_2 \simeq \frac{x_2 - x_1}{a} \quad \text{and} \quad \sin\theta_3 \simeq \frac{x_2}{a}.$$

Also, the tensions T_1, T_2 and T_3 can be shown to be approximately equal to T. Hence

$$T_1 = c\left[\sqrt{a^2 + x_1^2} - a_0\right] = c\left[a(1 + x_1^2/a^2)^{1/2} - a_0\right]$$

$$= c(a - a_0) + 0(x_1^2/a^2) \simeq T,$$

with analogous results for T_2 and T_3.

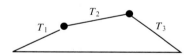

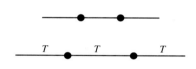

Fig. E.5 Figure for the solution of Exercise 4.5.

In this way, the linearized equations of motion may be written as:

$$\ddot{x}_1 = n^2[-x_1 + (x_2 - x_1)]$$

$$\ddot{x}_2 = n^2[-(x_2 - x_1) - x_2],$$

where

$$n^2 = T/ma.$$

As in Section 4.3.1, we substitute trial solutions:

$$X_1 = A\cos(\omega t + \phi)$$

$$X_2 = B\cos(\omega t + \phi),$$

to find (in the same way) solutions corresponding to $\omega = n$ and $\omega = \sqrt{3}n$. The normal modes of vibration are qualitatively the same as those sketched in Figure 4.11.

4.6 Referring to Figure E.6, the tensions are given by:

$$T_1 = c[(a + x_1) - a_0]$$

$$T_2 = c[(2a + x_2) - (a + x_1) - a_0] = c[x_1 + x_2 + a - a_0]$$

$$T_3 = c[(a - x_2) - a_0],$$

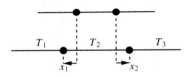

Fig. E.6 Figure for the solution of Exercise 4.6.

with the equations of motion from N2 as:

$$m\ddot{x}_1 = T_2 - T_1 = c(x_2 - 2x_1)$$

$$m\ddot{x}_2 = T_3 - T_2 = c(x_1 - 2x_2).$$

It is easily seen that the normal mode solutions must be the same as in the previous case, and consist of 'in phase' and 'anti-phase' longitudinal oscillations.

4.7 In this case, N2 leads to exact equations of motion as:

$$m\ddot{x}_1 = -T_1\sin\theta_1 + T_2\sin\theta_2 - 2mk\dot{x}_1$$

$$m\ddot{x}_2 = -T_2\sin\theta_2 - T_3\sin\theta_3 - 2mk\dot{x}_2,$$

just as in Exercise 4.5, but with frictional forces included. The analysis for small displacements is just a generalization of that in Exercise 4.5, with normal mode frequencies given by:

$$\omega_1 = n(1 - k^2/n^2)^{1/2}$$

$$\omega_2 = \sqrt{3}n(1 - k^2/3n^2)^{1/2},$$

along with a damping factor e^{-ht}. The form of the normal modes is the same as in Exercise 4.5.

4.8 This is a central force problem.

$$\mathbf{F}(\mathbf{x}) = f(r)\mathbf{e}_r = -Kr\mathbf{e}_r,$$

where

$$f_x = -Kx \qquad \text{and} \qquad f_y = -Ky.$$

Thus from N2, the equations of motion in the x and y directions are

$$m\ddot{x} = -Kx, \qquad \text{and} \qquad m\ddot{y} = -Ky.$$

Thus the orbital motion is the resultant of two SHMs of the same frequency at right angles and is therefore an ellipse. In other words, the simplest of the Lissajous figures.

4.9 Case 1 (as in Section 4.4.6),

$$f_L = \left(1 \pm \frac{V_L}{u}\right)f.$$

Case 2 (as in Section 4.4.6),

$$f' = \left(\frac{1}{1 \mp V_s/u}\right)f$$

$$\simeq [1 - (\mp)V_s/u]f$$

$$= [1 \pm V_s/u]f,$$

where we have used the binomial expansion and stopped at first-order for $V_s \ll u$.

Without loss of generality, consider source and listener approaching each other:

Case 1: $f_L = (1 + V_L/u)f$
Case 2: $f_L = (1 - V_s/u)f.$

Why are they not equivalent and therefore the same effect? Because the air through which the sound waves propagate provides a preferred reference frame. Consider taking $V_L = u$ and $V_s = u$:

Case 1: $f_L = 2f$ when $V_L = u.$
Case 2: $f_L = f/0 = \infty$ when $V_s = u,$

i.e. in the latter case, the source is moving at the speed of sound.

Chapter 5

5.1 Consider a subsystem of particles labelled by s, and consisting of masses $m_{1s}, m_{2s}, \ldots$ situated at $x_{1s}, x_{2s}, \ldots$ respectively.

The centre of mass R of the total system is given by

$$MR = \sum_{i,s} m_{is}\, x_{is},$$

where the sum runs over all particles in the system and M is the total mass of the system.

Noting that the centre of mass of the sth subsystem is given by

$$M_s R_s = \sum_i m_{is}\, x_{is},$$

we can rearrange the preceding expression as

$$MR = \sum_s \left(\sum_i m_{is}\, x_{is} \right)$$

$$= \sum_s M_s R_s,$$

as required.

5.2 This may be seen as an application of the result of the preceding exercise, with the cone being one subsystem and the hemisphere the other. Take the axis of symmetry of the composite body to be the z-axis, and take the origin of coordinates at the vertex of the cone.

From Section 5.6.1, we have:

$$\text{mass of hemisphere} = 2\pi a^3 \rho/3$$

and the position of the centre of mass is $3a/8$ from its centre and so is $11a/8$ from the vertex of the cone. Accordingly the coordinates of the centre of mass of the hemisphere are:

$$X = Y = 0 \qquad Z = 11a/8.$$

Similarly, one can obtain for the cone that:

$$\text{mass of cone} = \pi a^3 \rho/3,$$

and its centre of mass has coordinates:

$$X = Y = 0 \qquad Z = 3a/4.$$

Using the result of the previous exercise, we have for the composite system:

$$\left(\frac{2\pi a^3 \rho}{3} + \frac{\pi a^3 \rho}{3} \right) Z = \frac{2\pi a^3 \rho}{3} \cdot \frac{11a}{8} + \frac{\pi a^3 \rho}{3} \cdot \frac{3a}{4}$$

hence

$$\pi a^3 \rho \cdot Z = \frac{\pi a^4 \rho}{3} \left(\frac{22}{8} + \frac{6}{8} \right)$$

and

$$Z = 7a/6 \qquad \text{while} \qquad X = Y = 0.$$

5.3 If the point O is the existing reference point, consider the behaviour of the system with respect to a new reference point O', with $\mathbf{x}_o$ the position vector of O' relative to O. Taking the position of the sth particle to be given by $\mathbf{x}_s$ relative to O and $\mathbf{x}'_s$ relative to O', we may write:

$$\mathbf{x}_s = \mathbf{x}_o + \mathbf{x}'_s.$$

From equation (5.15) the angular momentum relative to O is:

$$\mathbf{L} = \sum_s \mathbf{x}_s \times \mathbf{p}_s = \sum_s m_s \mathbf{s} \times \dot{\mathbf{x}}_s$$

and similarly that relative to O' is:

$$\mathbf{L}' = \sum_s \mathbf{x}'_s \times \mathbf{p}_s = \sum_s m_s \mathbf{x}'_s \times \dot{\mathbf{x}}'_s.$$

Now substitute for $\mathbf{x}_s$ in the first expression, and noting that $\dot{\mathbf{x}}_o = O$, we have:

$$\mathbf{L} = \sum_s m_s \mathbf{x}_o \times \dot{\mathbf{x}}'_s + \sum_s m_s \mathbf{x}'_s \times \dot{\mathbf{x}}'_s$$

$$= \sum_s m_s \mathbf{x}_o \times \dot{\mathbf{x}}'_s + \mathbf{L}' = \mathbf{x}_o \times \sum_s m_s \dot{\mathbf{x}}'_s + \mathbf{L}'.$$

Now

$$\sum_s m_s \dot{\mathbf{x}}'_s = \frac{\mathrm{d}}{\mathrm{d}t} \sum_s m_s \mathbf{x}_s = 0,$$

if the centre of gravity is at rest, and hence

$$\mathbf{L} = \mathbf{L}'.$$

5.4 Referring to Figure E.7, the Cartesian coordinates of points A and B respectively are:

$$A = (a \cos \alpha, a \sin \alpha)$$

$$B = (a \cos \alpha + b \cos(\alpha + \beta), a \sin \alpha + b \sin(\alpha + \beta)).$$

Let the coordinates of the centre of mass be X and Y. Then, using equation (5.1), we have:

$$(m_A + m_B)X = m_A a \cos \alpha + m_B a \cos \alpha + m_B b \cos(\alpha + \beta)$$

$$(m_A + m_B)Y = m_A a \sin \alpha + m_B a \sin \alpha + m_B b \sin(\alpha + \beta).$$

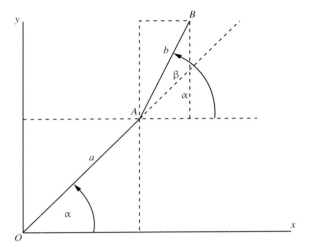

Dividing across by $m_A + m_B$ and introducing

$$\lambda = \frac{m_B}{m_A + m_B},$$

we have

$$X = a\cos\alpha + \lambda b\cos(\alpha + \beta)$$

$$Y = a\sin\alpha + \lambda b\sin(\alpha + \beta),$$

as required.

5.5 Introducing the relative coordinate $\mathbf{r}_s$ for particle s relative to the centre of mass through equation (2.28):

$$\mathbf{r}_s = \mathbf{x}_s - \mathbf{R},$$

it is readily shown that if $\mathbf{r}_A \equiv (r_{Ax}, r_{Ay})$, then:

$$r_{Ax} = -\lambda b\cos(\alpha + \beta) \qquad \text{and} \qquad r_{Ay} = -\lambda b\sin(\alpha + \beta)$$

and $\mathbf{r}_B \equiv (r_{Bx}, r_{By})$, then:

$$r_{Bx} = (1 - \lambda)b\cos(\alpha + \beta) \qquad \text{and} \qquad r_{By} = (1 - \lambda)b\sin(\alpha + \beta)$$

where λ is as defined in Exercise 5.4.

The total kinetic energy of the system is:

$$2T = (m_A + m_B)\dot{R}^2 + m_A\dot{r}_A^2 + m_B\dot{r}_B^2,$$

where the dot denotes differentiation with respect to time.

Now,

$$\dot{R}^2 = \dot{X}^2 + \dot{Y}^2$$

$$= [a^2 + \lambda^2 b^2 + 2ab\lambda \cos \beta]\dot{\alpha}^2 + (\lambda^2 b^2 + 2ab\lambda \cos \beta)\dot{\alpha}\dot{\beta} + \lambda^2 b^2 \dot{\beta}^2,$$

where we have used the usual double-angle formulae to expand out $\cos(\alpha + \beta)$ and $\sin(\alpha + \beta)$. Similarly,

$$\dot{r}_A^2 = \dot{r}_{Ax}^2 + \dot{r}_{Ay}^2 = \lambda^2 b^2 \dot{\alpha}^2 + 2\lambda^2 b^2 \dot{\alpha}\dot{\beta} + \lambda^2 b^2 \dot{\beta}^2,$$

and

$$\dot{r}_B^2 = \dot{r}_{Bx}^2 + \dot{r}_{By}^2$$

$$= (1 - 2\lambda + \lambda^2)b^2 \dot{\alpha}^2 + 2(1 - 2\lambda + \lambda^2)b^2 \dot{\alpha}\dot{\beta} + (1 - 2\lambda + \lambda^2)b^2 \dot{\beta}^2.$$

Write

$$2T = F\dot{\alpha}^2 + G\dot{\alpha}\dot{\beta} + H\dot{\beta}^2$$

and the three coefficients can be obtained by comparison with the previous expression for $2T$.

Chapter 6

6.1 The analysis given in Section 6.2.3 for a slender rod of length l should be applied to these cases. It is then possible to verify the answer by an appropriate substitution in equation (6.16), as follows:

(a) set $h = l/2$, $I = \dfrac{1}{12} M l^2$;

(b) set $h = 0$, $I = \dfrac{1}{3} M l^2$.

6.2 Using the analysis and results given in Section 6.2.6, we have:

(a) $I = \dfrac{1}{12} M(a^2 + b^2)$;

(b) $I = \dfrac{1}{12} M a^2$.

6.3 The analysis of the solid cylinder given in Section 6.2.4 can be adapted to the present case by taking the integral which leads to equation (6.17) to have limits $r = A$ and $r = B$, leading to

$$I = \frac{\pi \rho l}{2}[A^4 - B^4] = \frac{1}{2} M[A^2 + B^2],$$

where M is the mass of the cylinder.

6.4 Take spherical polar coordinates with the origin at the centre of the sphere and consider the moment of inertia about the z-axis. Assume that the sphere is of uniform density ρ and radius a.

The spherical polar coordinates (r, θ, ϕ) have the usual relationship to (x, y, z), viz.,

$$x = r \sin \theta \cos \phi$$

$$y = r \sin \theta \sin \phi$$

$$z = r \cos \phi,$$

where $0 \leq \theta \leq \pi$ and $0 \leq \phi \leq 2\pi$.

The element of volume is

$$\mathrm{d}r = r^2 \, \mathrm{d}r \sin \theta \, \mathrm{d}\theta \, \mathrm{d}\phi,$$

and it is easily verified that the volume of a sphere of radius a is:

$$V = \int_0^a r^2 \, \mathrm{d}r \int_0^\pi \sin \theta \, \mathrm{d}\theta \int_0^{2\pi} \mathrm{d}\phi.$$

Making the usual transformation to $\mu = \cos \theta$, $\mathrm{d}\mu = -\sin \theta \, \mathrm{d}\theta$, we have

$$V = \int_0^a r^2 \, \mathrm{d}r \int_{-1}^1 \mathrm{d}\mu \int_0^{2\pi} \mathrm{d}\phi = \frac{a^3}{3} \cdot 2 \cdot 2\pi = \frac{4\pi a^3}{3}.$$

In order to work out the moment of inertia we invoke equation (6.9), which we rewrite as

$$I = \rho \int_V r_d^2 \, \mathrm{d}V,$$

where r_d is the distance of element $\mathrm{d}r$ from the z-axis and is given by

$$r_d = r \sin \theta.$$

Hence,

$$I = \rho \int_0^a r^4 \, \mathrm{d}r \int_0^\pi \sin^3 \theta \, \mathrm{d}\theta \int_0^{2\pi} \mathrm{d}\phi$$

$$= 2\pi\rho \cdot \frac{a^5}{5} \cdot \int_{-1}^1 (1 - \mu^2) \mathrm{d}\mu$$

$$= \frac{4\pi\rho}{3} \cdot \frac{2a^5}{5} = \frac{2}{5} Ma^2.$$

For the second case, we use the parallel axes theorem. From equation (6.18)

$$I_\rho = I_{CM} + Ma^2 = \frac{2Ma^2}{5} + Ma^2 = \frac{7Ma^2}{5}.$$

6.5 There are two parts to the kinetic energy.

A. Rotational:

$$T_{\text{rot}} = \frac{1}{2} I \omega^2 = \frac{1}{5} M a^2 \omega^2$$

from the preceding exercise, and if we take the speed v to be along the slope (i.e. $v = \mathrm{d}l/\mathrm{d}t$), the rolling condition gives:

$$\frac{\mathrm{d}l}{\mathrm{d}t} = a\omega = v,$$

thus

$$T_{\text{rot}} = \frac{1}{5} M v^2.$$

B. Translational:

$$T_{\text{trans}} = \frac{1}{2} M v^2$$

By conservation of energy:

$$T = \left(\frac{1}{5} M + \frac{1}{2} M \right) v^2 = M g l \sin \alpha$$

and hence

$$v^2 = \frac{10}{7} g l \sin \alpha.$$

If the sphere were to slide without rotation, then the factor of 10/7 would be replaced by a factor of 2.

6.6 Note that in this case, with a smooth surface, the translational and rotational motions are independent.

Hence

$$T = \frac{1}{2} M v^2 + \frac{1}{2} I \omega^2$$

$$= \frac{1}{2} M v^2 + \frac{1}{5} M a^2 \omega^2.$$

(but note that $a\omega \neq v$ in this case!).

6.7 Let I be the moment of inertia of the fly-wheel and its attached pulley about the axis of rotation. As the mass m falls with speed v, the fly-wheel rotates with angular speed ω.

From the angular form of N2, we have for the motion of the flywheel:

$$I \frac{\mathrm{d}\omega}{\mathrm{d}t} = Tb,$$

where T is the tension in the string and clearly Tb is the applied torque. Applying the linear form of N2, we have for the motion of m:

$$m\frac{dv}{dt} = mg - T.$$

Also, the equivalent of the 'rolling condition' in Exercise 6.5 is now:

$$v = b\omega.$$

Eliminating ω from the angular form of N2 gives (with some rearrangement)

$$\frac{I}{b^2}\frac{dv}{dt} = T,$$

and substituting this for T in the linear form of N2:

$$\left(\frac{I}{b^2} + m\right)\frac{dv}{dt} = mg,$$

gives the acceleration of the particle as:

$$\frac{dv}{dt} = g \cdot \frac{mb^2}{I + mb^2},$$

and substituting back for dv/dt leads to the required result.

6.8 We use the angular form of N2:

$$I\dot{\omega} = \Gamma$$

where Γ is the torque and ω is the angular speed. From Section 6.2.4,

$$I = \frac{1}{2}Ma^2,$$

where a is the radius of the disk.

For a constant force, we can let the required time be T and write:

$$\frac{\omega - 0}{T} = \frac{\Gamma}{I} = \frac{aF}{I},$$

where F is the required force.

Thus

$$F = \frac{Ma\omega}{2T}.$$

Converting ω:

$$\omega = \frac{100}{60} \times 2\pi \text{ radian s}^{-1},$$

and

$$F = 5.24 \times 10^{-2}\,\text{N}.$$

6.9 This problem is an example of the conservation of angular momentum. We assume that λ is small and we note that the mass of the planet does not change. Hence we may write:

$$I\omega = I'\omega',$$

where the primed quantities refer to the time when cooling has taken place, and then

$$\frac{1}{5}Ma^2\omega = \frac{1}{5}Ma^2(1-\lambda)^2\omega'.$$

For small λ,

$$\omega' \simeq (1+2\lambda)\omega,$$

and

$$T' = \frac{2\pi}{\omega'} \simeq (1-2\lambda)T,$$

as required.

6.10 This is an example of a compound pendulum and we follow the procedure given in Section 6.2.8. In this case the cylinder oscillates by rolling about an axis which is one of its generators at the point of contact O (say) with the rough plane. At equilibrium, the centre of mass (CM) is vertically above O, and if we rotate the cylinder so that the radius through the CM moves through an angle θ, the subsequent oscillatory motion will satisfy the conservation of energy in the form:

$$\frac{1}{2}I_O\dot{\theta}^2 - Mgh\cos\theta = E,$$

where I_O is the moment of inertia of the cylinder about the axis of rotation. Making a comparison with the simple pendulum, as in Section 6.2.8, we may identify the equivalent length as

$$L = \frac{I_O}{Mh}$$

and the period as

$$T = 2\bar{u}\sqrt{\frac{L}{g}} = 2\pi\sqrt{\frac{I_O}{Mgh}}.$$

Then, by the parallel axes theorem.

$$I_O = I_{\text{CM}} + M(a-h)^2,$$

and the required result follows.

6.11 Referring to Section 6.4.1 for the details, the final calculation becomes:

$$I_{11} = \sum_\alpha m^\alpha x_2^2 = 12\mu a^2$$

$$I_{22} = \sum_\alpha m^\alpha x_1^2 = 12\mu a^2$$

$$I_{33} = \sum_\alpha m^\alpha (x_1^2 + x_2^2) = 24\mu a^2$$

$$I_{12} = I_{21} = -\sum_\alpha m^\alpha x_1 x_2 = -4\mu a^2$$

$$I_{13} = I_{31} = -\sum_\alpha m^\alpha x_1 x_3 = 0$$

$$I_{23} = I_{32} = -\sum_\alpha m^\alpha x_2 x_3 = 0.$$

6.12 From equations (6.40), we have the inertia tensor for the four particles:

$$I_{ij} = \sum_{\alpha=1}^4 m^{(\alpha)} \left[(x^{(\alpha)})^2 \delta_{ij} - x_i^{(\alpha)} x_j^{(\alpha)} \right],$$

where we have written

$$x_h x_h = x^2,$$

using the summation convention that repeated indices are summed.

We shall do this exercise in a different way, by working out the contributions to the inertia tensor for each particle and then adding the results. We begin by noting that δ_{ij} is the unit matrix and that for each particle $x_i x_j$ is also a matrix. Then for particle 1:

$$\mathbf{x} = a(1, -1, 1), \quad x^2 = 3a^2$$

$$x_i x_j = \begin{pmatrix} 1 & -1 & 1 \\ -1 & 1 & -1 \\ 1 & -1 & 1 \end{pmatrix}$$

and the contribution from particle 1 to the inertia tensor is:

$$I_{ij}^{(1)} = 3ma^2 \begin{pmatrix} 1 & 0 & 0 \\ 0 & 1 & 0 \\ 0 & 0 & 1 \end{pmatrix} - ma^2 \begin{pmatrix} 1 & -1 & 1 \\ -1 & 1 & -1 \\ 1 & -1 & 1 \end{pmatrix}$$

and

$$I_{ij}^{(1)} = ma^2 \begin{pmatrix} 2 & 1 & -1 \\ 1 & 2 & 1 \\ -1 & 1 & 2 \end{pmatrix}.$$

The results for the other three particles are:

$$I_{ij}^{(2)} = ma^2 \begin{pmatrix} 2 & -1 & 1 \\ -1 & 2 & 1 \\ 1 & 1 & 2 \end{pmatrix}$$

$$I_{ij}^{(3)} = ma^2 \begin{pmatrix} 2 & 1 & 1 \\ 1 & 2 & -1 \\ 1 & -1 & 2 \end{pmatrix}$$

$$I_{ij}^{(4)} = ma^2 \begin{pmatrix} 2 & -1 & -1 \\ -1 & 2 & -1 \\ -1 & -1 & 2 \end{pmatrix}.$$

Add up:

$$I_{ij} = I_{ij}^{(1)} + I_{ij}^{(2)} + I_{ij}^{(3)} + I_{ij}^{(4)},$$

and the required result follows.

6.13 Either method can be used but we now have a continuous distribution of mass with density ρ (say). For example,

$$I_{33} = \rho \int_0^{2a} \int_0^{2b} (x^2 + y^2) \, dV,$$

where

$$dV = 2c \, dx \, dy.$$

Hence

$$I_{33} = 2\rho c \int_0^{2a} dx \int_0^{2b} dy \, (x^2 + y^2) = \frac{4}{3} M(a^2 + b^2),$$

where $M = 8\rho abc$ is the mass of the block. Similarly,

$$I_{11} = \frac{4}{3} M(b^2 + c^2)$$

$$I_{22} = \frac{4}{3} M(c^2 + a^2).$$

For the off-diagonal terms,

$$I_{12} = \rho \int xy \, dV = 2c\rho \int_0^{2a} \int_0^{2b} xy \, dx \, dy.$$

$$= 8\rho ca^2 b^2 = Mab.$$

Similarly,

$$I_{23} = Mbc;$$

$$I_{13} = Mac.$$

Chapter 7

7.1 By considering the forces acting, it is easily seen that the required angle is $\tan^{-1}(a/g)$. The inertial observer will see the mass accelerate in the direction of motion due to the force supplied by the horizontal component of the tension in the string.

The observer in the car is non-inertial, because the car is accelerating, and will consider the pendulum bob to experience two forces: the weight mg acting downwards and an inertial force ma in the direction opposite to the motion.

7.2 This is easily calculated, provided we know the acceleration a. From the result of the previous exercise, we have:

$$a/g = \tan \theta,$$

where $\theta = 3$ degrees. Hence,

$$a/g = 0.05,$$

and

$$a = 0.05 \times 9.81 \, \mathrm{m \, s^{-2}}.$$

Thus

$$t_f = \frac{v(0)}{a} = 61.16 \, \mathrm{s}.$$

7.3 For a simple pendulum

$$T = 2\pi \sqrt{L/g}.$$

In the accelerating spaceship,

$$T' = 2\pi \sqrt{L/9g} = \frac{1}{3} T.$$

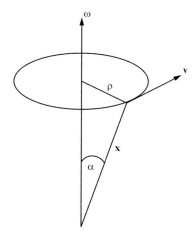

Fig. E.8 Figure for the solution to Exercise 7.4.

7.4 The situation is as shown in Figure E.8. The particle is moving in a circle of radius ρ with angular speed ω and so its linear speed is:

$$v = \omega\rho = \omega r \sin\alpha = |\boldsymbol{\omega} \times \mathbf{x}|$$

where $r = |\mathbf{x}|$.

In addition, the direction of $\mathbf{v}$ is perpendicular to the plane containing the axis and $\mathbf{x}$ and is therefore in the direction of $\boldsymbol{\omega} \times \mathbf{x}$.

7.5 It is easy to generalise the Galilean transformation of Section 1.6.3 to a vector form:

$$\mathbf{x} = \mathbf{x}' + \mathbf{V}t.$$

Then differentiating with respect to time:

$$\dot{\mathbf{x}} = \dot{\mathbf{x}}' + \mathbf{V}.$$

From the previous exercise, we can regard the **linear** velocity of a point fixed to the rotating frame as

$$\mathbf{V} = \boldsymbol{\omega} \times \mathbf{x},$$

which instantaneously points in a constant direction. Accordingly the above Galilean transformation can be written as:

$$\dot{\mathbf{x}} = \dot{\mathbf{x}}' + \boldsymbol{\omega} \times \mathbf{x},$$

at one instant and obviously generalizes to the required result which holds at all times.

7.6 This is a straightforward application of the analysis given in Section 7.2.5. From equation (7.38), we have that a particle released at $t = 0$ at a height h, will hit the ground when

$$t = \sqrt{2h/g}.$$

From (7.41), we then obtain the deviation as:

$$y' = \frac{1}{3}\omega g t^3 \cos\phi = \frac{1}{3}\omega\left(\frac{8h^3}{g}\right)^{1/2}\cos\phi,$$

when we substitute the above expression for t.

7.7 To the non-inertial observer on the rotating Earth, it seems that the particle experiences an inertial force which deflects it in an easterly direction. To a nearby inertial observer, it seems that the particle drops in a straight line, while the surface of the Earth rotates underneath it.

7.8 We use equations (7.33)–(7.35):

$$\ddot{x}' = 2\omega\sin\phi\dot{y}'$$

$$\ddot{y}' = -2\omega\sin\phi\dot{x}' - 2\omega\cos\phi\dot{z}'$$

$$\ddot{z}' = -g + 2\omega\cos\phi\dot{y}',$$

with initial conditions:

$$\dot{\mathbf{x}}'(0) = 0, \qquad \dot{\mathbf{x}}'(\epsilon) = (V \cos \beta \mathbf{i}' + V \sin \beta \mathbf{k}').$$

Integrate the equations of motion with respect to time:

$$\dot{x}' = 2\omega \sin \phi y' + A_1$$

$$\dot{y}' = -2\omega \sin \phi x' - 2\omega \cos \phi z' + A_2$$

$$\dot{z}' = -gt + 2\omega \cos \phi y' A_3,$$

and from the initial conditions:

$$A_1 = V \cos \beta, \qquad A_2 = 0, \qquad A_3 = V \sin \beta.$$

Thus:

$$\dot{x}' = V \cos \beta + 0(\omega)$$

$$\dot{y}' = 0(\omega)$$

$$\dot{z}' = -gt + V \sin \beta + 0(\omega).$$

Now substitute back:

$$\ddot{x}' = 0(\omega^2)$$

$$\ddot{y}' = -2\omega V \cos \beta \sin \phi - 2\omega V \sin \beta \cos \phi + 2\omega g t \cos \phi$$

$$\ddot{z}' = -g + 0(\omega^2).$$

Integrate again with respect to time:

$$\dot{x}' = B_1$$

$$\dot{y}' = -2\omega V t \cos \beta \sin \phi - 2\omega V t \sin \beta \cos \phi + \omega g t^2 \cos \phi + B_2$$

$$\dot{z}' = -gt + B_3.$$

Apply initial conditions again:

$$B_1 = V \cos \beta, \qquad B_2 = 0, \qquad B_3 = V \sin \beta.$$

The velocity equations to $O(\omega)$ become:

$$\dot{x}' = V \cos \beta$$

$$\dot{y}' = -2\omega V t (\cos \beta \sin \phi + \sin \beta \cos \phi) + g\omega t^2 \cos \phi$$

$$\dot{z}' = -gt + V \sin \beta.$$

Integrate again:

$$x' = Vt\cos\beta + C_1$$

$$y' = -\omega Vt^2(\cos\beta\sin\phi + \sin\beta\cos\phi) + \frac{g\omega t^3}{3}\cos\phi + C_2$$

$$z' = -\frac{1}{2}gt^2 + Vt\sin\beta + C_3.$$

Apply the initial conditions again:

$$C_1 = C_2 = C_3 = 0.$$

The projectile reaches the ground again when $z' = 0$ and $t = T$ (say). Hence

$$-\frac{1}{2}gT^2 + VT\sin\beta = 0,$$

and

$$T = \frac{2V\sin\beta}{g}.$$

The range is given by:

$$x'(T) = VT\cos\phi = \frac{2V^2\cos\beta\sin\beta}{g}.$$

Deviation:

$$y'(T) = -\frac{4\omega V^2\sin^2\beta}{g^2}(\cos\beta\sin\phi + \sin\beta\cos\phi)$$

$$+\frac{8\omega V^3\sin^2\beta\cos\phi}{3g^2}$$

$$-\frac{4\omega V^3}{3g^2}\sin^2\beta(\sin\beta\cos\phi + 3\cos\beta\sin\phi).$$

and as y' is positive in an easterly direction, a negative deviation is westerly by the above amount.

Chapter 8

8.1 Particle projected upwards to height h, with initial speed V_0. From equation (8.8),

$$m\dot{v} = -mg - v\dot{m},$$

for this problem, along with the given information

$$\dot{m} = km.$$

Thus

$$\frac{dv}{dt} = -g - kv.$$

Separating variables and integrating,

$$\int_{V_0}^{v} \frac{dv}{g + kv} = -\int_{0}^{t} dt,$$

thus

$$[\ln(g + kv)]_{V_0}^{v} = -kt,$$

and

$$\frac{g + kv}{g + kV_0} = e^{-kt},$$

and with some rearrangement

$$v = -\frac{g}{k}\left(1 - e^{-kt}\right) + V_0 e^{-kt}.$$

Expanding out the exponential, and dropping terms of order k^2 and higher:

$$v = V_0 - gt - V_0 kt$$

$$= V_0(1 - kt) - gt = V_0 - (k + g)t;$$

when $v = 0$ at maximum height,

$$t_f = \frac{V_0}{k + g}.$$

8.2 A gun of mass M fires a shell of mass m. Momentum is conserved in the absence of external forces and

$$mv = MU,$$

where U is the speed of recoil of the gun.

Conservation of energy yields:

$$\frac{1}{2}mu^2 + \frac{1}{2}MU^2 = E.$$

Eliminating U between the two equations

$$2E = mu^2 + M(mv/M)^2$$

$$= mu^2(1 + m/M)$$

$$= mu^2\left(\frac{m + M}{M}\right),$$

and rearranging, and taking the square root of both sides:

$$u = \sqrt{\frac{2ME}{M(M+M)}}.$$

8.3 The rocket equation is (8.14). In this problem, the force is due to gravity and

$$\mathbf{F} = (0, -mg, 0),$$

while the rocket's initial velocity is

$$\mathbf{v} = (v\cos\theta, v\sin\theta, 0)$$

and its initial exhaust velocity is

$$\mathbf{V} = (-V\cos\theta, -V\sin\theta, 0).$$

Equation (8.14) gives for the equations of motion in the x and y directions:

$$\frac{dv_1}{dt} \equiv \frac{dv\cos\theta}{dt} = -V\cos\theta\,\frac{\dot{m}}{m}$$

$$\frac{dv_2}{dt} \equiv \frac{dV\sin\theta}{dt} = -mg - V\sin\theta\,\frac{\dot{m}}{m}.$$

Integrating over the interval 0 to t produces the changes in velocity:

$$\Delta v_1 = V\sin\theta\ln(m_0/m)$$

$$\Delta v_2 = -gt + V\cos\theta\ln(m_0/m).$$

At time t:

$$\tan\alpha = \frac{\Delta U_2}{\Delta U_1} = \frac{-gt + V\cos\theta\ln(m_0/m)}{V\sin\theta\ln(m_0/m)}$$

$$= \frac{V\cos\theta\ln(m_0/m)}{V\sin\theta\ln(m_0 m)}\left[1 - \frac{gt}{V\cos\theta\ln(m_0/m)}\right]$$

$$= \tan\theta\left[1 - \frac{gt}{V\cos\theta\ln(m_0/m)}\right].$$

8.4 Equation (8.14) reduces to:

$$m\frac{dv}{dt} = -kv + VR.$$

Rearranging and integrating

$$\int_0^v \frac{dv}{RV - kv} = \frac{1}{m} \int_0^t dt$$

with initial conditions $t = 0$ and $v = 0$. Hence

$$\left[-\frac{1}{k} \ln(RV - kv) \right]_0^v = [\ln m]_{m_0}^m$$

and with some further manipulation the required result follows.

8.5 From the analysis given in Section 8.3.4, we have

$$v_f = V\left[\ln(M + m) + \ln(M_2 + m) - \ln(M + m - \varepsilon M + \varepsilon m_2)\right.$$

$$\left. - \ln(M_2 + m) - \varepsilon M_2\right].$$

The condition for a maximum is:

$$dv_f/dM_2 = 0,$$

and so, differentiating the right-hand side of the preceding equation with respect to M_2, and setting the result equal to zero gives the condition for a maximum as

$$V\left[\frac{1}{M_2 + m} - \frac{\varepsilon}{M + m - \varepsilon M + \varepsilon M_2} - \frac{(1 - \varepsilon)}{M_2 + m - \varepsilon M_2}\right] = 0.$$

Then, taking the first and third terms in the square brackets over to the right-hand side and putting them on a common denominator yields

$$\frac{\varepsilon}{M + m - \varepsilon M + \varepsilon M_2} = \frac{M_2 + m - \varepsilon M_2 - (1 - \varepsilon)(M_2 + m)}{(M_2 + m)(M_2 + m - \varepsilon M_2)}$$

$$= \frac{\varepsilon}{(M_2 + m)(M_2 + m - \varepsilon M_2)},$$

hence

$$(M_2 + m)^2 - \varepsilon(M_2 + m)M_2 = mM + m_2 + \varepsilon(M_2 m - mM),$$

and so

$$[M_2^2 + 2mM_2 - mM](1 - \varepsilon) = 0.$$

But $(1 - \varepsilon) \neq 0$, therefore v_f is a maximum for M_2 satisfying the equation

$$M_2^2 + 2mM_2 - mM = 0.$$

Taking only the positive solution,

$$M_2 = \sqrt{(M + m)m} - m,$$

and so substitution for M_2 back into the equation for v_f yields

$$v_f = V \ln 16 \simeq 2.77V.$$

Chapter 9

9.1 Conservation of momentum:

$$m_1 \mathbf{v}_{1i} = m_1 \mathbf{v}_{1f} + m_2 \mathbf{v}_{2f}.$$

Conservation of energy:

$$\frac{1}{2} m_1 v_{1i}^2 = \frac{1}{2} m_1 v_{1f}^2 + \frac{1}{2} m_2 v_{2f}^2.$$

Rearrange the energy equation to give an expression for $m_2^2 v_{2f}^2$:

$$m_2^2 v_{2f}^2 = m_1 m_2 v_{1i}^2 - m_1 m_2 v_{1f}^2.$$

Rearrange the momentum equation and square both sides to get:

$$m_2^2 v_{2f}^2 = m_1^2 v_{1i}^2 + m_1^2 v_{1f}^2 - 2 m_1^2 v_{1i} v_{1f} \cos \theta_1.$$

From the latter pair of equations:

$$m_1 m_2 v_{1i}^2 - m_1 m_2 v_{1f}^2 = m_1^2 v_{1i}^2 + m_1^2 v_{1f}^2 - 2 m_1^2 v_{1i} v_{1f} \cos \theta_1.$$

Rearrange:

$$(m_1^2 + m_1 m_2) v_{1f}^2 - 2 m_1^2 v_{1i} v_{1f} \cos \theta_1 + (m_1^2 - m_1 m_2) v_{1i}^2 = 0.$$

Divide across by $v_{1i}^2 (m_1^2 + m_1 m_2)$, to obtain:

$$(v_{1f}/v_{1i})^2 - \frac{2m_1}{m_1 + Fm_2} (v_{1f}/v_{1i}) \cos \theta_1 + \frac{m_1 - m_2}{m_1 + m_2} = 0.$$

For $m_1 = m_2$,

$$(v_{1f}/v_{1i})^2 - (v_{1f}/v_{1i}) \cos \theta_1 = 0,$$

therefore

$$\frac{v_{1f}}{v_{1i}} = \cos \theta_1 \quad \text{or} \quad 0.$$

For $\theta_1 = \pi/2$, $v_{1f} = 0$ for finite $v_{1i} \neq 0$ and all the energy is given to the recoiling particle.

9.2 Equation (9.30) is $\theta_1 = \theta'/2$, hence $\theta' = 2\theta_1$. Also from equation (9.25),

$$\theta_2 = \frac{\pi - \theta'}{2} = \frac{\pi}{2} - \frac{\theta'}{2} = \frac{\pi}{2} - \frac{2\theta_1}{2},$$

from the previous result, and is

$$\theta_2 = \frac{\pi}{2} - \theta_1,$$

or

$$\theta_1 + \theta_2 = \pi/2,$$

as required.

9.3 From equation (9.28),

$$\tan\theta_1 = \frac{\sin\theta'}{\cos\theta' + m_1/m_2}.$$

In the limit $m_2 \to \infty, m_1/m_2 \to 0,$ and hence

$$\lim_{m_1/m_2 \to 0} \tan\theta_1 = \tan\theta'.$$

Thus

$$\theta_1 = \theta'.$$

The physical interpretation is that the very massive target particle does not recoil and the **CM** and **LAB** frames become identical.

9.4 The magnitude of the scattering angle is determined by the distance between the centres of the incident and scattering particles at the distance of closest approach. This is in fact the impact parameter, as defined in Section 10.2.

9.5 Conservation of momentum:

$$m_1\dot{y}_1 + m_2\dot{y}_2 = m_1\dot{y}_1' + m_2\dot{y}_2',$$

and rearranging gives:

$$m_1(\dot{y}_1 - \dot{y}_1') = -m_2(\dot{y}_2 - \dot{y}_2').$$

Conservation of energy:

$$\frac{1}{2}m_1\dot{y}_1^2 + \frac{1}{2}m_2\dot{y}_2^2 = \frac{1}{2}m_1(\dot{y}_1')^2 + \frac{1}{2}m_2(\dot{y}_2')^2,$$

and rearranging and cancelling factor of $1/2,$

$$m_1\left[\dot{y}_1^2 - (\dot{y}_1')^2\right] = -m_2\left[\dot{y}_2^2 - (\dot{y}_2')^2\right],$$

and expanding:

$$m_1(\dot{y}_1 - \dot{y}_1')(\dot{y}_1 + \dot{y}_1') = -m_2(\dot{y}_2 - \dot{y}_2')(\dot{y}_2 + \dot{y}_2').$$

Dividing this on both sides by both sides of the relation obtained from conservation of momentum,

$$(\dot{y}_1 + \dot{y}_1') = (\dot{y}_2 + \dot{y}_2')$$

or

$$\dot{y}_1 - \dot{y}_2 = -(\dot{y}_2' - \dot{y}_1'),$$

as required.

9.6 The ball is projected horizontally with speed U and thus speed is constant as only the vertical component of velocity is affected by the bounces. Hence the distance travelled before the ball stops bouncing is

$$x_m = TU,$$

where T is the time taken for the ball to stop bouncing.

If the ball leaves the plane with speed V in the vertical direction, it will return in a time

$$t_b = 2V/g,$$

and will bounce again with speed eV.

Since it falls initially (at $t = 0$) from height h, it will reach the plane for the first time at $t = \sqrt{2h/g}$ and with speed $V = \sqrt{2gh}$, rebounding with speed $eV = e\sqrt{2gh}$; and so on.

The total time for all the bounces is given by the infinite series:

$$T = \sqrt{2h/g} + 2e\sqrt{2h/g} + 2e^2\sqrt{2h/g} + \cdots$$

$$= \sqrt{2hg}[1 + 2e + 2e^2 + \cdots]$$

$$= \sqrt{2gh}[2 + 2e + 2e^2 + \cdots - 1]$$

$$= 2\sqrt{2gh}[1 + e + e^2 + \cdots - 1/2]$$

$$= 2\sqrt{2gh}\left[\frac{1}{1-e} - \frac{1}{2}\right]$$

$$= \left(\frac{2h}{g}\right)^{1/2}\left(\frac{1+e}{1-e}\right),$$

and hence

$$x_m = \left(\frac{2h}{g}\right)^{1/2}\left(\frac{1+e}{1-e}\right)U.$$

Chapter 10

10.1 In Section 10.4.1, we found that

$$\sigma_T = \pi R^2,$$

which is readily interpreted as the cross-sectional area of the target sphere. In Section 10.4.2, the analysis was extended to incident particles also of finite radius R. In this case a collision will take place provided the projected area of the incident particle lies anywhere within a circle of radius $2R$ which is centred on the target particle. Hence on these grounds one would expect that in this case

$$\sigma_T = \pi(2R)^2 = 4\pi R^2.$$

10.2 In Sections 10.4.2 and 10.4.3, we found that

$$\sigma_T = 4\pi R^2,$$

irrespective of whether the target particle was free to recoil or was fixed.

In physical terms, the total cross-section is just the probability of the particle being scattered and this is unaffected by the target particle being free to recoil or not.

10.3 Using the terminology and analysis of Section 10.4.2, we can write the expression for the impact parameter b as

$$b = (R + r) \sin i,$$

which leads on to the result for the differential cross-section:

$$\sigma(\theta') = \frac{R^2}{4} \left(1 + \frac{r}{R}\right)^2,$$

and a total cross-section of

$$\sigma_T = \pi R^2 \left(1 + \frac{r}{R}\right)^2 = \pi(R + r)^2.$$

This result might have been predicted on physical grounds and reduces to the result of Section 10.4.1 for $r \to 0$; and to that of Section 10.4.2 for $r \to R$.

10.4 For a fixed centre of force, CM and LAB coordinates are the same. Given that the scattering angle is

$$\theta_1 = \pi \left[1 - \frac{1}{\sqrt{1 + k/mb^2 V^2}}\right],$$

we have (for $\theta' = \theta_1$)

$$\sigma(\theta_1) = \frac{b}{\sin \theta_1} \left|\frac{db}{d\theta_1}\right| = \frac{-1}{2 \sin \theta_1} \frac{db^2}{d\theta_1}.$$

Rearranging the expression for θ_1 gives

$$1 + \frac{K}{mb^2 V^2} = \frac{\pi^2}{(\pi - \theta_1)^2} \quad \Rightarrow \quad b^2 = \frac{K}{mV^2} \frac{(\pi - \theta_1)^2}{(2\pi\theta_1 - \theta_1)}.$$

Thus

$$\frac{db^2}{d\theta_1} = -\frac{K}{mV^2} \left[\frac{2(\pi - \theta_1)}{(2\pi - \theta_1)\theta_1} + \frac{(\pi - \theta_1^2)(2\pi - 2\theta_1)}{(2\pi - \theta_1)^2 \theta_1^2} \right]$$

$$= \frac{-K}{mV^2} \frac{\pi^2(\pi - \theta_1)}{(2\pi - \theta_1)^2 \theta_1^2},$$

and so

$$\sigma(\theta_1) = \frac{\pi^2 K(\pi - \theta_1)}{mV^2(2\pi - \theta_1)^2 \theta_1^2 \sin \theta_1}.$$

10.5 (a) In this case, the sphere captures all particles within the perimeter of its projected area and the capture cross-section is just

$$\sigma = \pi R^2.$$

(b) In the second case, the situation is like that in Section 10.6.1, but with the replacement of the gravitational force by the electrostatic force. If the electostatic force is written as

$$f = -\lambda/r^2,$$

then an identical analysis leads to

$$\sigma = \pi R^2 \left(1 + \frac{2\lambda}{mRV^2} \right).$$

Chapter 11

11.1 The student should carry out this task in order to gain an appreciation of the 'shape' of the γ-factor. No further comment will be made here.

11.2 From (11.14) and (11.15), the non-trivial Lorentz transformations are:

$$x' = \gamma(x - Vt) \qquad t' = \gamma(t - Vx/c^2).$$

We may write these as:

$$x' = \gamma x - (V/c)\gamma ct \qquad ct' = \gamma ct - (V/c)\gamma x.$$

Let

$$\beta = V/c,$$

then

$$x' = \gamma x - \gamma \beta ct$$
$$ct' = \gamma ct - \gamma \beta x,$$

and, with some rearrangement,

$$x' = \gamma(x - \beta ct)$$
$$ct' = \gamma(-\beta x + ct).$$

These equations may then be written as:

$$\begin{pmatrix} x' \\ ct' \end{pmatrix} = \gamma \begin{pmatrix} 1 & -\beta \\ -\beta & 1 \end{pmatrix} \begin{pmatrix} x \\ ct \end{pmatrix},$$

and the transformation matrix is as required.

11.3 For the outward journey B is in S, A is in S'_+.

For the inward journey B is in S, A is in S'_-.

For transformation of time intervals we have

$$\Delta t = \frac{\Delta t'}{\sqrt{1 - V^2/c^2}},$$

which does not depend on the sign of V.

Thus, for the outward trip, B records an interval $\frac{1}{2}T_B$, which is related to A's interval $\frac{1}{2}T_A$, by

$$\frac{1}{2}T_B = \frac{1/2T_a}{\sqrt{1 - V^2/c^2}},$$

and for the return trip $\gamma(V) = \gamma(-V)$ and so we get the same result, and in all:

$$T_B = \frac{T_A}{\sqrt{1 - V^2/c^2}}.$$

11.4 Anticipate that V will be close to c, and hence the time taken to make the journey from Earth to star as measured on Earth will be approximately equal to 10^5 years.

Take Earth's frame as S, spaceship's frame as S'.

In S: $T = 10^5$ years for journey to star.

In S': $T' = \gamma T$.

It would be reasonable to have a journey time of 10 years in S', so we require a γ-factor such that:

$$\gamma = T'/T = 10/10^5 = 10^{-4},$$

or

$$1 - V^2/c^2 = 10^{-8}.$$

Set $V = (1 - \delta)c$ and hence

$$2\delta + O(\delta^2) \simeq 10^{-8},$$

so that

$$\delta \simeq 5 \times 10^{-9}.$$

Thus confirming that $V \simeq c$!

11.5 The essential point here is that the observer in S' makes two successive comparisons with clocks fixed in S using a single clock which is moving in S. Hence, if it is in agreement with the first clock in S, then it must be slow when compared with the second clock in S, as moving clocks run slow.

In S, let the interval between the two events be Δt and clearly

$$\Delta t = L/V.$$

In S', let the interval between these events be $\Delta t'$ and as only a single clock is involved, this is a proper time interval and $\Delta t' \equiv \Delta t_0$ (say).

Then, from equation (11.26), the intervals in the two frames are related by

$$\Delta t = \gamma \Delta t_0,$$

and the difference between the two intervals is

$$\delta = \Delta t_0 - \Delta t = \Delta t \left(\frac{1}{\gamma} - 1 \right)$$

$$= \frac{L}{V} [(1 - V^2/c^2)^{1/2} - 1]$$

$$\simeq \frac{L}{V} \left(\frac{-V^2}{2c^2} \right) = \frac{-LV}{2c^2},$$

for $V \ll c$.

11.6 Let S be the reference frame of the Earth and S' be the reference frame of the spaceship.

From equation (11.15), the Lorentz transformation of time between the two frames is:

$$t' = \gamma(t - Vx/c^2).$$

An event on the Moon takes place at coordinates $x = 3.84 \times 10^8$ m and $t = 0.5$ s in S.

The event as recorded in S' takes place at x' and t'. We only need consider t', which is given by the above Lorentz transformation. With the values given, $\gamma = 1.25$ and $x/c = 1.28$ s, while $Vx/c^2 = 0.768$ s.

Hence,

$$t' = 1.25(0.5 - 0.768) = -0.34 \text{ s}.$$

Thus the observer on the spaceship records the event on the Moon as happening before the spaceship passed the Earth.

In order to answer the second part of the question, we must allow for the time taken for the light to travel from Moon to Earth so that the event can be **seen** on Earth. Hence the event is seen on Earth at time

$$T = t + x/c = 0.5 \text{ s} + 1.28 \text{ s} = 1.78 \text{ s}.$$

For the last part, we work in frame S', and the light from the event will be seen in the spaceship at time

$$T' = t' + x'/c = -0.34 \text{ s} + x'/c.$$

Now we have to find the space coordinate of the event (i.e. x') in S'.

From the Lorentz transform as given by (11.14) we have:

$$x = \gamma x' + \gamma V t',$$

and with some rearrangements:

$$\frac{x'}{c} + \frac{V}{c}t' = \gamma^{-1}\frac{x}{c},$$

Now $x/c = 1.28$ s, $V/c = 3/5$ and $\gamma^{-1} = 4/5$, thus

$$\frac{x'}{c} = \frac{4}{5} \times 1.28 \text{ s} - \frac{3}{5}t',$$

and for $t' = -0.34$ s

$$\frac{x'}{c} = 1.024 + \frac{3}{5}(-0.34) = 1.224 \text{ s}.$$

Substituting this value for x'/L in the formula for T' then yields

$$T' = 1.224 - 0.34 = 0.884 \text{ s}.$$

Note that the times T and T' of the event being **seen** in S and S' are **both** positive. In practice one would work backwards from these times, making allowance for the time taken for the light signal to travel from event to observer, to obtain the actual times t and t' of the event in both frames.

11.7 This is an example of the popular 'pole and barn' class of paradoxes. In my view, these are really too unphysical to qualify as 'thought experiments'. For instance, a typical form of words could be paraphased, as follows: 'An exceptional athlete carrying a pole runs into a barn with a speed of the same order as the speed of light and once the pole is in the barn comes to an immediate halt'.

As they are popular, I feel that one should be included. The present one—if fanciful—is at least a little more feasible than some. The resolution of the paradox is just the same as for the 'clock paradox' or 'twins paradox', as discussed in Section 11.6. The apparent symmetry is broken by the fact that the arrow accelerated and the barn did not. Accordingly, the arrow undergoes Lorentz–FitzGerald contraction and fits into the barn, at least momentarily.

It would be interesting to estimate, on the basis of plausible assumptions about the materials involved, the temperature rise of the target when an arrow with $\gamma = 2$ dissipates all its kinetic energy in it!

Chapter 12

12.1 In S, we have

$$U_A = 0 \qquad \text{and} \qquad U_B = u.$$

In S' we require:

$$U'_A + U'_B = 0 \qquad \text{hence} \quad U'_A = -U'_B.$$

By equation (12.18),

$$U'_A = \frac{U_A - V}{1 - VU_A/c^2} = -V$$

$$U'_B = \frac{U_B - V}{1 - VU_B/c^2} = \frac{U - V}{1 - VU/c^2}.$$

Equating

$$U'_B = -U'_A \qquad \frac{U - V}{1 - VU/c^2} = V,$$

thus

$$U - V = V(1 - VU/c^2).$$

Hence

$$U + \frac{V^2 U}{c^2} = 2V \text{ and so } U_2(1 + V^2/c^2) = 2V,$$

and finally:

$$U = \frac{2V}{1 + V^2/c^2}.$$

The Galilean result follows from taking $c \to \infty$, hence $V^2/c^2 \to 0$ and

$$U = 2V,$$

as one would expect.

12.2 Let S be the Earth's frame of reference and S' that of the spaceship.

In S, the lifeboat takes off at an angle θ to the direction of motion of the spaceship, which we take to be the x-axis.

Thus the velocity of the lifeboat is:

$$\mathbf{U} \equiv (U_x, U_y, U_z) \equiv (U_1, U_2, U_3) = (U\cos\theta_1, U\sin\theta_1, 0).$$

In S', we are given that the velocity of the lifeboat is

$$\mathbf{U}' \equiv (U_1', U_2', U_3') = (0, c/3, 0).$$

From equations (12.15) and (12.16),

$$U_1 = \frac{U_1' + V}{1 + VU_1'/c^2} = \frac{0 + c/2}{1 + 0} = c/2$$

$$U_2' = \frac{U_2'}{\gamma(V)(1 + VU_1'/c^2)} = \frac{c/3}{2/\sqrt{3}(1 + 0)} = \frac{c}{2\sqrt{3}},$$

where $V = c/2$ is the speed of the spaceship relative to the Earth and $\gamma(V) = 2/\sqrt{3}$. Hence:

$$U_1 \equiv U\cos\theta = c/2$$

$$U_2 \equiv U\sin\theta = c/2\sqrt{3},$$

and

$$U = \sqrt{(c/2)^2 + (c/2\sqrt{3})^2} = c/\sqrt{3},$$

while

$$\tan\theta = \frac{c}{2\sqrt{3}} \bigg/ \frac{c}{2} = \frac{1}{\sqrt{3}},$$

and so

$$\theta = 30 \text{ degrees.}$$

12.3 The velocity component of the flash along the x'-axis is

$$U' = c\cos\theta',$$

and, by equation (12.15),

$$U = \frac{U' + V}{1 + VU'/c^2} = \frac{c\cos\theta' + V}{c + Vc\cos\theta'}.$$

In S we must also have

$$U = c\cos\theta,$$

and equating the two expressions for U yields:

$$\cos\theta = \frac{c\cos\theta' + V}{c + Vc\cos\theta'}.$$

Light emitted into the forward hemisphere in S' has $\theta' \leq 90$ degrees. When $\theta' = 90$ degrees, $\cos\theta' = 0$, and the above expression gives us:

$$\cos\theta = \frac{V}{c}.$$

Thus in S, the light is emitted into a cone with semi-angle $\cos^{-1} V/c$.
    For the last part,

$$\sin^2\theta = 1 - \cos^2\theta = 1 - V^2/c^2,$$

hence

$$\sin\theta = \sqrt{1 - V^2/c^2} = 1/\gamma(V),$$

and, for small angles,

$$\theta \simeq \sin\theta = 1/\gamma(V).$$

12.4 Use the given identity in the form:

$$\tan\theta'/2 = \frac{\sin\theta'}{1 + \cos\theta'}$$

and substitute for $\sin\theta'$ and $\cos\theta'$ from equations (12.28) and (12.27), respectively:

$$\tan\theta'/2 = \frac{\sin\theta}{\gamma(1 + V\cos\theta/c)} \cdot \frac{1}{1 + (\cos\theta + V/c)(1 + V\cos\theta/2)^{-1}}.$$

Now multiply above and below by $(1 + V\cos\theta/c)$:

$$\tan\theta'/2 = \sin\theta\frac{\gamma^{-1}}{(1 + V/c)(1 + \cos\theta)}$$

$$= \frac{\sin\theta}{1 + \cos\theta} \cdot \frac{\sqrt{(1 + V/c)(1 - V/c)}}{1 + V/c}$$

$$= \frac{\sin\theta}{1 + \cos\theta} \cdot \sqrt{\frac{1 - V/c}{1 + V/c}}$$

$$= \left(\frac{c - V}{c + V}\right)^{1/2} \tan\theta^{1/2},$$

where we have invoked the given identity again.

12.5 From equation (12.34), for an approaching source,

$$\frac{\nu_0}{\nu} = \frac{(1 + U_r/c)}{\sqrt{1 - u^2/c^2}}.$$

In this case, there is no transverse component of velocity and we set

$$u = U_r \equiv V,$$

for simplicity. Hence:

$$\frac{\nu}{\nu_0} = \frac{1 + V/c}{\sqrt{1 - V^2/c^2}} = \sqrt{\frac{1 + V/c}{1 - V/c}}.$$

This leads to:

$$V/c = \frac{(\nu/\nu_0)^2 - 1}{(\nu/\nu_0)^2 + 1} \simeq \frac{0.44}{2.44} = 0.18.$$

Hence the car was going at a speed of $V = 0.18c$!

12.6 U is speed in S'.

In S_1 assuming Galilean transformation, the speed is W_G, given by

$$W_G = U + V.$$

In S_1 assuming Lorentz transformation, the speed is W_L, given by:

$$W_L = \frac{u + V}{1 + UV/c^2}.$$

Note: $W_L \rightarrow W_G$ as $c \rightarrow \infty$.

From the definition of rapidity, we have

$$V/c = \tanh \alpha_V,$$

and so on. Dividing both sides of the above Lorentzian velocity composition law by c,

$$\frac{W}{c} = \frac{U/c + V/c}{1 + UV/c^2}$$

implies

$$\tanh \alpha_W = \frac{\tanh \alpha_U + \tanh \alpha_V}{1 + \tanh \alpha_U \tanh \alpha_V}.$$

But, by the given identity, the right-hand side may be written as:

$$\tanh(\alpha_U + \alpha_V)$$

and so

$$\tanh \alpha_W = \tanh(\alpha_U + \alpha_V)$$

which implies

$$\alpha_W = \alpha_U + \alpha_V.$$

12.7
$$\frac{u}{c} = \tanh \alpha = \tanh\left(\frac{aT}{c}\right).$$

From equation (11.26),

$$dt = \gamma \, dT.$$

$$\gamma^{-1} = \sqrt{1 - u^2/c^2} = \sqrt{1 - \tanh^2 \alpha} = \operatorname{sech} \alpha,$$

thus

$$\gamma = \cosh \alpha.$$

Hence

$$\frac{dt}{dT} = \cosh(aT/c),$$

and integrating with $t = 0$ at $T = 0$,

$$t = \frac{c}{a} \sinh\left(\frac{aT}{c}\right),$$

as required.

Now to find the distance travelled, we write

$$u = \frac{dx}{dt} = c \tanh\left(\frac{aT}{c}\right),$$

so

$$\frac{dx}{dT} = u\gamma(u) = c \sinh\left(\frac{aT}{c}\right).$$

Next, integrate with initial conditions $x = 0$ at $T = 0$:

$$x = \frac{c^2}{a}\left[\cosh\left(\frac{aT}{c}\right) - 1\right],$$

as required.

Lastly, expand out cosh for small (aT/c) as $c \to \infty$, thus:

$$x = \frac{c^2}{a}\left[1 + \frac{1}{2}\frac{a^2 T^2}{c^2} + \cdots - 1\right]$$

$$= \frac{1}{2}a^2 T^2 + \cdots.$$

Chapter 13

13.1 We use equation (13.8) and the criteria which follow it. Then simple arithmetic serves to establish the following:

– E_1 and E_2 can be related.
– E_1 and E_3 cannot be related.
– E_2 and E_3 can only be related by a light signal.

13.2 Work only with the non-trivial coordinates: (ct, x).

In S: $E_1 \equiv (0,0)$ and $E_2 \equiv (1,2)$.

In S': $E_1 \equiv (0,0)$ and $E_2 = (\gamma[1 - 2V/c], \gamma[2 - V])$, where the coordinates of E_1 and E_2 in S' are obtained from those in S by means of Lorentz transformation.

(a) The two events are simultaneous in S' if $t_2' = t_1'$ and hence

$$\gamma(1 - 2V/c) = 0,$$

hence

$$V = c/2.$$

(b) E_2 precedes E_1 by $1/c$ in S', thus

$$t_2' = t_1' - 1/c,$$

and so

$$\gamma(V)(1 - 2V/c) = -1,$$

or

$$(1 - 2V/c) = -\sqrt{1 - V^2/c^2}.$$

Squaring both sides:

$$1 - 4V/c + 4V^2/c^2 = 1 - V^2/c^2,$$

$$(5V/c - 4)(V/c) = 0,$$

thus $V/c = 0$ or $4/5$. Hence, ignoring the trivial solution,

$$V = 4c/5.$$

13.3 From equation (13.29),

$$U = \gamma(u)(c, \mathbf{u}) = [\gamma(u)c, \gamma(u)\mathbf{u}].$$

Inner product gives (13.30):

$$U^2 = r^2(c^2 - u^2) = \gamma^2 c^2 - \gamma^2 u^2$$

$$= \frac{c^2}{1 - u^2/c^2} - \frac{u^2}{1 - u^2/c^2}$$

$$= \frac{c^2 - u^2}{1 - u^2/c^2} = \frac{c^2(c^2 - u^2)}{c^2 - u^2} = c^2,$$

as required.

13.4 From equations (13.18) and (13.19), we have:

$$A^\mu \equiv (A^0, \mathbf{a})$$

$$A_\mu \equiv (A^0, -\mathbf{a}).$$

From (13.49) we have the Minkowski metric tensor as:

$$g_{\mu\nu} = \mathrm{diag}(1, -1, -1, -1).$$

From the properties of matrix multiplication, the required result follows by inspection of the following equations:

$$\begin{pmatrix} A^0 \\ -a_1 \\ -a_2 \\ -a_3 \end{pmatrix} = \begin{pmatrix} 1 & 0 & 0 & 0 \\ 0 & -1 & 0 & 0 \\ 0 & 0 & -1 & 0 \\ 0 & 0 & 0 & -1 \end{pmatrix} \begin{pmatrix} A^0 \\ a_1 \\ a_2 \\ a_3 \end{pmatrix}.$$

13.5 From equation (13.48),

$$g_{\mu\nu}\, dX^\mu\, dX^\nu = ds^2.$$

Divide through by $d\tau^2$ and take the limit $d\tau \to 0$. Assuming constant metric tensor:

$$g_{\mu\nu} \lim_{d\tau \to 0} \left(\frac{dX^\mu}{d\tau}\right)\left(\frac{dX^\nu}{d\tau}\right) = \frac{ds^2}{d\tau^2} = c^2,$$

where the least step follows from the definition of proper time in equation (13.23). Hence taking the limits

$$g_{\mu\nu}\dot{X}^\mu \dot{X}^\nu = c^2,$$

as required.

From equation (13.27),

$$U^\mu = \frac{dX^\mu}{d\tau} = \dot{X}^\mu,$$

hence our previous result becomes

$$g_{\mu\nu} U^\mu U^\nu = c^2,$$

and by the raising/lowering properties of $g_{\mu\nu}$,

$$U_\mu U_\nu \equiv U^2 = c^2,$$

in agreement with equation (13.32).

13.6 In the usual notation,

$$X^0 \equiv ct, \qquad X' \equiv R,$$

$$X^2 \equiv R\cos\theta = 0 \quad (\text{for } \theta = \pi/2).$$

and at any time t,

$$X^3 \equiv R \sin \theta \, \omega t = R\omega t \quad (\text{for } \theta = \pi/2).$$

Now we may introduce the proper time τ through the relationship

$$t = \gamma \tau,$$

where

$$\gamma = (1 - u^2/c^2)^{-1/2},$$

and

$$U = R\omega.$$

Hence, in all, we have:

$$X^\mu = (rc\tau, R, 0, \gamma R\omega\tau),$$

with the 4-velocity as:

$$U^\mu = \frac{dX^\mu}{d\tau} = (\gamma c, 0, 0, \gamma R\omega) = \gamma(c, 0, 0, R\omega).$$

13.7 An ideal clock is defined as one which is unaffected by acceleration. From equation (13.26), the proper time interval of a moving clock is given by

$$\gamma(u) \, d\tau = dt$$

and hence

$$d\tau = [1 - U^2(t)/c^2]^{1/2} \, dt$$

and so

$$\tau = \int_{t_i}^{t_f} [1 - U^2(t)/c^2]^{1/2} \, dt.$$

13.8 We take S and S' to be in our usual standard configuration and so without loss of generality we put

$$\mathbf{U} = (U_1, 0, 0),$$

and consider the transformation laws for 4-vectors in equations (13.43) and (13.45), as applied to the 4-acceleration A^μ, thus:

$$(A^1)' = \gamma(A' - U_1 A^0/c)$$

$$(A^2)' = A^2$$

$$(A^3)' = A^3.$$

The 4-acceleration is given by (13.34) and is:

$$A^\mu = \gamma \left(c \frac{d\gamma}{dt}, \frac{d}{dt}(\gamma\mathbf{u}) + \gamma\mathbf{a} \right).$$

Begin with the relationship between a_2' and a_2.

In S'

$$(A^2)' = a_2' \qquad \text{as } \gamma = 1 \text{ for the comoving frame.}$$

In S_1

$$A^2 = \gamma^2 a_2.$$

Then invoking the transformation

$$(A^2)' = A^2,$$

it follows that

$$a_2' = \gamma^2 a_2,$$

and

$$a_3' = \gamma^2 a_3,$$

follows by similar reasoning. Now turn to the relationship between a_1' and a_1.

In S',

$$(A^1)' = a_1', \qquad \text{as } \gamma = 1 \text{ and } U_1 = 0 \text{ in the comoving frame.}$$

In S,

$$A^1 = \gamma \left(U_1 \frac{dr}{dt} + \gamma a_1 \right).$$

Now invoke the transformation:

$$(A^1)' = \gamma A^1 - \gamma \frac{U_1}{c} A^0,$$

which implies:

$$a_1' = \gamma^2 \left(U_1 \frac{d\gamma}{dt} + \gamma a_1 \right) - \frac{\gamma U_1}{c} \left(\gamma c \frac{d\gamma}{dt} \right)$$
$$= \gamma^3 a_1,$$

as required, where we note the cancellation of the first and third terms.

13.9 From equation (12.44), proper acceleration in the x-direction is given by

$$\frac{du'}{dt'} = \gamma^3 \frac{du}{dt} = \frac{d}{dt}[u\gamma(u)].$$

For constant proper acceleration a_0, we put

$$a_0 = \frac{\mathrm{d}}{\mathrm{d}t}[u\gamma(u)],$$

and integrate with respect to time to obtain

$$a_0 t = u\gamma(u) = \frac{u}{\sqrt{1 - u^2/c^2}},$$

with initial conditions:

$$x = 0, \qquad u = 0 \qquad \text{at } t = 0.$$

Squaring both sides of this result,

$$a_0^2 t^2 = \frac{u^2}{1 - u^2/c^2},$$

and rearranging gives

$$u^2 = \frac{a_0^2 t^2}{1 + a_0^2 t^2/c^2},$$

or

$$u = \frac{a_0 t}{\sqrt{1 + a_0^2 t^2/c^2}} \equiv \frac{\mathrm{d}x}{\mathrm{d}t}.$$

We integrate again with respect to time, to obtain

$$x(t) = \frac{c^2}{a_0}\left[\left(1 + \frac{a_0^2 t^2}{c^2}\right)^{1/2} - 1\right],$$

where we again invoked the initial conditions to fix the constant of integration.

Lastly, with some rearrangement, the required result follows.

13.10 For large x, the hyperbola of the previous exercise tends asymptotically to a straight line. Rearrange the equation for the world line of the spaceship as

$$c^2 t^2 = x^2 + 2c^2/a_0 = x^2(1 + 2c^2/a_0^2 x^2).$$

Taking the square roots of both sides, and using the binomial expansion, we find

$$ct = x(1 + 2c^2/a_0^2 x^2) \simeq x\left(1 + \frac{c^2}{a_0 x}\right) = x = +c^2/a_0.$$

This world line corresponds to a light ray emitted at a time $t = c/a_0$.

Chapter 14

14.1 Four-momentum P^μ is given by equation (14.3), along with (14.4) for the relativistic 3-momentum **p**. For a photon, we use Q^μ, as given by (14.26). Let the proper mass of the atom after the collision be M and its speed be u. We shall invoke the principle of conservation of 4-momentum to equate 4-momentum of initial (I) and final (F) states, which are:

$$P_I = \frac{h\nu}{c}(1, n, 0, 0) + [m\gamma(0)c, 0, 0, 0]$$

$$P_F = [M\gamma(u)c, M\gamma(u)u, 0, 0].$$

Equating the non-trivial components yields:

$$\frac{h\nu}{c} + mc = M\gamma(u)c;$$

$$\frac{h\nu}{c} = m\gamma(u)u.$$

The first of these implies

$$M = \frac{h\nu + mc^2}{\gamma(u)c^2};$$

while the second gives us:

$$u = \frac{\lambda\nu}{c}\frac{1}{M\gamma(u)} = \frac{ch\nu}{h\nu + mc^2},$$

the last step resulting upon substitution of the preceding result for M.

Now that we have an expression for u, we may substitute back into the previous result for M in order to evaluate the γ-factor. Then, with some algebra:

$$M = m\left(1 + \frac{2h\nu}{mc^2}\right)^{1/2}.$$

14.2 From equation (14.12), the kinetic energy is

$$Em\gamma(u)c^2.$$

We now consider the behaviour of the gamma factor for $u > c$. Let us put $u = \lambda c$, for $\lambda > 1$.
Then,

$$\gamma(u) = \frac{1}{\lambda\sqrt{-1}(1 - 1/\lambda^2)^{1/2}} \simeq \frac{1}{i\lambda}$$

for large λ, where $i = \sqrt{-1}$.

Under these circumstances, the energy becomes:

$$E = \frac{mc^2}{i\lambda} = \left(\frac{m}{i}\right)\frac{c^2}{\lambda}.$$

Hence, if E is real, the mass m must be imaginary.

14.3 For the case where $\mathbf{E} = 0$, the Lorentz force, as given by equation (1.17), may be written as:

$$\mathbf{F} = \mathbf{u} \times \mathbf{B},$$

with a change of notation for the velocity. As $\mathbf{E} = 0$, there is no acceleration and the relativistic generalization of N2 becomes:

$$m\frac{d^2\mathbf{x}}{d\tau^2} = \gamma(u)q\mathbf{u} \times \mathbf{B},$$

where we have invoked equations (13.29), (14.34) and (14.38). Note that the differentiation is with respect to proper time τ.

Now we choose our coordinate system such that

$$\mathbf{B} = (0, 0, B),$$

and the equations of motion in component form are:

$$m\ddot{x} = qB\dot{y}$$

$$m\ddot{y} = -qB\dot{x}$$

$$m\ddot{z} = 0,$$

where the dots all denote differentiations with respect to proper time. As $\dot{z}(0) = 0$, we have $\dot{z}(\tau) = 0$, and the motion is restricted to the xy-plane. We note that the equations of motion are identical to those for classical mechanics (albeit the time parameter is now τ rather than t) and hence we obtain the same result that the motion is in a circle.

At this stage, it is convenient to change to plane polar coordinates (r, ϕ) and, using equations (3.51) and (3.52), we have:

$$m[\ddot{r} - r\dot{\phi}^2] = qBr\dot{\phi}$$

$$\frac{1}{r}\frac{d}{d\tau}(r^2\dot{\phi}) = 0,$$

with the single change that all time differentiations are with respect to proper time τ. As the motion is circular under a constant central force, $\dot{r} = \ddot{r} = 0$. Then, the equation of radial motion yields:

$$\dot{\phi} = -qB/m,$$

and

$$u\gamma(u) = |r\dot\phi| = r(qB/m).$$

Hence

$$r = \frac{m}{qB} U\gamma(u).$$

14.4
$$K^\alpha = (K^0, \mathbf{k}) = \frac{q}{c}\left[\frac{\partial}{\partial X^\alpha}(U_\beta \Phi^\beta) - \frac{\partial \Phi^\beta}{\partial \tau}\right].$$

Consider the components in turn:

$$K^0 = \frac{q}{c}\frac{\partial}{\partial X^0}(\gamma c\phi - \gamma u_j A_j) = 0,$$

for time-independent potentials. Also,

$$K_i = \frac{q}{c}\frac{\partial}{\gamma x_i}(\gamma c\phi + \gamma U_j A_j) = \gamma q\mathbf{E} + \gamma\frac{q}{c}\cdot\frac{\partial}{\partial x_i}u_j A_j.$$

Note:

1. $\mathbf{E} = -\nabla\phi$.
2. The vector identity:

$$\mathbf{X}\times\mathbf{Y}\times\mathbf{Z} = \mathbf{Y}(\mathbf{X}\cdot\mathbf{Z}) - \mathbf{Z}(\mathbf{X}\cdot\mathbf{Y}).$$

Now, from (1.17)

$$\mathbf{F} = q(\mathbf{E} + \mathbf{u}\times\mathbf{B}).$$

Substituting $\mathbf{B} = \nabla\times\mathbf{A}$ and using the above identity,

$$\mathbf{F} = q(\mathbf{E} + \mathbf{u}\times\nabla\times\mathbf{A})$$
$$= q\mathbf{E} + q\nabla(\mathbf{u}\cdot\mathbf{A}) - q\mathbf{A}(\nabla\cdot\mathbf{u}).$$

Comparing the expressions for k_i and F_i shows that

$$k_i \rightarrow F_i,$$

as the particle velocity $u \rightarrow 0$ and $\gamma(u) \rightarrow 1$.

14.5 Take inner product of K^α with U^α:

$$U_\alpha K^\alpha = U_\alpha \frac{\mathrm{d}}{\mathrm{d}\tau}mU^\alpha = \frac{1}{2}\frac{\mathrm{d}}{\mathrm{d}\tau}mU_\alpha U^\alpha.$$

But, from (13.32),

$$U_\alpha U^\alpha = c^2,$$

hence

$$U_\alpha K^\alpha = 0.$$

Now decompose into components:

$$U_\alpha = (\gamma c, \gamma \mathbf{u}) \equiv (U^0, \gamma \mathbf{u}),$$

while

$$K^\alpha \equiv (K^0, \mathbf{k}) \equiv (K^0, \gamma \mathbf{F}).$$

Thus

$$U_\alpha K^\alpha = 0 \quad \Rightarrow \quad U^0 K^0 - \gamma^2 \mathbf{F} \cdot \mathbf{u} = 0$$

and

$$\gamma c K^0 = \gamma^2 \mathbf{F} \cdot \mathbf{u},$$

or

$$K^0 = \frac{\gamma}{c} \mathbf{F} \cdot \mathbf{u}.$$

Now go back to the defining relationship:

$$K^0 = \frac{\mathrm{d}}{\mathrm{d}\tau} m U^0 = \frac{\gamma}{c} \mathbf{F} \cdot \mathbf{u},$$

thus

$$\frac{\mathrm{d}t}{\mathrm{d}\tau} \frac{\mathrm{d}}{\mathrm{d}t} m \gamma c = \frac{\partial}{c} \mathbf{F} \cdot u,$$

and, as

$$\frac{\mathrm{d}t}{\mathrm{d}\tau} = \gamma,$$

it follows that:

$$\gamma \frac{\mathrm{d}}{\mathrm{d}t} m \gamma c = \gamma \frac{\mathbf{F} \cdot \mathbf{u}}{c}.$$

Cancel the gamma factors and divide across by c:

$$\frac{\mathrm{d}}{\mathrm{d}t} (m \gamma c^2) = \mathbf{F} \cdot \mathbf{u}.$$

In Newtonian terms, $\mathbf{F} \cdot \mathbf{u}$ is the rate at which the force does work on the particle and must equal the rate of change of energy. Hence if we put

$$E = m \gamma c^2,$$

then the above equation becomes:

$$\frac{dE}{dt} = \mathbf{F} \cdot \mathbf{u}.$$

14.6 From (14.35),

$$k_i = \frac{d}{dt} m\gamma u U_i = m \frac{d}{dt} U_i \left(1 - \frac{u^2}{c^2}\right)^{-1/2}.$$

Differentiate and let

$$\frac{du_i}{dt} = a_i,$$

to find

$$k_i = m\gamma(u)a_i + \frac{m\gamma^3}{c^2} U_i(a_j u_j).$$

(1) **a** perpendicular to **u**:
   Hence

$$\mathbf{a} \cdot \mathbf{u} = a_j u_j = 0,$$

and the above equation reduces to:

$$k_i = m\gamma(u)a_i,$$

as required.

(2) **a** parallel to **u**:
   Set $a_1 = a\mathbf{n}_i$ and $u = u\mathbf{n}_i$. Thus the above equation becomes

$$k_i = m\gamma(u)a\mathbf{n}_i \left(1 + \frac{u^2}{c^2}\gamma^2\right),$$

as $\mathbf{n}_j\mathbf{n}_j = 1$, and

$$k_i = m\gamma^3(u)a_i.$$

Chapter 15

15.1 Observers S and S'_{rot} agree on the diameter D of the turntable. Also, S says the circumference is L, and hence finds:

$$\pi = L/D.$$

However, S'_{rot} has a measuring stick which suffers a Lorentz–Fitzgerald contraction relative to S, by a factor $\gamma(u)$, where $u = D\omega/2$. Hence S'_{rot}

measures the circumference as having a greater number of units of length than in S_1 and accordingly finds:

$$L' = \gamma L.$$

Thus, in S'_{rot},

$$\pi' = \frac{L'}{D} = \frac{\gamma L}{D} = \gamma \pi = \pi \left(1 - \frac{D^2 \omega^2}{2c^2}\right)^{1/2}.$$

15.2 Usual formula for the distance a 'particle' falls under gravity is:

$$d = \frac{1}{2}gt^2, \qquad g = 9.81 \text{ m s}^{-2}.$$

$$t = \frac{30 \text{ m}}{3 \times 10^8 \text{ m s}^{-1}} = 10^{-7} \text{ s}.$$

Hence

$$d \simeq 5 \times 10^{-14} \text{ m}.$$

15.3 $\quad g_{\text{sun}} = \frac{GM}{R^2} = \frac{6.72 \times 10^{-11} \times 2 \times 10^{30}}{(7 \times 10^8)^2} \sim 3 \times 10^2 \text{ m s}^{-2}.$

Transit time $\sim$ solar diameter$/c \sim ((14 \times 10^8)/(3 \times 10^8)) \sim 5$s.
Deflection $= \frac{1}{2}g_{\text{sun}}t^2 \sim \frac{1}{2} \times 3 \times 10^2 \times 2.5 \times 10 \sim 4 \times 10^3$ m.

15.4 Differentiate the given curve:

$$2y\frac{dy}{dx} - 2a\frac{dy}{dx} + 2x = 0.$$

At $x = 0, y = 0 \Rightarrow dy/dx = 0$.
   Differentiate again:

$$2\left(\frac{dy}{dx}\right)^2 + 2y\left(\frac{d^2y}{dx^{-2}}\right) - 2a\frac{d^2y}{dx^2} + 2 = 0$$

and at $x = 0, y = 0, dy/dx = 0$,

$$-2a\frac{d^2y}{dx^2} = -2.$$

Hence

$$\frac{1}{\rho} = \frac{1}{a}$$

and so

$$\rho = a,$$

as required.

15.5
$$x_1 = a \cos t, \qquad x_2 = a \sin t, \qquad x_3 = ct$$
$$\mathbf{x} \equiv (x_1, x_2, x_3) = (a \cos t, a \sin t, ct)$$
$$\dot{\mathbf{x}} = (-a \sin t, a \cos t, c)$$
$$\ddot{\mathbf{x}} = (-a \cos t, -a \sin t, 0)$$
$$|\dot{\mathbf{x}}|^3 = (a^2 + c^2)^{3/2}.$$
$$|\dot{\mathbf{x}} \times \ddot{\mathbf{x}}| = a(c^2 + a^2)^{1/2}.$$

Hence, using the given formula.

$$\rho = \frac{a^2 + c^2}{a}.$$

15.6 From equation (15.34),

$$\Lambda = GM/c^2$$

$$c \sim 3 \times 10^8 \, \text{m s}^{-1}, \qquad G \sim 6.672 \times 10^{-11} \, \text{Nm}^2 \, \text{kg}^{-2}$$

$$M(\text{Earth}) = 5.98 \times 10^{24} \, \text{kg}.$$

$$\Lambda \simeq 4 \times 10^{-3} \, \text{m}.$$

15.7 For $(1 + 3)$D space, equation (13.48) is the definition of the metric tensor:

$$ds^2 = g_{\mu\nu} \, dX^\mu \, dX^\nu,$$

and (15.31) gives the Scharzschild metric as:

$$dS^2 = c^2 \left(1 - \frac{2\Lambda}{r}\right) dt^2 - \left(1 - \frac{2\Lambda}{r}\right)^{-1} dr^2 - r^2 d\theta^2 - r^2 \sin^2 \theta \, d\phi^2.$$

Comparison indicates that if (13.48) is to reduce to (15.33), it must be diagonal such that it takes the intermediate form (as in Section13.10.2):

$$dS^2 = g_{00} \, dX^0 \, dX^0 + g_{11} \, dX^1 \, dX^1 + g_{22} \, dX^2 \, dX^2 + g_{33} \, dX^3 \, dX^3.$$

Comparison of the two forms yields:

$$g_{00} = \left(1 - \frac{\Lambda}{r}\right)$$

$$g_{11} = -\left(1 - \frac{\Lambda}{r}\right)^{-1}$$

$$g_{22} = -1$$

$$g_{33} = -1.$$

and the required result follows.

15.8 From Exercise (13.6),

$$\dot{X}^\mu = (\gamma c, 0, 0, \gamma r \omega).$$

Also, from (15.7),

$$g_{\mu\nu} = \text{diag}\left[\left(1 - \frac{\Lambda}{r}\right), -\left(1 - \frac{\Lambda}{r}\right)^{-1}, -1, -1\right].$$

Then,

$$g_{\mu\nu}\dot{X}^\mu \dot{X}^\nu = \left(1 - \frac{\Lambda}{r}\right)\gamma^2 c^2 - \gamma^2 R^2 \omega^2 = c^2,$$

so that

$$\gamma^2 c^2 \left(1 - \frac{\Lambda}{r} - \frac{R^2 \omega^2}{c^2}\right) = c^2$$

and so

$$\gamma^{-2} = 1 - \frac{\Lambda}{r} - \frac{R^2 \omega^2}{c^2},$$

and the required result follows.

15.9 Gravitational redshift reduces the frequency and hence the energy of the climbing photon.

15.10 Escape velocity for a particle at the surface of a planet or a star is given by:

$$U_0^2 = \frac{2GM}{R}.$$

If the particle is a photon, then its velocity must be c, and hence for a photon:

$$c^2 = \frac{2GM}{R}.$$

Inverting this, the critical radius for a photon to escape the gravitational pull of the star is

$$R = \frac{2GM}{c^2},$$

where we assume that the mass M remains constant. From (15.34), it is clear that for $R < 2\Lambda$ no light will escape.

Index